Process Design and Modeling of Low-Carbon Energy Systems

Process Design and Modeling of Low-Carbon Energy Systems

Guest Editors

Chenyu Wu
Zhongkai Yi
Chenhui Lin

MDPI

Basel • Beijing • Wuhan • Barcelona • Belgrade • Novi Sad • Cluj • Manchester

Guest Editors
Chenyu Wu
College of Energy and Electrical Engineering
Hohai University
Nanjing
China

Zhongkai Yi
School of Electrical Engineering and Automation
Harbin Institute of Technology
Harbin
China

Chenhui Lin
Department of Electrical Engineering
Tsinghua University
Beijing
China

Editorial Office
MDPI AG
Grosspeteranlage 5
4052 Basel, Switzerland

This is a reprint of the Special Issue, published open access by the journal *Processes* (ISSN 2227-9717), freely accessible at: https://www.mdpi.com/journal/processes/special_issues/VK1RMZU786.

For citation purposes, cite each article independently as indicated on the article page online and as indicated below:

Lastname, A.A.; Lastname, B.B. Article Title. *Journal Name* **Year**, *Volume Number*, Page Range.

ISBN 978-3-7258-4125-7 (Hbk)
ISBN 978-3-7258-4126-4 (PDF)
https://doi.org/10.3390/books978-3-7258-4126-4

Contents

About the Editors

Chenyu Wu

Chenyu Wu, a professor in Jiangsu Province, is currently employed at the College of Energy and Electrical Engineering, Hohai University. With extensive experience in academia and industry, he has published over 20 SCI papers (Q1) and participated in editing two team standards. Wu has chaired five projects, including the National Natural Science Foundation, the State Grid Corporation of China's Headquarters Science and Technology Project, the China Southern Power Grid Company's Science and Technology Project, and basic research projects of central universities. He has also participated in over 10 projects as a technical backbone, such as the National Science and Technology Support Program, the National Key Research and Development Program, and technology projects for the State Grid and China Southern Power Grid. The products developed based on his research in the field of power systems/power markets have been successfully demonstrated in power grids in Jilin, Yunnan, Myanmar, and other regions. He has received accolades such as the First Prize of Jiangsu Province Science and Technology Award, the National Innovation and Entrepreneurship Outstanding Postdoctoral Fellow title, and recognition for an outstanding doctoral dissertation from the China Simulation Society. He has also served as a Guest Editor for journals like *Processes* and *Energies* and has been appointed as a reviewer for IEEE PES Transactions, Applied Energy, etc. His primary research focuses on the optimal operation of integrated energy systems, virtual power plants, active distribution networks, and electricity market trading and bidding.

Zhongkai Yi

Zhongkai Yi is an Associate Professor and Doctoral Supervisor at the School of Electrical Engineering and Automation, Harbin Institute of Technology, and he serves as the Deputy Director of the Power Systems Research Institute. His research interests lie in the application of machine learning and optimization theory in power systems and the operation and control of virtual power plants. Dr. Yi has authored more than over 40 SCI-indexed papers in JCR Q1 journals, with over 1,200 citations to date, and he holds over 30 granted international and national invention patents. He has led or participated in 12 major scientific research projects, including those funded by the National Natural Science Foundation of China, the National Key R&D Program, and science and technology programs supported by the State Grid Corporation of China. Dr. Yi's research interests include the application of machine learning and optimization theory in power system operation and control. Dr. Yi's outstanding contributions have been recognized with several prestigious awards, including the Global Championship in the NeurIPS City Learn Challenge, the Special Award in the "Silk Road" Ecological Development Competition of the China Electrotechnical Society, the First Prize for Electric Power Technology Innovation, the First Prize for Science and Technology Progress in Heilongjiang Province, the Excellent Doctoral Dissertation Award in Beijing, and the Outstanding Doctoral Graduate Award from Tsinghua University.

Chenhui Lin

Chenhui Lin received his bachelor's degree in 2015 and his Ph.D. degree in 2020, both in Electrical Engineering, from Tsinghua University. He is currently a Research Assistant Professor in the Department of Electrical Engineering, Tsinghua University. His main research interests include the operation and control of low-carbon power systems, as well as the application of machine learning in energy systems. Dr. Lin has been selected for the Youth Talent Support Programme by the China Association for Science and Technology. He has received a number of awards and research funds, including the first prize of the Beijing Science and Technology Award, the first prize of the China Electricity Council Innovation Award, the Outstanding Doctoral Dissertation Award of Tsinghua University, the Outstanding Technology Paper Award from the China Association for Science and Technology, and the Golden Award at the Geneva International Invention Exhibition (twice). He is currently serving as Secretary of the Active Distribution Network and Distributed Generation Committee of the Chinese Electrotechnical Society. He is a member of the IEEE Energy Internet Coordinating Committee and is an editorial board member of the journal *IET Renewable Power Generation*. Dr. Lin has published over 40 academic papers and filed more than 60 invention patents, with a total of over 1,200 citations to date. Dr. Lin's work has also led to the development of coordination technologies for high-penetration renewable energy into power systems, active distribution network control systems, and cluster control equipment, which have been implemented in several provincial power grids in China.

Preface

The transition to low-carbon energy systems stands as one of the most pressing challenges of our time, driven by the urgent need to mitigate climate change while ensuring energy security, affordability, and equity. This Special Issue of *Processes*, titled "Process Design and Modeling of Low-Carbon Energy Systems", brings together cutting-edge research at the nexus of engineering, economics, and policy to address the multifaceted barriers hindering global decarbonization. This collection of 14 studies seeks to bridge the gap between theoretical innovation and real-world implementation.

Motivated by escalating climate crises and the complexity of integrating renewable energy, storage, grid resilience, and cross-sector coupling, this reprint emphasizes interdisciplinary solutions. It explores AI-driven renewable forecasting models, hybrid storage systems, carbon-internalized market mechanisms, resilient multi-energy networks, and nanomaterials for thermal applications. By synthesizing advancements in process optimization, system integration, and policy coherence, the issue aims to empower policymakers, industry leaders, and academic researchers with actionable insights for scalable, equitable energy transitions.

We extend our gratitude to the reviewers, the editorial team of *Processes*, and institutions supporting this work. Their expertise and dedication ensured the rigor and relevance of the contributions. As the energy landscape evolves, we hope this reprint inspires collaborative efforts to accelerate the global shift toward sustainable, low-carbon futures.

Chenyu Wu, Zhongkai Yi, and Chenhui Lin
Guest Editors

processes

Editorial

Process Design and Modeling of Low-Carbon Energy Systems

Chenyu Wu [1,*], Zhongkai Yi [2] and Chenhui Lin [3]

1 College of Electrical and Power Engineering, Hohai University, Nanjing 211100, China
2 School of Electrical Engineering, Harbin Institute of Technology, Harbin 150001, China; yzk_article@163.com
3 Department of Electrical Engineering, Tsinghua University, Beijing 100084, China; linch11@yeah.net
* Correspondence: wcy@hhu.edu.cn

The need to transition toward low-carbon energy systems has never been more urgent [1]. Amid escalating climate crises, the global community faces a dual challenge: to decarbonize energy production while meeting growing demand for affordable and reliable energy [2]. Low-carbon energy systems are inherently complex, spanning combined heat and power generation [3], renewable generation [4], energy storage [5], telemeters [6], and electricity–gas–thermal coupling networks [7]. However, their deployment is hindered by multifaceted challenges. Technologically, the intermittent nature of solar and wind energy necessitates advanced forecasting [8] and storage solutions to ensure grid stability. Economically, the high capital costs of emerging technologies, such as hydrogen electrolyzers [9] and carbon capture systems [10], need innovative financing mechanisms. Politically, fragmented regulatory frameworks and misaligned incentives often slow the pace of adoption. Moreover, the socio-environmental dimensions (such as community engagement, land-use conflicts, and lifecycle environmental impacts) add layers of complexity to the energy transition.

This Special Issue of Processes (ISSN: 2227-9717), titled "Process Design and Modeling of Low Carbon Energy Systems", responds to these challenges by curating cutting-edge research at the intersection of engineering, economics, and environmental science. Our goal is to bridge the gap between theoretical advancements and real-world implementation, offering actionable insights for policymakers, industry stakeholders, and university researchers. By focusing on process optimization, system integration, and policy coherence, this collection highlights how interdisciplinary approaches can accelerate decarbonization while addressing equity and scalability. This Special Issue, thus, serves as a platform to showcase innovations that not only enhance technical performance but also align with the United Nations Sustainable Development Goals.

The following sections synthesize the key contributions of the 14 papers published in this collection, spanning renewable energy prediction, carbon market mechanisms, energy storage optimization, and socio-technical analyses.

Citation: Wu, C.; Yi, Z.; Lin, C. Process Design and Modeling of Low-Carbon Energy Systems. *Processes* **2025**, *13*, 1119. https://doi.org/10.3390/pr13041119

Received: 8 March 2025
Accepted: 3 April 2025
Published: 8 April 2025

Citation: Wu, C.; Yi, Z.; Lin, C. Process Design and Modeling of Low-Carbon Energy Systems. *Processes* **2025**, *13*, 1119. https://doi.org/10.3390/pr13041119

1. Renewable Energy Prediction and Optimization

Integrating renewable energy into power systems hinges on accurate prediction and intelligent optimization [11]. This Special Issue showcases groundbreaking methodologies that address the variability and uncertainty inherent in solar and wind while balancing economic and operational constraints.

The IAO-LSTM model [12] represents an advancement in solar prediction accuracy. By integrating the Improved Aquila Optimization (IAO) algorithm with Long Short-Term Memory networks, this framework dynamically adjusts hyperparameters to minimize prediction errors caused by cloud cover and seasonal irradiance fluctuations. A patch time series Transformer-based non-parametric model [13] combines the non-parametric

Huberized composite quantile regression method to predict voltage fluctuations in low-voltage grids with distributed energy storage. Unlike conventional Gaussian assumption-based methods, this approach quantiles uncertainty bounds without prior distribution knowledge. In addition to directly improving the prediction method, [14] applies maximum power point tracking techniques to further reduce the uncertainty of photovoltaic power.

The VMD-AOA-GRU hybrid model [15] tackles the non-stationarity of wind signals through a two-stage decomposition–optimization approach. First, Variational Mode Decomposition separates raw wind speed data into intrinsic mode functions, reducing noise interference. Then, the arithmetic optimization algorithm (AOA) is employed to optimize the hyperparameters of the model of the gated recurrent unit (GRU), including the number of hidden neurons, training epochs, learning rate, learning rate decay period, and training data temporal length, thereby constructing a high-precision AOA-GRU forecasting model.

The bi-level inverse robust optimization model [16] bridges wind variability and grid demand through pumped storage hydropower. The upper layer minimizes total generation costs, while the lower layer enforces an Optimal Inverse Robustness Index to ensure stability against wind forecast errors. A robust power grid dispatching technology is proposed in [17,18] that integrates deep learning-based forecasting, reinforcement learning, and optimization techniques. This technology is capable of forecasting future electricity demand and solar power generation. References [19,20] make charging/discharging decisions for energy-storage devices based on current grid conditions. Moreover, this technology is effective in optimizing the configuration of circuit breakers and switches to improve the reliability of power systems [21].

2. Carbon Trading and Multi-Energy System

Carbon trading and multi-energy system are currently highly researched topics in the fields of environment and economy [22], especially against the backdrop of global climate change and the increasing significance of renewable energy. This research direction focuses on internalizing the cost of carbon emissions through market mechanisms and multi-energy complementary mode [23].

A thermoeconomic modeling approach is presented in [24] to incorporate carbon credits into the analysis of multiproduct systems. The study uses a gas turbine cogeneration system as a case study to demonstrate how carbon market dynamics can be integrated into thermoeconomic models. The authors develop a methodology to allocate carbon-related costs to final products, considering both revenue and expenses associated with carbon credits.

Reference [25] explores the resilience enhancement of electric and natural gas networks against extreme events such as windstorms and wildfires. The study proposes a novel integrated energy system planning strategy that combines deep learning-based forecasting, reinforcement learning, and optimization techniques. By integrating these approaches, the authors demonstrate a robust framework for improving the resilience of energy systems.

A centralized regional integrated market structure is developed in [26] involving industrial users, carbon capture, utilization, storage facilities, and carbon market operators. The authors formulate a generalized Nash equilibrium model to analyze the trading behaviors of different entities and their impacts on system operations. This research highlights the significance of market structures and equilibrium analysis in optimizing the performance of integrated energy systems with carbon trading.

3. Thermal Transmission and Nanomaterials

Heat transfer is an important subject in engineering, and it involves the process of heat transfer from one object or region to another [27]. Improving heat transfer efficiency is a key factor in many practical applications, such as in compression equipment and air conditioning systems [28]. With the development of nanotechnology, nanomaterials have shown great potential in improving heat transfer properties.

The insufficient heat dissipation capacity of the gas head cover can lead to overheating, resulting in safety issues and increased operating costs. Ref. [29] investigates the heat dissipation issue of the gas head cover in a diaphragm compressor. The study analyzes the structure and heat transfer characteristics of the gas head cover, establishing a finite element simulation model for temperature distribution. Additionally, based on the temperature field distribution characteristics, two enhanced heat transfer gas head cover structures are proposed, and both simulation and experimental verifications are conducted.

Nanoporous alumina sheets have been widely applied in air conditioning heat exchangers. Ref. [30] focuses on the ability of nanoporous alumina sheets to inhibit frost layer growth in low-temperature environments. The researchers prepare nanoporous alumina sheets with various pore diameters using the anodic oxidation method and conduct an in-depth analysis of their anti-frosting properties. The results reveal that compared to conventional polished aluminum sheets, the nanoporous alumina sheets exhibited excellent anti-frosting performance. Notably, the porous alumina sheet with a 100 nm pore diameter demonstrated strong anti-frosting properties under low-temperature and high-humidity conditions.

4. Conclusions and Future Directions

The research presented in this Special Issue underscores the transformative potential of process design and modeling in advancing low-carbon energy systems. Key innovations, such as AI-driven prediction models, hybrid storage systems, nanomaterials, and carbon-internalized economic frameworks, demonstrate pathways to mitigate technical and economic barriers.

However, challenges remain. Future studies should prioritize the following:

(1) Cross-Sector Integration: Deeper coupling of electricity, hydrogen, and thermal networks to maximize resource synergy;
(2) Scalability: Translating laboratory-scale innovations (e.g., nanomaterials) into industrial applications;
(3) Policy Alignment: Developing adaptive regulatory frameworks to incentivize low-carbon investments and community participation.

Conflicts of Interest: The authors declare no conflict of interest.

References

1. Saraji, M.K.; Streimikiene, D. Challenges to the Low Carbon Energy Transition: A Systematic Literature Review and Research Agenda. *Energy Strategy Rev.* **2023**, *49*, 101163.
2. Belaid, F.; Al-Sarihi, A.; Al-Mestneer, R. Balancing climate mitigation and energy security goals amid converging global energy crises: The role of green investments. *Renew. Energy* **2023**, *205*, 534–542.
3. Wu, C.; Gu, W.; Jiang, P.; Li, Z.; Cai, H.; Li, B. Combined Economic Dispatch Considering the Time-Delay of District Heating Network and Multi-Regional Indoor Temperature Control. *IEEE Trans. Sustain. Energy* **2018**, *9*, 118–127.
4. Shi, Y.; Li, W.; Fan, G.; Zhang, L.; Yang, F. Low-Carbon Dispatch of an Integrated Energy System Considering Confidence Intervals for Renewable Energy Generation. *Energy Eng.* **2024**, *121*, 461–482. [CrossRef]
5. Li, J.; Chen, H.; Li, J.; Zhang, Y.; Pan, P.; Bian, J.; Yu, Z. Bi-level optimization model of hydrogen-blended gas units and multi-type energy storage system considering low-carbon operation. *Energy* **2025**, *314*, 134162. [CrossRef]
6. Qiu, M.; Su, H.; Chen, M.; Ming, Z.; Yang, L.T. Balance of security strength and energy for a PMU monitoring system in smart grid. *IEEE Commun. Mag.* **2012**, *50*, 142–149.
7. Zhu, M.; Xu, C.; Dong, S.; Tang, K.; Gu, C. An Integrated Multi-Energy Flow Calculation Method for Electricity-Gas-Thermal Integrated Energy Systems. *Prot. Control Mod. Power Syst.* **2021**, *6*, 1–12. [CrossRef]

8. Ren, Y.; Suganthan, P.N.; Srikanth, N. Ensemble Methods for Wind and Solar Power Forecasting-A State-of-the-Art Review. *Renew. Sustain. Energy Rev.* **2015**, *50*, 82–91. [CrossRef]
9. Lei, J.; Ma, H.; Qin, G.; Guo, Z.; Xia, P.; Hao, C. A Comprehensive Review on the Power Supply System of Hydrogen Production Electrolyzers for Future Integrated Energy Systems. *Energies* **2024**, *17*, 935. [CrossRef]
10. Lockwood, T. A Compararitive Review of Next-generation Carbon Capture Technologies for Coal-fired Power Plant. *Energy Procedia* **2017**, *114*, 2658–2670.
11. Wu, C.; Gu, W.; Yi, Z. Competitive Equilibrium Analysis for Renewables Integration in Dynamic Combined Heat and Power Trading Market. *IEEE Trans. Power Syst.* **2024**, *39*, 53–65.
12. Liu, L.; Li, Y. Research on a Photovoltaic Power Prediction Model Based on an IAO-LSTM Optimization Algorithm. *Processes* **2023**, *11*, 1957. [CrossRef]
13. Chen, X.; Ge, Y.; Zhang, Y.; Qian, T. Leveraging Transformer-Based Non-Parametric Probabilistic Prediction Model for Distributed Energy Storage System Dispatch. *Processes* **2023**, *12*, 779.
14. Fan, J.X.; Li, S.; Liu, S.J.; Deng, X.Q.; Zhu, X.P. Maximum Power Point Tracking Constraint Conditions and Two Control Methods for Isolated Photovoltaic Systems. *Processes* **2023**, *11*, 3245. [CrossRef]
15. Yang, J.; Pang, F.; Xiang, H.; Li, D.; Gu, B. A Novel Hybrid Deep Learning Model for Forecasting Ultra-Short-Term Time Series Wind Speeds for Wind Turbines. *Processes* **2023**, *11*, 3247. [CrossRef]
16. Jing, X.; Ji, L.; Xie, H. Bi-Level Inverse Robust Optimization Dispatch of Wind Power and Pumped Storage Hydropower Complementary Systems. *Processes* **2023**, *12*, 729.
17. Yang, C.; Zhang, J.; Jiang, W.; Wang, L.; Zhang, H.; Yi, Z.; Lin, F. Reinforcement Learning and Stochastic Optimization with Deep Learning-Based Forecasting on Power Grid Scheduling. *Processes* **2023**, *11*, 3188. [CrossRef]
18. Yi, Z.K.; Xu, Y.; Wu, C. Model-Free Economic Dispatch for Virtual Power Plants: An Adversarial Safe Reinforcement Learning Approach. *IEEE Trans. Power Syst.* **2023**, *39*, 3153–3168.
19. Zhu, M.; Sun, Y.; Lu, Y.; Sang, L.; Yi, Z.K.; Xu, Y.; Ma, K. The Value of Energy Storage in Facilitating Renewables: A Northeast Area Analysis. *Processes* **2023**, *11*, 3449. [CrossRef]
20. Yang, M.; Zhang, Y.H.; Liu, J.H.; Yin, S.; Chen, X.; She, L.H.; Fu, Z.X.; Liu, H.M. Distributed Shared Energy Storage Double-Layer Optimal Configuration for Source-Grid Co-Optimization. *Processes* **2023**, *11*, 2194. [CrossRef]
21. Huang, G.; Zhou, Y.; Yang, C.; Zhu, Q.; Zhou, L.; Dong, X.F.; Li, J.T.; Zhu, J.P. Optimal Allocation Method of Circuit Breakers and Switches in Distribution Networks Considering Load Level Variation. *Processes* **2023**, *11*, 2235. [CrossRef]
22. Zhou, Y. Worldwide Carbon Neutrality Transition? Energy Efficiency, Renewable, Carbon Trading and Advanced Energy Policies. *Energy Rev.* **2023**, *2*, 100026.
23. Wang, G.; Zhang, Z.; Lin, J. Multi-energy complementary power systems based on solar energy: A review. *Renew. Sustain. Energy Rev.* **2024**, *199*, 114464. [CrossRef]
24. Santos, J.J.C.S.; Faria, P.R.d.; Belisario, I.C.; Dos Santos, R.G.; Barone, M.A. Thermoeconomic Modeling as a Tool for Internalizing Carbon Credits into Multiproduct System Analysis. *Processes* **2024**, *12*, 705. [CrossRef]
25. Pan, W.; Li, Y.; Guo, Z.; Zhang, Y. Interdependent Expansion Planning for Resilient Electricity and Natural Gas Networks. *Processes* **2024**, *12*, 775. [CrossRef]
26. Yang, D.; Wang, S.; Wang, W.; Zhang, W.; Yu, P.; Kong, W. Regional Operation of Electricity-Hythane Integrated Energy System Considering Coupled Energy and Carbon Trading. *Processes* **2024**, *12*, 2245. [CrossRef]
27. She, X.; Cong, L.; Nie, B.; Leng, G.; Peng, H.; Chen, Y.; Zhang, X.; Wen, T.; Yang, H.; Luo, Y. Energy-efficient and-economic technologies for air conditioning with vapor compression refrigeration: A comprehensive review. *Appl. Energy* **2018**, *232*, 157–186. [CrossRef]
28. Vakiloroaya, V.; Samali, B.; Pishghadam, K. A comparative study on the effect of different strategies for energy saving of air-cooled vapor compression air conditioning systems. *Energy Build.* **2014**, *74*, 163–172. [CrossRef]
29. Ren, S.D.; Jia, X.H.; Zhang, J.T.; Xin, D.B.; Peng, X.Y. Analysis of Heat Transfer of the Gas Head Cover of Diaphragm Compressors for Hydrogen Refueling Stations. *Processes* **2024**, *11*, 2274. [CrossRef]
30. He, S.; Liu, H.Y.; Zhang, Y.; Liu, H.L.; Chen, W. Investigation of the Interface Effects and Frosting Mechanism of Nanoporous Alumina Sheets. *Processes* **2023**, *11*, 2019. [CrossRef]

processes

Article

Research on a Photovoltaic Power Prediction Model Based on an IAO-LSTM Optimization Algorithm

Liqun Liu * and Yang Li

College of Electronic and Information, Taiyuan University of Science & Technology, Taiyuan 030024, China; liyang@tyust.edu.cn
* Correspondence: llqd2004@163.com; Tel.: +86-035-1699-8245

Abstract: With the rapid popularization and development of renewable energy, solar photovoltaic power generation systems have become an important energy choice. Convolutional neural network (CNN) models have been widely used in photovoltaic power forecasting, with research focused on problems such as long training times, forecasting accuracy and insufficient speed, etc. Using the advantages of swarm intelligence algorithms such as global optimization, strong adaptability and fast convergence, the improved Aquila optimization algorithm (AO) is used to optimize the structure of neural networks, and the optimal solution is chosen as the structure of neural networks used for subsequent prediction. However, its performance in processing sequence data with time characteristics is not good, so this paper introduces a Long Short-Term Memory (LSTM) neural network which has obvious advantages in time-series analysis. The Cauchy variational strategy is used to improve the model, and then the improved Aquila optimization algorithm (IAO) is used to optimize the parameters of the LSTM neural network to establish a model for predicting the actual photovoltaic power. The experimental results show that the proposed IAO-LSTM photovoltaic power prediction model has less error, and its overall quality and performance are significantly improved compared with the previously proposed AO-CNN model.

Keywords: Aquila optimization algorithm; PV power prediction; neural networks

Citation: Liu, L.; Li, Y. Research on a Photovoltaic Power Prediction Model Based on an IAO-LSTM Optimization Algorithm. *Processes* **2023**, *11*, 1957. https://doi.org/10.3390/pr11071957

Academic Editors: Chenyu Wu, Zhongkai Yi and Chenhui Lin

Received: 4 June 2023
Revised: 21 June 2023
Accepted: 24 June 2023
Published: 28 June 2023

1. Introduction

With global energy demand increasing and the problem of climate change worsening, photovoltaic power generation as an environmentally friendly, renewable and reliable new energy source, is increasingly applied on a global scale. It is well known that photovoltaic power generation systems are greatly affected by environmental conditions, such as light intensity, temperature, wind speed, and so on. They also have impacts and challenges for the power system and bring greater security risks. Therefore, how to accurately predict the power generation of photovoltaic power generation systems is the key to ensuring their stable operation. At the same time, it can also help power system operators make real-time dispatch decisions, reduce the security risk of power grids, improve the quality of power supply and provide economic benefits [1–3].

By predicting the output power of photovoltaic power generation, the optimal dispatching of power grids can be realized, the stability level of power systems can be effectively improved, and the potential safety problems in power systems can be eliminated. It can also effectively reduce the output limit of photovoltaic power generation systems and increase the rate of return on investment, thus increasing the economic benefits and operation management level of photovoltaic power generation systems. At present, the commonly used methods for PV power prediction include physical methods [3], statistical methods [4–6], meta-heuristic learning methods [7,8] and combination methods [9], etc.

Swarm intelligence algorithms are a kind of intelligent optimization method that solve practical problems by simulating the swarm intelligence behaviors of natural organisms.

Neural networks are a network structure composed of many neurons, which can learn complex nonlinear relationships adaptively, and have good performance in prediction and classification. In recent years, more and more scholars combine swarm intelligence algorithms with neural networks for PV power prediction. For example, Wang et al. proposed a deterministic and probabilistic prediction of photovoltaic power based on deep convolutional neural networks, which can improve prediction accuracy [10]. In [11], an artificial neural network is used to reduce the complexity of a PV power prediction model and improve its prediction accuracy. The advantage of a hybrid method is that it can make full use of the advantages of both methods and improve prediction accuracy and efficiency.

Various artificial intelligence technologies with adaptive and self-learning abilities have been developed and are gradually becoming more widely used in the field of electric power. Through a comparative study of various methods, this paper adopts the method combining LSTM and optimization methods to realize photovoltaic output power prediction. Compared with regular CNN networks, an LSTM network is more suitable for processing classification or prediction of time-series data. By introducing a gate structure, an LSTM neural network has greater selectivity compared with traditional recursive neural networks [12–23]. In this paper, a neural network model based on the Aquila optimization algorithm combined with a neural network prediction model is proposed to speed up the prediction speed of the neural network and improve the prediction accuracy and speed of photovoltaic power systems. Then, the Aquila optimization algorithm (IAO) is used to optimize the parameters of the LSTM neural network to establish a model for predicting the actual photovoltaic power. The proposed IAO-LSTM photovoltaic power prediction model reduces error and its overall quality and performance are significantly improved compared with AO-CNN models.

2. Photovoltaic Power Data Preprocessing

Domestic and foreign research studies generally choose irradiance, temperature, wind direction and wind speed as the main influencing factors of photovoltaic power generation. In different distributed photovoltaic power stations, considering that the direct installation angle and position of the photovoltaic array are not exactly the same, different effects will gradually form. However, according to the connection between meteorological factors and the rated output power of photovoltaic new energy, determining more reasonable meteorological factors will help to further improve the accuracy of power prediction. For specific analysis of relevant data, the internal relationship between different functions can be used to judge the correlation between different functions through a curve, so as to understand the degree of correlation of independent variables and dependent variables according to the data. The correlation coefficient refers to the correlation between the variable and parameters of the function. If the value is regular, it is considered as positive correlation, and if the value is less than zero, it is considered a negative correlation. From the correlation analysis, it can be found that for photovoltaic arrays, it is reasonable to select the influence of irradiance, temperature, wind speed and wind direction.

The problem of missing or abnormal data caused by abnormal local function is inevitable, and photovoltaic power generation and various meteorological factors, such as light intensity, ambient temperature, wind speed and other unit function relationships are not the same. Therefore, the original data must be preprocessed to obtain data that can be directly applied in model training and prediction. In the data preprocessing, data outliers are corrected, missing data are completed, and the data are normalized.

(1) There is a strong correlation between PV power and meteorological data. Missing values are replaced with the mean of values before and after the missing value. If a large amount of data are missing during the day, the data for that day are deleted to prevent human influence. Replacement of missing data can be performed using the following formula:

$$data_j = \frac{data_{j-5} + data_{j-4} + \ldots data_{j-1} + data_{j+1} + data_{j+4} - data_{j+5}}{10} \quad (1)$$

The data between five moments before and after the missing data are selected to calculate the mean value to supplement the missing data, and Formula (1) shows the value between the left and right five moments of the missing data, respectively, as can be seen from (1).

(2) If there is no significant change in radiance or other meteorological data but the data on photovoltaic power generation have changed significantly, this value needs to be removed. In addition, if the photoelectric energy is negative, then in the case of very low radiation or zero, 0 is used instead of the negative value.

(3) The resolution frame rate of the database data is changed. The data interval needed to predict actual PV power over a short period of time is between 15 min and 1 h. Given the short time span of minute-level database data, the application of the original 1 min resolution data is not common and even less in production practice. The data collected are, therefore, converted into 15 min resolution.

(4) Data normalization is necessary. Because meteorological factors such as solar radiation have different dimensions, directly introducing them into the model reduces the accuracy of power prediction. Normalization of data can speed up model training and improve prediction accuracy. In general, maximum and minimum principles are used in combination for data normalization, and the formula is as follows:

$$data_{\mathrm{i}} = \frac{data_{\mathrm{i}} - data_{\mathrm{max}}}{data_{\mathrm{min}} - data_{\mathrm{max}}} \quad (2)$$

where $data_{\mathrm{i}}$ represents the value to be normalized, $data_{\mathrm{min}}$ and $data_{\mathrm{max}}$ represent the minimum value and the maximum value, respectively.

3. Principle of Convolutional Neural Networks

A convolutional neural network (CNN) is a relatively simple neural network used to solve prediction and classification problems. However, this algorithm requires much data to predict the model and its structural parameters directly affect the accuracy and generalization ability of the model, which also makes it difficult to determine the parameters. In order to obtain better results, the accuracy of CNN prediction models must be further improved.

In a CNN, the neural model refers to the convolutional nucleus (also known as a filter) in the convolutional layer. The filter is a small two-dimensional weight matrix, usually much smaller than the input image. In a CNN, the filter performs convolutional operations on different local regions of the input data to extract local features of the image. Each filter has a set of learnable weight parameters that are gradually adjusted by backpropagation and gradient-descent optimization. In forward propagation, the convolutional core sliding window scans different positions of the input image, convolves each position and generates a new feature map. In this process, each weight of the convolutional nucleus is equivalent to the weight of the neuron, which is used to control the response of the different information in the input data to the convolutional nucleus.

The neural network model is the basic structure of a deep neural network and has wide applicability. The advantage of this method is that it can effectively extract the external resources needed for global reinforcement training and complete the final classification task by using the external features of the local organization of the database data. The CNN architecture comprises a multi-layered feedforward neural network of convolutional layers, pooled layers, and fully connected layers. After a series of convolutional operations on the input data, the neural network extracts information from the simplest features and gradually becomes more complex until it can uniquely define the target's eigenvalues.

As shown in Figure 1, in the CNN model, the input layer is the point layer, which preprocesses the input display image. Preprocessing standardizes the display image, maintains its balance, rotates and moves it in parallel, converts the image into a mirror file, and converts perspective. Then, the input image data are transformed into a digital vector, and the image range is reduced to a numerical region suitable for the activation function. A hidden layer is a convolutional layer that includes one or more convolutional layers and one or more fully connected layers. The convolutional layer is the CNN core layer that performs most of the calculations. This layer convolves the input data with the filter and passes the result to the next layer. Convolution is a linear operation, similar to traditional neural networks. he operation is ordered, multiplying the input data one by one with the filter, and calculating the sum of the products at each spatial location. The convolution layer contains a set of filters, each convolving the input matrix. In this operation, the filter slides along the vertical and horizontal directions of the input matrix to compute the sum of the products at each spatial position. The green region is the region corresponding to the filter selected from the input data. The yellow area is the filter, and the blue area is the output data corresponding to the selected filter. The convolution process is shown in Figure 2.

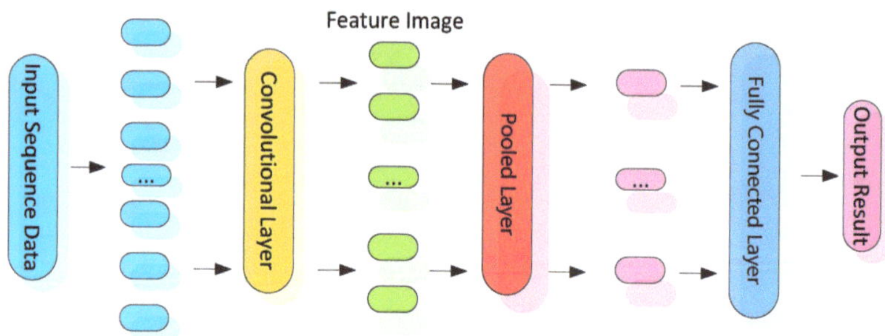

Figure 1. Schematic diagram of convolutional neural network model.

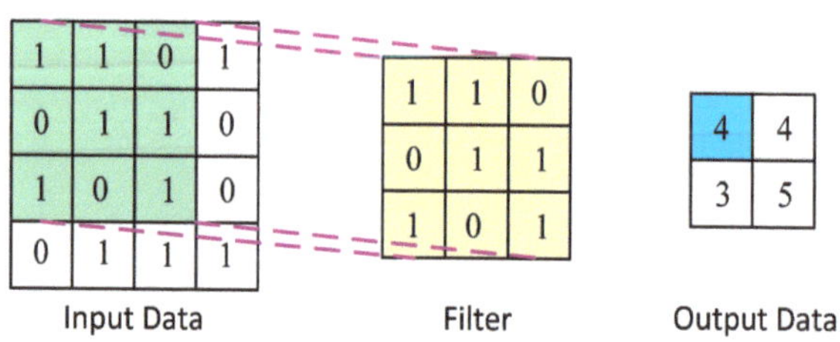

Figure 2. Schematic diagram of convolutional calculation.

In the training process, the weight in the filter is adjusted by back propagation and gradient descent to minimize the model's loss function. The convolution calculation process is given by the formula below. The mathematical expression of neurons in the convolution layer is as follows:

$$x_m^y = f(\sum_{n \in X} x_n^{y-1} * k_{mn}^y + b_m^y) \tag{3}$$

where x_m^y is the eigen graph matrix corresponding to the n-th feature graph of the $(y-1)$ layer, k_{mn}^y is the weight matrix corresponding to the mn-th feature graph of the y layer, * is the convolution operator, b_m^y is the deviation value of the m feature graph of the n layer, and f is the activation function. After convolution, the Eigen map matrix x_m^y is composed of the neurons of the m eigen map of the y layer.

$$x_m^y = f_1(\sum_{n \in X} x_n^{y-1} * k_{mn}^y + b_m^y) \tag{4}$$

There are two common pooling methods in the pooling layer. One is the maximum function pool, the other is the average function pool. Maximum pooling selects the maximum value of an area as the output of the response, and average pooling selects the arithmetic mean of an area as the output response. The mathematical expression for the pool layer is:

$$x_m^y = f_1(a_m^y pool(x_m^{y-1}) + b_m^y) \tag{5}$$

where pool is the pooling function, and a and b are the deviation values of each feature graph, respectively. Excitation function convolution is performed on the linear transformation layer. Therefore, when joining several hidden layers, the input and output show linear correlation. As a result, its performance is limited by a certain level of approximation. In practice, convolutional neural networks are not just linear operations.

4. Aquila Optimization Algorithm

The Aquila optimization algorithm is a new intelligent optimization algorithm [20–23]. It is mainly used to solve real number optimization problems. This algorithm has many exploration and development strategies. Compared with other meta-heuristic algorithms, the Aquila optimization algorithm has obvious advantages. The algorithm was inspired by four swarm behaviors of Northern Aquila birds during predation: (1) expanding the search area by soaring vertically and hunting birds in flight; (2) flying by contours of short gliding attacks and attacking prey in low-level air near the ground; (3) attacking prey gradually by low-flying and slow descent; and (4) walking and catching prey on land by diving. The initialization process is as follows: first, it randomly initializes the population position matrix as follows:

$$X_{ij} = rand \times (UB_j - LB_j) + LB_j, i = 1, 2, \ldots, Dim \tag{6}$$

where $rand$ is a random vector, LB_j represents the j-th lower bound for a given problem, and UB_j represents the j-th upper bound for a given problem.

(1) Expand the exploration is the first stage when the Aquila is hunting birds in the air. The birds use vertical glide height to expand the search scope. Its mathematical formula is:

$$X_1(t+1) = X_{best}(t) \times (1 - \frac{t}{T}) + (X_M(t) - X_{best}(t) \times rand) \tag{7}$$

$$X_M(t) = \frac{1}{N}\sum_{i=1}^{N} X_i(t), \forall j = 1, 2, \ldots, Dim \tag{8}$$

where $X(t)$ and $X(t+1)$ represent the individual position of the AO algorithm in the t iteration and the t + 1 iteration, respectively, $X_{best}(t)$ represents the optimal individual position obtained by the algorithm at the t iteration, $X_M(t)$ is the average position of the population in the t iteration, and T is the maximum number of iterations.

(2) Downsizing is the second stage when the Aquila flock finds its prey from high in the air. It chooses to spiral over the target, prepares to land, and then attacks. The mathematical expression can be shown as:

$$X_2(t+1) = X_{best}(t) \times \text{Levy}(D) + X_R(t) + (y - x) \times rand \tag{9}$$

where Levy (D) is the Levis strategy, s is a constant with the value 0.01, and u and v are random numbers between 0 and 1.

$$\text{Levy}(D) = \text{s} \times \frac{u \times \sigma}{|v|^{\frac{1}{\beta}}} \tag{10}$$

$$\sigma = \left[\frac{\tau(1+\beta) \times \sin e(\frac{\pi\beta}{2})}{\tau(\frac{1+\beta}{2}) \times \beta \times 2(\frac{\beta-1}{2})} \right] \tag{11}$$

Here, $\tau(x)$ is a Gamma function and β is a constant with a fixed value of 1.5. x and y represent the shape of a spiral flight. r is the search step, the radius of the helix. D_1 is the integer matrix from 1 to the length of the search space. θ is the helix angle, and θ_1 is the initial helix angle.

$$y = r \times \cos(\theta) \tag{12}$$

$$x = r \times \sin(\theta) \tag{13}$$

$$r = r_1 + U \times D_1 \tag{14}$$

$$\theta = -\omega \times D_1 + \theta_1 \tag{15}$$

$$\theta_1 = \frac{3 \times \pi}{2} \tag{16}$$

where r_1 ranges from 1 to 20, U takes the value 0.00565, and ω takes the value 0.005.

(3) To expand the development phase in the third stage, when the Aquila birds are in the hunting area, ready for landing and attack, they generally adopt the vertical drop method. The mathematical formula is:

$$X_3(t+1) = (X_{best}(t) - X_M(t)) \times \alpha\text{-}rand + ((UB - LB) \times rand + LB) \times \sigma \tag{17}$$

In the formula, α and β represent the development adjustment parameters, which are smaller than 0.1.

(4) To reduce the development in this stage when the Aquila bird is close to its prey, there is a certain randomness due to attack on the prey, and walking and capturing the prey. This is expressed in the mathematical formula:

$$X_4(t+1) = QF \times X_{best}(t) - (G_1 \times X(t) \times rand) - G_2 \times Levy(D) + rand \times G_1 \tag{18}$$

where QF represents the average search strategy of the mass function, G_1 represents the various trajectories of the movement of the Aquila bird during the escape of the prey, and G_2 represents the decreasing value from 2 to 0 during the escape of the prey when the Aquila bird follows the slope of the prey from the first position to the final position.

$$QF(t) = t\frac{2 \times rand() - 1}{(1-T)^2} \tag{19}$$

$$G_1 = 2 \times rand() - 1 \tag{20}$$

$$G_2 = 2 \times (1 - \frac{t}{T}) \tag{21}$$

The Aquila optimization algorithm is a method that can obtain the best result for a complex multi-objective problem. First, each index is evaluated by an initial group. At this stage, the algorithm, based on the existing best results, generates a new group, and each individual is given a new parameter. On this basis, the algorithm continuously searches for new optimal solutions according to the best individuals in the current population. In the current population, it makes a series of choices based on the best individual. When the number of optimal solutions is insufficient, the optimal strategy can be chosen by means of a rotary table. By solving a new group, the method updates all groups, and updates the existing optimization scheme according to the status of the existing groups. After finding the optimal solution, the method terminates the search and returns to the initial stage. Finally, the algorithm sorts the best schemes and selects the best schemes from them. In this way, we can find the optimal solutions, and then put the optimal solutions in a certain order so as to achieve better results.

5. AO-CNN Short-Term Photovoltaic Power Prediction Model

CNN is a common method of automatically learning PV power prediction. The filter, as an important parameter of the CNN, directly affects its accuracy and generalization ability. Improving the filter selection to improve model performance has become the focus of many researchers. In order to optimize the prediction performance of the CNN, the convolutional step size is inputted into the AO algorithm to optimize the convolutional kernel of the convolutional neural network.

In order to verify which prediction model is better, the measurement error evaluation index system of the prediction model is used. The evaluation mechanisms used in this paper are root mean square error (RMSE), absolute mean error (MAE) and average absolute error rate (MAPE). The mathematical formulas for RMSE, MAE and MAPE are as follows:

$$\mathrm{RMSE} = \sqrt{\frac{1}{M}\sum_{i=1}^{M}(P_j - \hat{P}_j)}^{2} \tag{22}$$

$$\mathrm{MAE} = \frac{1}{M}\sum_{j=1}^{M}\left|P_j - \hat{P}_j\right| \tag{23}$$

$$\mathrm{MAPE} = \frac{100\%}{M}\sum_{j=1}^{M}\left|\frac{P_j - \hat{P}_j}{\hat{P}_j}\right| \tag{24}$$

where P_j is the estimated photovoltaic output power at j-time, $\hat{P}_j$ is the actual photovoltaic output power at j-time, and M is the length of the PV power data series.

After training and testing the deep learning model, simulation tests can predict the actual photovoltaic power. The rated output power of PV is predicted by the AO-CNN and its CNN comparison chart, and the prediction effect is reflected by the chart. The smaller the error value, the better the prediction model.

Figure 3 shows the spring model of the fitting graph, which is used to show the fitting effect of the CNN and AO-CNN models. The vertical axis on the left shows the difference between the true and predicted values. The horizontal axis represents photovoltaic power data in hours. The blue dotted line represents the predicted value of the AO-CNN model. The green dotted line represents the predicted value of the original CNN model, and the red line represents the true value. As can be seen from the figure, the two models are less affected by spring weather and both models fit the real value well, but the predicted value of the AO-CNN model is closer to the real value, and the difference is smaller. The AO-CNN model has a large deviation in the prediction value from the sample time of 8–9 h, which may be caused by data abnormality or model inadequacy.

The summer model fitting diagram can be seen in Figure 4. It can be seen that the curve fluctuates greatly, and the two models are greatly affected by the weather in summer, which eventually leads to the green curve. The line deviates significantly from the original data, and the CNN model cannot properly match the true value. The purple curve basically coincides with the red curve. That is, the predicted value of the AO-CNN model proposed in this paper has a better fitting effect than that of the CNN model. The value is smaller. The predicted value of the AO-CNN model has a large deviation from 6–9 h in the sample time, which may be the actual number. It is also possible that the model was in error or did not fit, but overall, it is a better predictor of real data as there is better precision.

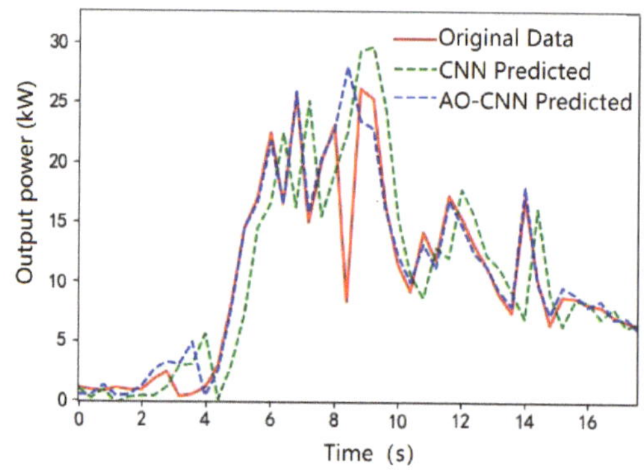

Figure 3. Spring forecast fit plot for each model with traditional methods.

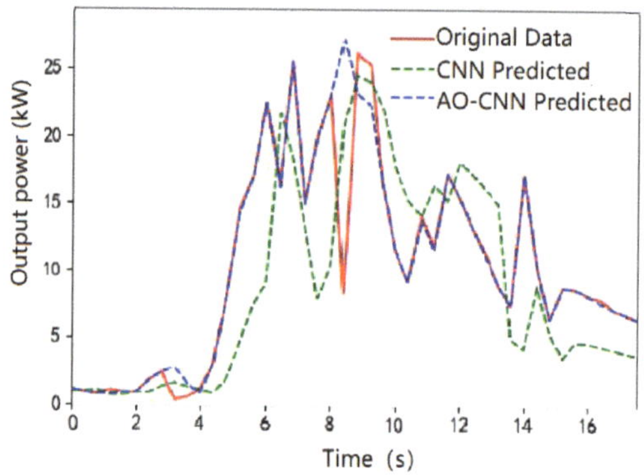

Figure 4. Summer forecast fit plot for each model with traditional methods.

Figure 5 shows the fitting diagram of the model in the fall. It can be observed that the weather in autumn has less effect. The two models discussed in this paper closely fit the real values. Blue is the AO-CNN prediction curve, which is similar to the original data. That is, the curve of the red line is fitted more comprehensively, and the predicted value of the proposed AO-CNN model is higher than that of the CNN model, indicating that the fitting effect is better.

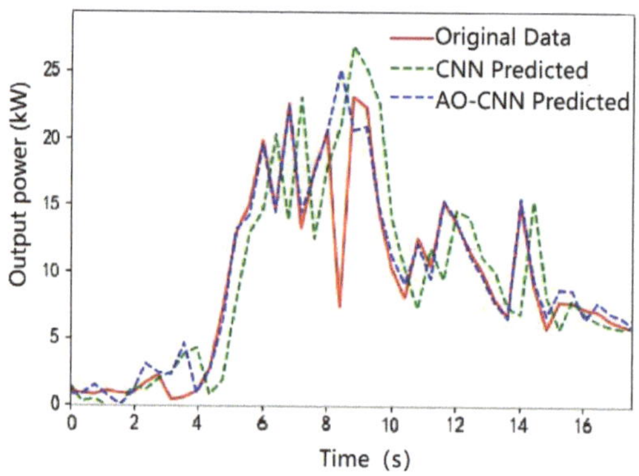

Figure 5. Autumn forecast fit plot for each model with traditional methods.

Figure 6 shows the winter fit of the model. It can be seen from the figure that the two models are affected by the winter weather, with the result that the CNN model does not match the real values well, but the forecast value of the AO-CNN model proposed in this paper is closer to the real values, with smaller differences. The predicted values of the AO-CNN model show a large deviation at the sample time from 8 to 9 h, which may be due to data anomalies or model underfitting.

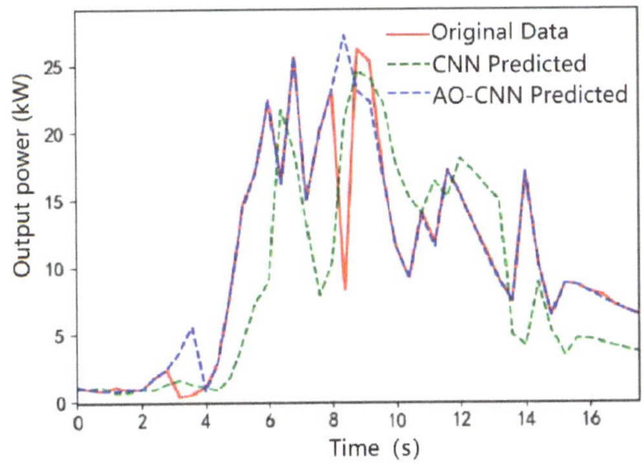

Figure 6. Winter forecast fit plot for each model with traditional methods.

Table 1 shows the differences between CNN and AO-CNN. It can be seen that the PV power prediction error of AO-CNN is smaller. Overall, the AO-CNN model performs better than the CNN model, demonstrating that the accuracy requirements of computer models can be further improved after the basic structure parameters of CNN model graphs are cleared by the deep learning algorithm based on group intelligence. The prediction effect of AO-CNN is better than that of the CNN model.

Table 1. Season forecast error table for each model with traditional methods.

Season	Error Type	CNN	AO-CNN
Spring	RMSE (%)	2.14	1.83
	MAE (%)	4.95	2.90
	MAPE(%)	1.25	3.84
Summer	RMSE (%)	2.00	1.75
	MAE (%)	3.07	2.68
	MAPE(%)	5.02	5.11
Autumn	RMSE (%)	1.49	1.25
	MAE (%)	2.76	2.24
	MAPE(%)	6.56	6.88
Winter	RMSE (%)	1.20	0.89
	MAE (%)	2.72	1.60
	MAPE(%)	6.91	8.39

6. Photovoltaic Power Prediction Based on the IAO-LSTM Network

6.1. LSTM Neural Network

Although CNN is often used in actual photovoltaic power prediction, its performance in processing sequence data with time characteristics is not advantageous, and the accuracy of photovoltaic power prediction still needs to be further improved. Therefore, a neural network called Long Short-Term Memory (LSTM) was introduced, which has obvious superiority in time-series analysis. At the same time, because the Aquila optimization algorithm appearing in the global index search process is insufficient, a new improved Aquila optimization algorithm (IAO) is proposed. This algorithm is used to optimize the parameters of LSTM neural networks, and finally, the photovoltaic power prediction model of IAO-LSTM network is constructed. It is then verified with the database-related data mentioned above.

The LSTM LAN comprises several LSTM units, and the neurons in each layer are the same as those in BP deep neural network. The LSTM network is similar to the training of many neural networks in that its learning mode is also a backward propagation of error calculation law. Since the LSTM network is a network with a repetitive structure, its learning algorithm is also called backward propagation algorithm. Its learning process mainly includes two aspects: forward transfer and backward transfer. In LSTM neural network learning, the optimal solution has a great influence, and the most common one is the gradient descent method.

6.2. Improved Aquila Optimization Algorithm

In view of the problems of the existing standard AO algorithm, such as insufficient overall search ability and the tendency to fall into local extreme values, we attempt to improve the standard AO algorithm with the Cauchy mutation and other methods, and test it using four standard functions.

In the particle swarm optimization, the Cauchy variation coefficient is added. Specifically, the Cauchy variation method is proposed for the first specific iteration process to ensure that the algorithm can jump out of the optimal solution of local organization and try to find an optimal algorithm. The complex density function of the Cauchy distribution in standard three-dimensional space is shown as follows:

$$f(x) = \frac{1}{\pi} \cdot \frac{1}{1+x^2} \quad -\infty < x < \infty \tag{25}$$

Since the characteristic of the density function of the Cauchy distribution is that it does not cross the X axis, the random numbers produced may leave the origin. The formula for generating random numbers using the Cauchy distribution is as follows:

$$Cauchy(0,1) = \tan[\pi \cdot (\xi - \frac{1}{2})] \tag{26}$$

where ξ is random, less than 1 and greater than 0. The formula for updating the individual position of the Aquila by Cauchy mutation method is as follows:

$$\overrightarrow{X^{q+1}} = \overrightarrow{X^q} \cdot (1 + Cauchy(0,1)) \tag{27}$$

where $\overrightarrow{X^q}$ is the individual before the mutation, and $\overrightarrow{X^{q+1}}$ is the individual after the mutation. Using the principle of survival of the fittest, the fittest values before and after mutation are compared to update the individual position.

Time complexity analysis usually includes three aspects: population initialization calculation fitness function, and updating the solution. It is assumed that the population number is N, the computational complexity of population initialization is $O(N)$, and the computational complexity of solution updating process is $O(T \times N) + O(T \times N \times D)$, where T is the total number of iterations and D is the dimension of the problem.

Therefore, the total time complexity of the standard AO algorithm is $O(N \times (T \times (D+1)))$. However, only the T distribution strategy $O(T \times N)$ is added to the IAO algorithm, without increasing the computational complexity, because the total time-interval complexity of the IAO algorithm is $O(N \times (T \times (D+1)+1))$. The IAO algorithm has the same complexity as the traditional AO algorithm and does not add any additional level of operation.

Four standard functions were selected, of which two were unimodal and two were multimodal, and their results were used to test the IAO performance. Table 2 presents the standard functions: F1 describes four standard functions, F1 to F2 are single mode functions, and F3 to F4 are multi-mode functions. For a single mode function, its search capability can be tested because it has a unique global optimal solution. The other part of the test composite function curve is shown in Figure 7.

Table 2. Four benchmark functional test functions.

Functions	Expression	Search Space	Dimension	Optimal Solution
F1	$F(x) = \sum_{k=1}^{D} x_k^4 + rand(0,1)$	[−200,200]	30	0
F2	$F(x) = \sum_{k=1}^{D} x_k^2$	[−2.28,2.28]	30	0
F3	$F(x) = \sum_{k=2}^{D} (10^6)^{(k-1)(D-1)} \cdot x_k^2$	[−22,22]	30	0
F4	$F(x) = \sum_{k=1}^{11} [a_k - \frac{x_1(b_k^2+b_k x_2)}{b_k^2+b_k x_3+x_4}]^2$	[−40,40]	30	0

Figure 7 shows a portion of the two-dimensional plane display images of the compound test function and the contraction change curves of various algorithms. Figure 7a–d display the digital F1, F2, F3 and F4 plane display images, respectively. In addition, F1 and F2 are single-peak composite functions, and there are only two optimal choices. F3 and F4 are complex multimodal functions with multiple optimal local fabric choices. It should be noted that the composite function of the selected cell is semicircular, but the smaller the final convergence, the faster the specific requirements are met.

The deep learning algorithm is used to compare the test results of each test compound function including the weighted average, probability distribution, optimal solution value, and difference value. The test results for the four benchmark functions are shown in Table 2. For all the combined F1-F2 functions of a single neural network, the AO algorithm performs better than the CPSO algorithm. The global optimal choice of the composite function for calculating F1-F2 elements is 0, but the optimal solution and weighted average of the AO and Marine Predators Algorithm (MPA) are not 0. The optimal solution values and weighted average of F1 and F2 are 0. For the composite function of a single neural network model, the test results of the F1-F2 test composite function all verify that the quality and performance of the proposed IAO are more competitive.

For composite functions F3 and F4, the final linear distance test of the CPSO algorithm is expected to have a large final large gap. When testing the composite function F4, IAO's optimal solution value is the best choice for calculating the closest test composite function in the deep learning algorithm. In both algorithms, the final value of IAO is closer to the expected value. Through the testing of F3 and F4, it is found that the optimization potential of IAO is relatively better than other specific methods. In general, often according to the above analysis, most single neural network model composite functions are still composite functions of natural language understanding. Compared with AO and its improved MPA, IAO is more competitive, stable and secure. Finally, it is shown that after the improvement of the computing unit AO with the help of the Cauchy mutation and other operational strategies, IAO has relatively effective computing and development potential, and IAO has stronger global optimization potential.

6.3. The Short-Term Photovoltaic Power Prediction Model of the IAO-LSTM Network

The LSTM network has a good ability to predict time series. However, the structure and modeling accuracy of LSTM networks depend on the selection of their hyperparameters, which directly affect the prediction effect of LSTM networks. At present, hyperparameter selection in the LSTM network mainly relies on prediction and many experiments. This is not only inefficient, but it is also difficult to obtain reasonable hyperparameter values. Here, in the context of machine learning, hyperparameters are parameters whose values are set before the learning begins, rather than parameter data obtained through training. Under normal circumstances, hyperparameters are optimized to select a group of optimal hyperparameters for the learning machine to improve the performance and effect of learning. Therefore, hyperparameters are derived from human experience and are subject to hardware constraints. On this basis, the IAO method is proposed to optimize the key parameters in the LSTM network and improve the accuracy of LSTM network modeling. Hyperparameters have a great influence on the prediction ability of LSTM networks. Improper selection leads to a decline in the prediction capacity of the LSTM network. In LSTM, the number of hidden layer neurons and the interval of reinforcement training batches are two indispensable related parameters which hinder the overall quality and performance of LSTM. Increasing the number of hidden layer neurons in LSTM neural networks can improve their fitting performance and prediction accuracy.

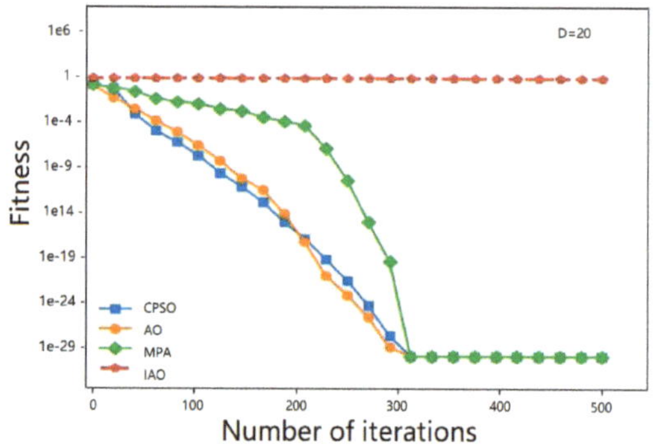

(**a**) F1

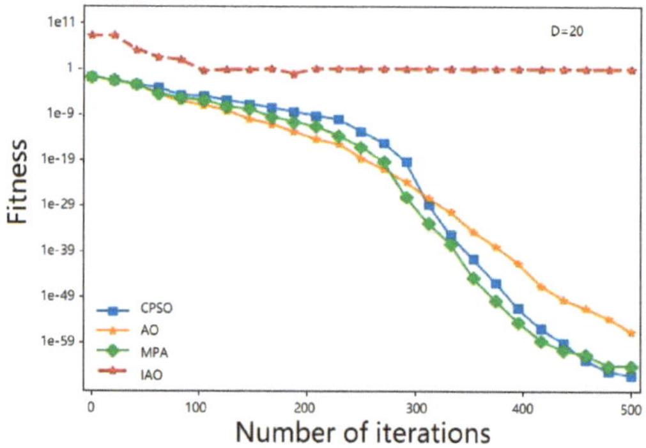

(**b**) F2

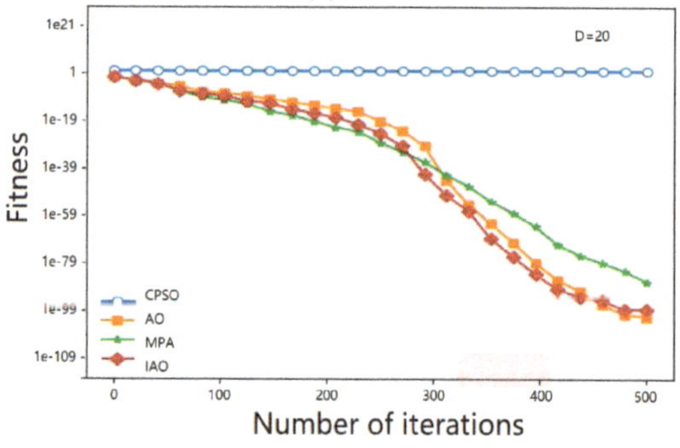

(**c**) F3

Figure 7. *Cont.*

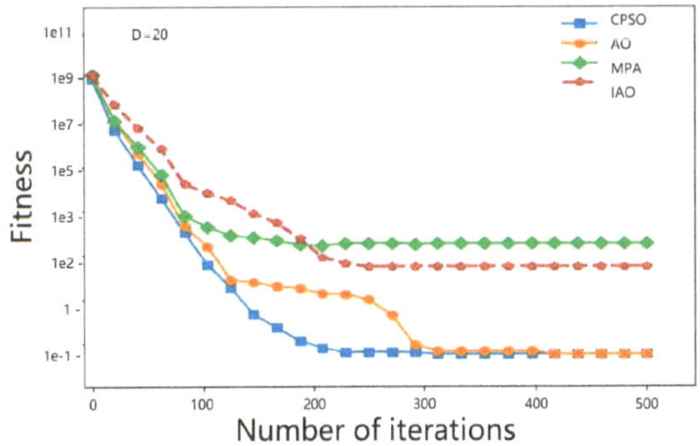

(**d**) F4

Figure 7. Two-dimensional function convergence curve.

As the number of neural networks increases, the need for computation also increases, and the learning speed of neural networks is affected to some extent. Each neural network can be used to describe the training batch and time interval of intensive training, but further improving the training batch time interval of neural network intensive training can effectively further improve computing power. It can also further improve the memory utilization efficiency of the operating system and shorten the product iteration cycle. However, in specific cases where the same training batch has improved significantly, database overflows, software program crashes, and so on can occur. If the appropriate parameters are selected, not only can the learning speed of motion be accelerated, but also the accuracy of 3D images can be further improved to avoid the possibility of generalization ability and other problems being difficult to solve.

In this paper, a new and improved IAO algorithm for Aquila deep learning is adopted. By testing the composite function, the aim was to verify the quality and performance of the IAO cell, thereby enabling the continuous improvement of the two new and current effective algorithm models. This testing indicates that IAO has good optimization potential. IAO was selected to optimize LSTM local area network (LAN) connectivity parameters, the number of hidden layer neurons, and the time interval of connection reinforcement training batches, in order to promote IAO automatic convergence in hyperparameters and optimize LSTM connectivity correlation. Thus, IAO can optimize the LSTM connection parameters.

Determining the number of hidden layer neurons enables us to optimize the LSTM LAN connection and further enhance the training batch time interval. The overall optimal choice is to allow the target compound function to obtain the corresponding value range of Aquila-related parameters, where the Aquila position in the vector space of the number of neurons in the time interval between the hidden layer and the reinforcement training batch corresponding to the LSTM neural network prediction model error is less than one. With IAO, the LSTM connection parameters can be automatically optimized to avoid measurement errors caused by manual selection of relevant parameters. IAO-LSTM makes basic connections based on LSTM neural networks. The optimal number of hidden layer neurons and the interval of intensive training batches were used as parameters for LSTM neural network connection.

The basic steps of the IAO-LSTM network model optimization process are:

Step 1: The relevant LSTM connection parameters are initialized. Then, the basic framework of LSTM connectivity and its related parameters are preliminarily preprocessed and inputted into the IAO-LSTM connectivity training set.

Step 2: The optimal IAO parameters are initialized. The population number, iteration times and other parameters are set. We then take the optimal LSTM network model as the optimal Skyhawk individual, and take the error function in the algorithm network model as the optimal fitness.

Step 3: Based on the number of training batches of the LSTM network and the number of neurons in the hidden layer, each Aquila is positioned and trained with the initial parameters to obtain the adaptive value of the Aquila (training error of the LSTM network), and then compared with the adaptive value of each individual to find the optimal search subject.

Step 4: The IAO correction formula is used to correct and determine the Aquila's position. On this basis, genetic algorithms are used to solve the algorithm. Otherwise, the number of optimal individuals and the size of the optimal adaptation value are maintained.

Step 5: When the maximum number of iterations is reached, the iteration is terminated to obtain the optimal solution. Otherwise, the model returns to step 4 and continues looking for the best individual.

Step 6: The relevant parameters in the IAO from the optimal individual scheme location are decoded. Then the relevant parameters considered most suitable for the LSTM connection are decoded.

Step 7: After garbage file cleaning, the LSTM neural network is intensively trained using the deep learning model. Then the test data set is selected to predict the nominal photovoltaic output power, and finally the prediction is recorded.

Through the construction and optimization of the IAO-LSTM network model, the photovoltaic power prediction mechanism is further understood. However, the existing IAO-LSTM neural network model cannot train and predict the entire system. Specifically, the actual photovoltaic power generation has a certain seasonal variation so it needs to be pretreated. According to the actual situation, the prediction model suitable for the data set is established. The distributed PV active power database data and their corresponding meteorological database data are determined. Then, the deep learning model is further divided into four parts. The deep learning model can be divided into spring, summer, autumn, and winter according to the physical characteristics of spring and autumn. After the deep learning model is preprocessed, the number of selected parameters in the IAO-LSTM sub-deep learning model is cleaned with the help of the enhanced training set, and then the actual power output data of photovoltaic power generation is predicted by continuing to select the test set.

6.4. Model Verification

The model evaluation indices used in this paper were the mean root square error (RMSE) and mean absolute error (MAE). When using the same database data as the AO-CNN prediction model above for model verification, the selected database data had to be relatively stable. Only on this basis could the results be compared. In order to make comparisons and draw differences, the predictive power of the proposed PV statistical model needs to be verified and the LSTM and AO-LSTM of the computing element compared.

The interval of intensive training for all images was set to 200 times. The image of LAN connection uses MSE as the activation function. The value of genetic diversity of the deep learning algorithms was set to 30, and the product iteration interval of deep learning algorithms was 20 times. Through the deep learning models and tests, the actual photovoltaic power could be predicted. Finally, AO-CNN and its comparison graphs were used to predict the rated photovoltaic output power, and the different prediction effects are shown in Figure 8a–d. The figure shows the prediction effects of the LSTM, AO-LSTM, and IAO-LSTM models after fitting the data for four seasons. The left vertical axis shows the difference between the true and predicted values, and the horizontal axis shows the PV power data time in hours. The green dotted line represents the prediction result of the LSTM model, the blue dotted line represents the prediction result of the original AO-LSTM model, the black dotted line represents the prediction result of the IAO-LSTM model, and the red solid line represents the true value.

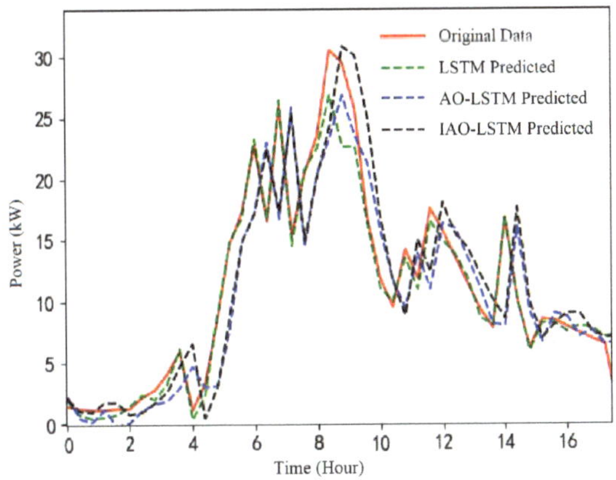

(**a**) Spring forecast fit plot for each model

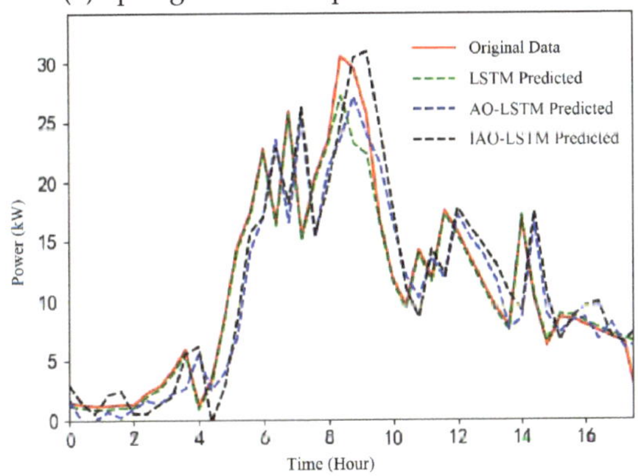

(**b**) Summer forecast fit plot for each model

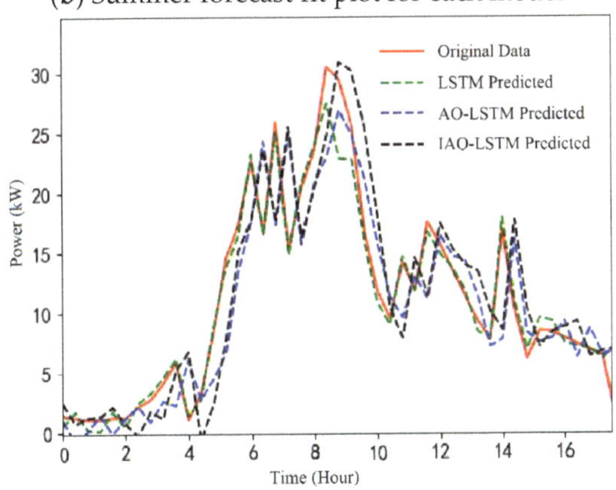

(**c**) Autumn forecast fit plot for each model

Figure 8. *Cont.*

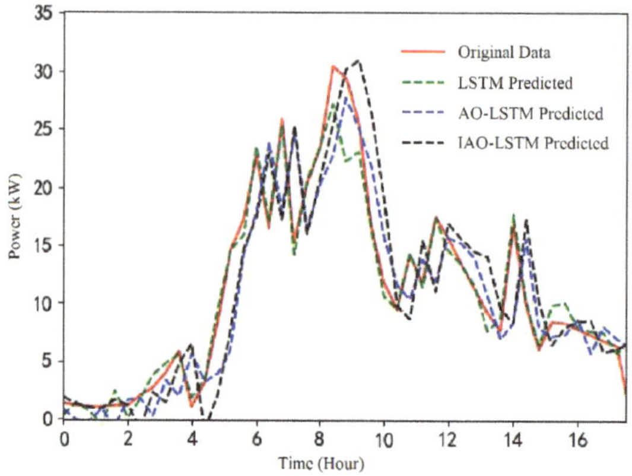

(**d**) Winter forecast fit plot for each model

Figure 8. Season forecast fit plot of each model.

As can be seen in Figure 8, the red and black lines tend to be more similar. Compared with the LSTM model and the AO-LSTM model, the IAO-LSTM model can better fit the true values, and the difference is small. There is a large deviation in the predicted values of the LSTM model in the 8–10 h of sample time range, which may be caused by abnormal data processing. Simply comparing the calculated PV active power curve with the reference curve cannot quantify the chart quality or visually evaluate its relative effect, so the error values are used to assess the results. Table 3 lists the values of the LSTM model, AO-LSTM model, and IAO-LSTM comparison model in terms of RMSE and MAE in four seasons. Comparing the error values of the LSTM, AO-LSTM and IAO-LSTM models in Table 3, it is clear that IAO-LSTM has higher accuracy in photovoltaic power prediction.

Table 3. Season forecast error table for each model.

Season	Error Type	LSTM	AO-LSTM	IAO-LSTM
Spring	RMSE (%)	1.87	1.48	1.38
	MAE (%)	2.67	1.76	1.91
	MAPE (%)	1.32	3.84	6.56
Summer	RMSE (%)	1.84	1.08	0.91
	MAE (%)	2.92	1.66	1.29
	MAPE (%)	4.61	6.38	6.31
Autumn	RMSE (%)	0.90	0.84	0.71
	MAE (%)	1.13	1.26	0.84
	MAPE (%)	7.76	6.88	7.47
Winter	RMSE (%)	1.01	0.71	0.61
	MAE (%)	1.43	1.15	0.91
	MAPE (%)	6.91	8.39	5.02

In conclusion, the experimental results show that the IAO-LSTM photovoltaic power prediction model proposed in this paper has smaller errors in all four seasons, and its overall quality and performance are better than that of LSTM and AO-LSTM. Compared with the AO-CNN model mentioned above, the performance of the IAO-LSTM model is also a significant improvement.

7. Conclusions

With the increase in the penetration rate of new energy, it is necessary to improve the ability of new energy to participate in the frequency regulation and voltage regulation of the power grid. However, the current energy storage technology is not mature in terms of safety, and the cost performance is low. To enable its participation in frequency regulation and voltage regulation of the power grid, it is necessary to accurately predict the output of the new energy field. This paper analyzes and studies the common methods of photovoltaic electric field output power prediction models, such as the CNN and AO algorithms, and the AO-CNN, LSTM and AO-LSTM models, etc. After comparing and analyzing the advantages and disadvantages of these models, we propose using IAO to optimize LSTM neural network parameters and establish a model for predicting actual photovoltaic power. The proposed IAO-LSTM model was applied in the field of photovoltaic power prediction. The experimental results show that the IAO-LSTM photovoltaic power prediction model has less error and that its overall quality and performance are better than the other abovementioned prediction models. In the future, if the proposed method can be combined with load forecasting, it will be able to accurately determine the frequency modulation region and adopt frequency modulation means to realize a supporting role in the power grid.

Author Contributions: Writing—review and editing, L.L.; writing—original draft preparation, Y.L.; All authors have read and agreed to the published version of the manuscript.

Funding: This work was financially supported by the Basic Research Surface Projects of Shanxi Province (202203021221153).

Data Availability Statement: Which will be found in Weipu net in future.

Acknowledgments: The authors acknowledged the financial support from Basic Research Surface Projects of Shanxi Province (202203021221153).

Conflicts of Interest: The authors declare no conflict of interest.

References

1. Yang, D.; Kileissl, J.; Gueymard, C.A.; Pedro, H.T.; Coimbra, C.F. History and trends in solar irradiance and PV power forecasting: A preliminary assessment and review using text mining. *Sol. Energy* **2018**, *168*, 60–101. [CrossRef]
2. Ma, T.; Yang, H.X.; Lin, L. Solar photovoltaic system modeling and performance prediction. *Renew. Sustain. Energy Rev.* **2014**, *31*, 75–83. [CrossRef]
3. Li, Y.T.; Su, Y.; Shu, L.J. An ARMAX model for forecasting the power output of a grid connected photovoltaic system. *Renew. Energy* **2014**, *66*, 820–823. [CrossRef]
4. Persson, C.; Bacher, P.; Shiga, T.; Madsen, H. Multi-site solar power forecasting using gradient boosted regression trees. *Sol. Energy* **2017**, *150*, 820–823. [CrossRef]
5. Ruby, N.; Jayabarathi, R. Predicting the Power Output of a Grid-Connected Solar Panel Using Multi-Input Support Vector Regression. *Procedia Comput. Sci.* **2017**, *115*, 75–83.
6. Li, Y.Z.; Niu, J.C.; Li, L. Forecast of Power Generation for Grid-Connected Photo-voltaic System Based on Grey Theory and Verification Model. *Energy Power Eng.* **2013**, *5*, 177–181. [CrossRef]
7. Wang, H.; Yi, H.; Peng, J.; Wang, G.; Liu, Y.; Jiang, H.; Liu, W. Deterministic and probabilistic forecasting of photovoltaic power based on deep convolutional neural network. *Energy Convers. Manag.* **2017**, *153*, 409–422. [CrossRef]
8. Rodríguez, F.; Fleetwood, A.; Galarza, A.; Fontán, L. Predicting solar energy generation through artificial neural networks using weather forecasts for microgrid control. *Renew. Energy* **2018**, *126*, 855–864. [CrossRef]
9. Yan, A.Y.; Gu, J.B.; Mu, Y.H.; Li, J.; Jin, S.; Wang, A. Research on photovoltaic ultra short-term power prediction algorithm based on attention and LSTM. *IOP Conf. Ser. Earth Environ. Sci.* **2021**, *675*, 151–158. [CrossRef]
10. Bruni, V.; Della Cioppa, L.; Vitulano, D. An automatic and parameter-free information-based method for sparse representation in wavelet bases. *Math. Comput. Simul.* **2020**, *176*, 73–95. [CrossRef]
11. Koster, D.; Minette, F.; Braun, C.; O'Nagy, O. Short-term and regionalized photovoltaic power forecasting, enhanced by reference systems, on the example of Luxembourg. *Renew. Energy* **2019**, *132*, 455–470. [CrossRef]
12. Hafeez, G.; Alimgeer, K.S.; Khan, I. Electric load forecasting based on deep learning and optimized by heuristic algorithm in smart grid. *Appl. Energy* **2020**, *269*, 114915. [CrossRef]
13. Ge, Q.B.; Guo, C.; Jiang, H.Y.; Lu, Z.; Yao, G.; Zhang, J.; Hua, Q. Industrial power load forecasting method based on reinforcement learning and Psolssvm. *IEEE Trans. Cybern.* **2022**, *52*, 1112–1124. [CrossRef] [PubMed]
14. Du, P.; Wang, J.Z.; Yang, W.D.; Niu, T. Multi-step ahead forecasting in electrical power system using a hybrid forecasting system. *Renew. Energy* **2018**, *122*, 533–550. [CrossRef]
15. Kong, W.C.; Dong, Z.Y.; Jia, Y.W.; Hill, D.J.; Xu, Y.; Zhang, Y. Short-term residential load forecasting based on LSTM recurrent neural network. *IEEE Trans. Smart Grid* **2019**, *10*, 841–851. [CrossRef]
16. Ozawa, A.; Furusato, R.; Yoshida, Y. Determining the relationship between a household's lifestyle and its electricity consumption in Japan by analyzing measured electric load profiles. *Energy Build.* **2016**, *119*, 200–210. [CrossRef]

17. Wang, H.; Qi, L.H.; Yan, L.; Li, Z. Load photo: A novel analysis method for load data. *IEEE Trans. Smart Grid* **2021**, *12*, 1394–1404. [CrossRef]
18. Ye, C.J.; Ding, Y.; Wang, P.; Lin, Z. A data-driven bottom-up approach for spatial and temporal electric load forecasting. *IEEE Trans. Power Syst.* **2019**, *34*, 1966–1979. [CrossRef]
19. Al-Wakeel, A.; Wu, J.Z.; Jenkins, N. K-means based load estimation of domestic smart meter measurements. *Appl. Energy* **2017**, *194*, 333–342. [CrossRef]
20. Muthu Kumar, B.; Maram, B. AACO: Aquila Anti-Coronavirus Optimization-Based Deep LSTM Network for Road Accident and Severity Detection. *Int. J. Pattern Recognit. Artif. Intell.* **2023**, *37*, 2252030.
21. Mohammed, A.A.A.; Ahmed, A.E.; Fan, H.; Ayman Mutahar, A.; Mohamed, A.E. Modified aquila optimizer for forecasting oil production. *Geo-Spat. Inf. Sci.* **2022**, *25*, 519–535.
22. Li, Z.; Luo, X.; Liu, M.; Cao, X.; Du, S.; Sun, H. Short-term prediction of the power of a new wind turbine based on IAO-LSTM. *Energy Rep.* **2022**, *8*, 9025–9037. [CrossRef]
23. Huang, Y.; Jiang, H. Soil Moisture Content Prediction Model for Tea Plantations Based on a Wireless Sensor Network. *J. Comput.* **2022**, *33*, 125–134. [CrossRef]

processes

Article

Leveraging Transformer-Based Non-Parametric Probabilistic Prediction Model for Distributed Energy Storage System Dispatch

Xinyi Chen [1], Yufan Ge [1], Yuanshi Zhang [1,2,*] and Tao Qian [1,2]

[1] School of Electrical Engineering, Southeast University, Nanjing 210096, China; chxy@seu.edu.cn (X.C.); taylorqian@seu.edu.cn (T.Q.)
[2] Jiangsu Provincial Key Laboratory of Smart Grid Technology and Equipment, Southeast University, Nanjing 210096, China
* Correspondence: yuanshizhang@seu.edu.cn

Abstract: In low-voltage distribution networks, distributed energy storage systems (DESSs) are widely used to manage load uncertainty and voltage stability. Accurate modeling and estimation of voltage fluctuations are crucial to informed DESS dispatch decisions. However, existing parametric probabilistic approaches have limitations in handling complex uncertainties, since they always rely on predefined distributions and complex inference processes. To address this, we integrate the patch time series Transformer model with the non-parametric Huberized composite quantile regression method to reliably predict voltage fluctuation without distribution assumptions. Comparative simulations on the IEEE 33-bus distribution network show that the proposed model reduces the DESS dispatch cost by 6.23% compared to state-of-the-art parametric models.

Keywords: low-voltage distribution networks; distributed energy storage system; chance-constrained programming; PatchTST; composite quantile regression; non-parametric probabilistic prediction

Citation: Chen, X.; Ge, Y.; Zhang, Y.; Qian, T. Leveraging Transformer-Based Non-Parametric Probabilistic Prediction Model for Distributed Energy Storage System Dispatch. *Processes* **2024**, *12*, 779. https://doi.org/10.3390/pr12040779

Academic Editor: Hsin-Jang Shieh

Received: 4 March 2024
Revised: 6 April 2024
Accepted: 9 April 2024
Published: 12 April 2024

1. Introduction

Effective voltage management is essential to ensure the safe and stable operation of low-voltage distribution networks [1,2]. However, the random nature of electrical loads presents a significant challenge in maintaining the bus voltage within the nominal range [3]. These uncertainties may result in voltage fluctuations or exceedances, thereby jeopardizing the stability and reliability of the power grid [4]. In recent decades, the distributed energy storage system (DESS) has emerged as a vital solution to manage this challenge and maintain voltage safety [5,6].

The accurate estimation of voltage fluctuation caused by the stochastic characteristics of loads [7] enables the optimal dispatch of DESSs. Existing techniques for handling uncertainties in distribution networks primarily include scenario-based stochastic programming [8,9], robust optimization [10,11], chance-constrained programming [12,13], etc. Among these, the chance-constrained programming approach is an effective approach that directly incorporates uncertainties into the optimization model by defining constraints that must be satisfied with a certain probability [14].

There are existing studies that investigate the formulations and solving methods for the chance-constrained economic dispatch (CCED) problem. Ref. [15] introduces a new probabilistic distribution model, called versatile distribution, to represent prediction errors in wind power. This probabilistic distribution model is incorporated into the CCED problem that includes wind power, with the aim of reducing the non-linearity and complexity of the problem. Ref. [16] also utilizes the versatile distribution to model the stochastic output of wind turbines, thus transforming the probabilistic constraints of wind power in the proposed decentralized CCED model into deterministic constraints. Although the fitting accuracy of the versatile distribution has been shown to outperform the Gussian and Bata distributions, the parametric probabilistic forecasting method may face limitations when dealing with complex uncertainties [14]. Therefore, some studies have begun exploring non-parametric probabilistic forecasting methods to better capture the uncertainties of renewable energy or electricity load. Ref. [17] combines extreme learning machines and quantile regression to efficiently produce non-parametric probability forecasts for wind power generation. Ref. [18] formulates a CCED model for DESS planning in active distribution networks, utilizing empirical probability density functions without parametric assumptions and a numerical convolution method to deal with uncertainties of various distributed energy resources (DERs). These studies validate that non-parametric probabilistic forecasting methods can accurately estimate various quantiles of random variables in the CCED problem without the need for any prior knowledge, statistical inference, or the assumption of specific probability distributions. This enables a more accurate and efficient conversion of the uncertain CCED problem into a linear programming problem.

Furthermore, with the applicability and extensibility of deep learning methods continuously verified [19–21], their capabilities in time series forecasting have received widespread attention [22–25]. Recently, the channel-independent patch time series Transformer (PatchTST) model has been proven to exhibit exceptional performance in time series prediction [26]. Its channel-independent processing, patching processing, and the Transformer architecture together enhance the model's deep understanding of both global trends and granularity in time-series data. Therefore, it is highly suitable to predict voltage fluctuations in distribution networks affected by random loads.

In this paper, we integrate a non-parametric probabilistic forecasting approach, Huberized Composite Quantile Regression (HuberCQR), into the Transformer-based PatchTST model, to address the uncertainty of random loads in the DESS CCED problem. HuberCQR is an effective technique that combines the robustness of Huber loss [27,28] with the flexibility of composite quantile regression (CQR) [29], enabling the model to generate accurate probabilistic forecasts even in the presence of noisy or outlier data. By integrating CQR into the PatchTST framework, we aim to leverage the Transformer's ability to capture complex temporal relationships while enhancing its prediction accuracy and efficiency across different quantiles. This integration allows for more reliable and robust forecasting of non-stationary voltage uncertainties, thereby facilitating more efficient and effective decision-making in the DESS CCED problem.

Overall, the contributions of this paper can be concluded as follows: (1) This paper leverages the non-parametric HuberCQR method to estimate composite quantiles of the uncertain voltage fluctuation caused by random loads, which is vital for transforming the original DESS CCED problem into linear form without complex mathematical derivations and predefined probabilistic assumptions. (2) The Transformer-based PatchTST forecasting framework integrated with the HuberCQR loss function is utilized to efficiently learn the uncertainties of bus voltage fluctuations. (3) The Transformer-based non-parametric probabilistic prediction framework demonstrates superior performance in providing accurate quantification of the voltage fluctuation range, which facilitates an effective trade-off between the DESS dispatch cost and the voltage violation risk.

The remainder of this paper is organized as follows. The problem formulation of the DESS CCED for voltage management in the distribution network is introduced in Section 2. Section 3 presents the Transformer-based PatchTST forecasting framework integrated with the HuberCQR loss function for composite quantile predictions of voltage fluctuation. In Section 4, comprehensive case studies are conducted to verify the effective and economical dispatch of DESS based on the proposed method. Finally, Section 5 concludes the paper.

2. Problem Formulation

In this section, we formulate a day-ahead DESS CCED problem for voltage management in a distribution network considering load uncertainty. Then we demonstrate how to transform chance constraints into deterministic constraints by introducing the cumulative distribution function (CDF) of a random variable and its inverse function: quantile function.

2.1. Linear DistFlow Model

Consider a radial distribution network denoted by $(\mathcal{N}, \mathcal{L})$, where $\mathcal{N} := \{1, \cdots, N\}$ represents the distribution network buses and $\mathcal{L} \subset \{\mathcal{N} \times \mathcal{N}\}$ represents the distribution lines. Let $\mathcal{L}_{i,1}$ represent the branch on the direct path from bus i to the reference bus, with $i \in \mathcal{N} \backslash 1$ denoting the non-reference bus. Define the set of descendants of bus m as $\mathcal{N}_m$, and each branch between two buses as (n, m). Take the IEEE 33-bus distribution network (shown in Figure 1) as an illustrative case, where bus #1 is the reference bus with a voltage magnitude of 1 p.u. The set $\mathcal{L}_{25,1}$ includes direct branches connecting bus #25 to the reference bus, i.e., $\{(1,2), (2,3), (3,23), (23,24), (24,25)\}$, and $\mathcal{N}_6$ refers to the descendants of bus #6, i.e., $\{7, 8, \ldots, 18, 26, 27, \ldots, 33\}$.

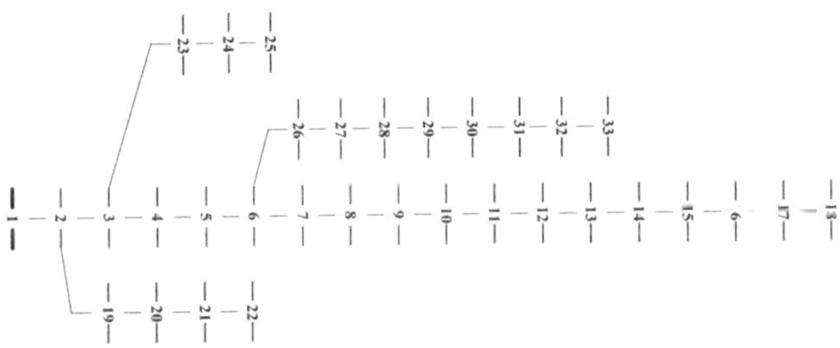

Figure 1. IEEE 33-bus radial distribution network [30].

In this paper, we adopt the Linear DistFlow model to describe the power flow in the distribution network [31,32], and assume a common scenario where the DESS only offers active power support [18]:

$$V_{i,t} = 1 - \sum_{nm \in \mathcal{L}_{i,1}} (2R_{nm}P_{nm,t} + 2X_{nm}Q_{nm,t}), \tag{1}$$

$$P_{nm,t} = \sum_{j \in \mathcal{N}_m} (P^{\text{cha}}_{j,t} - P^{\text{dis}}_{j,t} + \tilde{P}^L_{j,t}), \tag{2}$$

$$Q_{nm,t} = \sum_{j \in \mathcal{N}_m} \tilde{Q}^L_{j,t} \tag{3}$$

where $V_{i,t}$ represents the bus voltage squared, calculated based on the line resistance R_{nm} and reactance X_{nm}, considering the active and reactive power flows $P_{nm,t}$ and $Q_{nm,t}$. The active power flows $P_{nm,t}$ and $Q_{nm,t}$ are determined by adding the active power consumption of DESSs $(P^{\text{cha}}_{j,t} - P^{\text{dis}}_{j,t})$ and random loads $\tilde{P}^L_{j,t}$ on all descendants of the bus m, while reactive power flow $Q_{nm,t}$ is calculated as the sum of net reactive power consumption $\tilde{Q}^L_{j,t}$. This linear power flow model illustrates how the spatial distribution and electrical power of DESSs and random loads influence the bus voltage.

2.2. Objective and Constraints

This section introduces the objectives and constraints of the DESS CCED problem and presents the deterministic conversion of the chance constraints.

2.2.1. Objective

The objective of the DESS CCED problem in this paper is to minimize the total operational cost of all DESSs in the distribution network [33] as formulated in Equation (4):

$$\min \sum_{i \in \mathcal{N}_B} \sum_{t=1}^{T} \lambda(1 + \eta) \cdot (P^{\text{cha}}_{i,t} + P^{\text{dis}}_{i,t})\Delta t \tag{4}$$

where $\mathcal{N}_B$ is the set of all DESSs in the distribution network, T is the entire dispatch horizon, λ is the cost per unit of charging and discharging of a DESS (\$/MWh), η is the charging and discharging efficiency (%), $P_{i,t}^{\text{cha}}$ and $P_{i,t}^{\text{dis}}$ are the charging and discharging power of DESS i at time t (MW), and Δt is the dispatch time interval.

This formulation reflects the wear cost from battery degradation due to the charging/discharging operation. Under a reasonable depth of discharge, the overall capacity for charging/discharging of a DESS remains at a certain level. Therefore, the wear cost can be considered to be nearly proportional to the charging/discharging energy. Moreover, the term $1+\eta$ considers the energy loss during the charging and discharging process due to efficiency.

2.2.2. Constraints

The DESS CCED problem should be subject to the following constraints:

$$0 \le P_{i,t}^{\text{cha}} \le I_{i,t}^{\text{cha}} P_{i,\max}^{\text{cha}} \tag{5}$$

$$0 \le P_{i,t}^{\text{dis}} \le I_{i,t}^{\text{dis}} P_{i,\max}^{\text{dis}} \tag{6}$$

$$0 \le I_{i,t}^{\text{cha}} + I_{i,t}^{\text{dis}} \le 1, \tag{7}$$

$$\text{SOC}_{i,t} = \text{SOC}_{i,t-1} + (\eta P_{i,t}^{\text{cha}}/E_i^b - \frac{1}{\eta E_i^b} P_{i,t}^{\text{dis}})\Delta t \tag{8}$$

$$\text{SOC}_{i,\min} \le \text{SOC}_{i,t} \le \text{SOC}_{i,\max} \tag{9}$$

$$\Pr\{V_{i,t} \le V_{i,\min}\} \le \epsilon \tag{10}$$

$$\Pr\{V_{i,t} \ge V_{i,\max}\} \le \epsilon \tag{11}$$

where $I_{i,t}^{\text{cha}}$ and $I_{i,t}^{\text{dis}}$ represent the charging and discharging states of DESS i at time t, with a value of 1 indicating charging/discharging and 0 otherwise; $P_{i,\max}^{\text{cha}}$ and $P_{i,\max}^{\text{dis}}$ mean the maximum allowable charging/discharging power of DESS; SOC is the state of charge of the battery, indicating the ratio of the remaining capacity to the total capacity; $\text{SOC}_{i,\min}$ and $\text{SOC}_{i,\max}$ signify the minimum/maximum allowed SOC lower limits of DESS i; E_i^b represents the DESS capacity; $V_{i,t}$ is bus voltage at time t calculated by the Equation (1); $V_{i,\min}$ and $V_{i,\max}$ represent the lower/upper voltage thresholds of bus i; and ϵ symbolizes the allowed violation probability of the voltage constraint.

Equations (5)–(9) define the operational constraints of the DESS unit. More specifically, Equations (5) and (6) give the limits of the DESS charging and discharging power. Equation (7) indicates that DESS cannot be simultaneously in charging and discharging states. Equations (8) and (9) describe the energy balance and depth of discharge (DOD) limit of DESS. Equations (10) and (11) assume that the probability of bus voltage violation remains below a certain level, which is vital for the safe operation of the power system.

2.2.3. Deterministic Conversion of Chance Constraints

To solve the formulated DESS CCED model (4)–(11), the chance constraints (10) and (11) need to be transformed into deterministic constraints. Then, the optimal solution of the formulated model can be obtained directly by applying professional solvers to the resulting mixed-integer linear programming (MILP) problem. Next, the conversion of chance constraints into deterministic linear constraints will be explained.

To begin with, we substitute Equations (1)–(3) into (10) and obtain:

$$\Pr\left\{1 - \sum_{nm\in\mathcal{L}_{i,1}} \sum_{j\in\mathcal{N}_m} 2R_{nm}(P_{j,t}^{\text{cha}} - P_{j,t}^{\text{dis}}) - \sum_{nm\in\mathcal{L}_{i,1}} \sum_{j\in\mathcal{N}_m} (2R_{nm}\tilde{P}_{j,t}^L + 2X_{nm}\tilde{Q}_{j,t}^L) \le V_{i,\min}\right\} \le \epsilon \tag{12}$$

Then, a new random variable $\Xi_{i,t}$ is defined, which indicates the voltage drop caused by random loads compared with the reference voltage (1 p.u.) at the reference bus:

$$\Xi_{i,t} := \sum_{nm\in\mathcal{L}_{i,1}} \sum_{j\in\mathcal{N}_m} (2R_{nm}\tilde{P}^L_{j,t} + 2X_{nm}\tilde{Q}^L_{j,t}) \tag{13}$$

Thereafter, Equation (12) can be expressed as Equation (14) by substituting Equation (13). Equation (15) takes the complement of Equation (14). Next, Equation (16) substitutes the CDF definition for the probability term in Equation (15), which reflects the probability that the random variable $\Xi_{i,t}$ is less than or equal to a certain value. Finally, Equation (17) incorporates the inverse CDF term, $F_{\Xi}^{-1}(\cdot)$, resulting in an equivalent expression of the original chance constraint but now in a deterministic form:

$$\Pr\left\{ \Xi_{i,t} \geq 1 - \sum_{nm\in\mathcal{L}_{i,1}} \sum_{j\in\mathcal{N}_m} 2R_{nm}(P^{\text{cha}}_{j,t} - P^{\text{dis}}_{j,t}) - V_{i,\min} \right\} \leq \epsilon \tag{14}$$

$$\Rightarrow \Pr\left\{ \Xi_{i,t} \leq 1 - \sum_{nm\in\mathcal{L}_{i,1}} \sum_{j\in\mathcal{N}_m} 2R_{nm}(P^{\text{cha}}_{j,t} - P^{\text{dis}}_{j,t}) - V_{i,\min} \right\} \geq 1-\epsilon \tag{15}$$

$$\Rightarrow F_{\Xi}(1 - \sum_{nm\in\mathcal{L}_{i,1}} \sum_{j\in\mathcal{N}_m} 2R_{nm}(P^{\text{cha}}_{j,t} - P^{\text{dis}}_{j,t}) - V_{i,\min}) \geq 1-\epsilon \tag{16}$$

$$\Rightarrow 1 - \sum_{nm\in\mathcal{L}_{i,1}} \sum_{j\in\mathcal{N}_m} 2R_{nm}(P^{\text{cha}}_{j,t} - P^{\text{dis}}_{j,t}) - V_{i,\min} \geq F_{\Xi}^{-1}(1-\epsilon) \tag{17}$$

A similar transformation can be applied to Equation (11), resulting in:

$$1 - \sum_{nm\in\mathcal{L}_{i,1}} \sum_{j\in\mathcal{N}_m} 2R_{nm}(P^{\text{cha}}_{j,t} - P^{\text{dis}}_{j,t}) - V_{i,\max} \leq F_{\Xi}^{-1}(\epsilon) \tag{18}$$

In Equations (17) and (18), $F_{\Xi}^{-1}(1-\epsilon)$ and $F_{\Xi}^{-1}(\epsilon)$ can be interpreted as the quantiles of Ξ at level $1-\epsilon$ and ϵ, according to the inverse relationship between CDF and the quantile function. For simplicity of notation, we denote $F_{\Xi}^{-1}(1-\epsilon)$ and $F_{\Xi}^{-1}(\epsilon)$ by $q_{i,t}(1-\epsilon)$ and $q_{i,t}(\epsilon)$, respectively. In other words, the voltage violation probability ϵ is also the probability level that defines $q_{i,t}(\epsilon)$. Thus, the key to transforming the DESS CCED model into a directly solvable MILP problem is being able to accurately obtain quantiles of $\Xi_{i,t}$.

However, the probability distributions of bus voltage fluctuations are often complex and unknown, which is due to the network topology and random loads. In addition, analytical expressions of the quantile function may not be obtainable even though the distribution is known. Therefore, technique is needed to accurately predict the values of the voltage fluctuation probability distribution at composite quantiles without relying on the assumptions of underlying probability distribution and complex numerical derivation.

In Section 3, we will introduce a learning-driven prediction model, which leverages the strengths of the improved Transformer model for time series forecasting and the robust CQR method for multi-quantile output. The proposed prediction model can efficiently capture the bus voltage fluctuation patterns affected by temporal random loads.

3. Transformed-Based Non-Parametric Probabilistic Prediction Model

In this section, we introduce the PatchTST prediction model combined with the HuberCQR method for the estimation of $q_{i,t}(\cdot)$.

PatchTST is a deep learning model that excels at capturing complex temporal patterns in time series data. In our problem, we use the PatchTST model to learn the patterns of voltage fluctuations at each non-reference bus caused by random loads. CQR is a non-parametric statistical method that can simultaneously estimate composite quantiles of a variable. It extends traditional quantile regression, which estimates a single quantile at a time. HuberCQR is a robust improvement over the CQR method, where it introduces Huber loss to make CQR predictions less sensitive to outliers.

Specifically, in our framework, we utilize the HuberCQR method to define a loss function, which measures the difference between real data and predicted composite quantiles. The training process of the PatchTST model aims to minimize the average of this HuberCQR loss function over the entire training period. Overall, the integration of the PatchTST model and HuberCQR loss function forms a powerful framework that can capture temporal fluctuations of voltages at different buses in the distribution network, and generate composite predictions at specified probability levels.

3.1. PatchTST Prediction Model

This section introduces the framework of the PatchTST prediction model combined with the HuberCQR loss function, and then outlines its key components.

3.1.1. PatchTST Framework

The framework of the PatchTST prediction model is given in Figure 2. The model's inputs are the collection of historical data of $\Xi_{i,t}$, namely, the voltage drop fluctuations at each non-reference bus, calculated by Equation (13). The outputs are the predictions of $\Xi_{i,t}$ at different quantile levels, i.e., $q_{i,t}(\cdot)$.

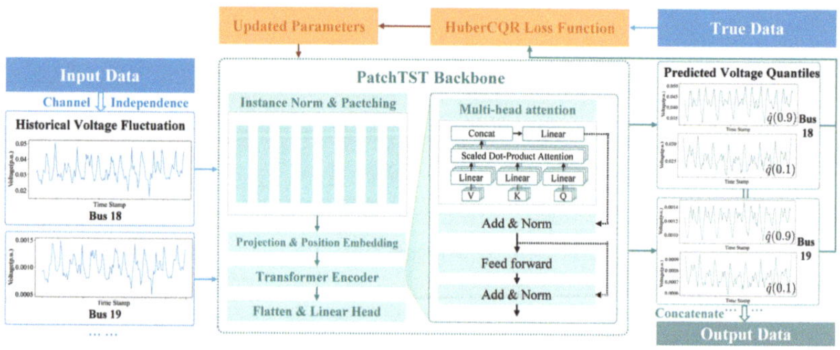

Figure 2. The framework of PatchTST.

Initially, PatchTST divides the input multi-bus voltage time series into separate channels. Then, the independent multi-bus voltage time series in each channel is normalized to ensure consistency across different scales. Following normalization, each time series is segmented into patches. After that, the projection and position embedding step projects each patch into a higher-dimensional space and integrates position embedding, to preserve the sequential context of the original time series. The Transformer Encoder, which is the essence of the classic Transformer frameworks, then analyzes patches and understands both overarching trends and fine-grained temporal dynamics in the voltage data. Further, the Flatten and Linear Head step combines the output of the Transformer Encoder by flattening it and using a linear transformation to produce accurate voltage predictions at various quantiles. Once the PatchTST backbone processes the data, the predicted voltage drops at different quantiles, and $q_{i,t}(\cdot)$ are compared against actual values using the HuberCQR loss function. By minimizing the HuberCQR loss, the model's parameters are updated, resulting in a trained PatchTST model. Finally, the well-trained PatchTST model outputs a concatenation of composite quantiles on different buses.

3.1.2. Core Components of PatchTST Backbone

PatchTST enhances prediction capabilities, mainly benefiting from three core components: channel independence process, patching process, and the multi-head attention mechanism of the Transformer backbone.

Channel Independence for Precision Analysis: Channel independence refers to the treatment of separating multivariate time series into individual channels which share the same embedding and Transformer weights. In the context of our problem, PatchTST separates the voltage fluctuation time series for each bus ($\Xi_{i,t}$) into distinct channels. This segregation allows for the generation of customized attention maps for each bus voltage, ensuring the accuracy of voltage predictions. The channel-independence model has several advantages over the channel-mixing model: (1) reducing computational complexity and improving processing speed, as the model can process each channel in parallel and a faster learning convergence rate can be achieved; (2) reducing risk of over-fitting, due to the smaller number of parameters for modeling complex interactions between different channels; and (3) increasing robustness to noise by preventing its propagation across mixed channels.

Efficient Data Segmentation with Patching: PatchTST employs a patching technique to effectively manage high-dimensional time series data. This technique divides a time series $x^{(i)} \in \mathbb{R}^{1\times L}$ into N patches of length P, denoted as $x_p^{(i)} \in \mathbb{R}^{P\times N}$, achieving a reduction in time and space complexity by a factor of stride: $N \approx L/S$. This technique not only captures the local information within each subsequence but also eases the computational and storage pressure when processing the entire series, enhancing model performance and efficiency.

Comprehensive Insight with Multi-Head Attention Mechanism: The multi-head attention mechanism in PatchTST is vital for analyzing complex dependencies in the input patches on various time scales. Specifically, the multi-head attention mechanism operates through the following steps:

1. Input Transformation: This step transforms each patch as a whole to capture different aspects of the data. For each attention head i, the entire patches represented by original queries (Q), keys (K), and values (V) are transformed by multiplying the respective weight matrices W_i^Q, W_i^K, and W_i^V. This transformation is expressed by Equation (19):

$$Q_i = QW_i^Q, \quad K_i = KW_i^K, \quad V_i = VW_i^V \tag{19}$$

Here, the transformation is applied at the patch level, treating each patch as an entity to grasp its unique characteristics and relationships with other patches.

2. Scaled Dot-Product Attention: This step assesses the relevance of each patch in relation to the others by calculating the similarity between queries and keys at the patch level. For each head i, the similarity between transformed queries Q_i and keys K_i is determined by dot products and scaling. The similarity scores for each head $\frac{Q_i K_i^T}{\sqrt{d_k}}$ are then converted into a probability distribution using the softmax function. A weighted summation is performed on the transformed values V_i based on this distribution as shown in Equation (20):

$$\text{head}_i\text{Attention}(Q_i, K_i, V_i) = \text{softmax}\left(\frac{Q_i K_i^T}{\sqrt{d_k}}\right) V_i \tag{20}$$

This process enables the model to prioritize patches based on their significance in predicting outcomes, emphasizing the importance of understanding interactions at the patch level.

3. Output Merging: By integrating insights from all heads, this step provides a comprehensive analysis that improves prediction accuracy through various temporal perspectives. The concatenated outputs of all heads are merged via an additional linear transformation W^O as illustrated by Equation (21):

$$\text{MultiHead}(Q, K, V) = \text{Concat}(\text{head}_1, \text{head}_2, \ldots, \text{head}_h)W^O, \tag{21}$$

where W^O is the weight matrix designed to combine the insights from individual patches.

3.2. HuberCQR Loss Function

As illustrated in Section 3.1.1, the loss function is the key to training the PatchTST model. In Ref. [26], where the PatchTST method was first introduced, the loss function of mean squared error was used to measure the discrepancy between the predicted values and actual values, but this design was only suitable for point predictions. In our case, we aim to obtain composite probabilistic predictions. In addition, it is significant to ensure the accuracy of the prediction throughout the distribution range for quantile predictions. Thus, we integrate the Huber loss function to provide a more reliable error metric that is less sensitive to extreme deviations. Next, we will provide a detailed description of the HuberCQR loss function.

First, the classic formula for quantile regression is [34]:

$$\min_{\beta_\epsilon} \left[\sum_{t=1}^{T} \rho_\epsilon (y_t - \hat{q}_t(\epsilon)) \right] \tag{22}$$

where t is the total training epochs, ϵ is the quantile level of interest, y_t is the true data, $\hat{q}_t(\epsilon)$ is the predicted quantile value, and $\rho_\epsilon(\cdot)$ is the quantile loss function, defined as:

$$\rho_\epsilon(u) = \begin{cases} \epsilon \cdot u & \text{if } u \geq 0 \\ (\epsilon - 1) \cdot u & \text{if } u < 0 \end{cases} \tag{23}$$

where $u := y_t - \hat{q}_t(\epsilon)$, which is the difference between the actual value y_t and the predicted quantile value $\hat{q}_t(\epsilon)$. Note that the quantile loss function applies penalties to residuals in a way that captures the asymmetry inherent in quantile estimation, giving different importance to underestimations and overestimations relative to the target quantile level.

However, in the classic quantile regression formula, the objective is to find the model parameters β_ϵ that minimize the overall loss. This allows the prediction model to estimate the single quantile of interest. CQR extends this original loss function by simultaneously estimating composite quantiles of interest. The loss function of CQR can be written as follows, which minimizes the average quantile loss [29]:

$$\min_{\beta_K} \frac{1}{KT} \sum_{k=1}^{K} \sum_{t=1}^{T} \rho_{\epsilon_k} (y_t - \hat{q}_t(\epsilon_k)) \tag{24}$$

where $\rho_{\epsilon_k}(\cdot)$ measures the quantile loss for the ϵ_k-th quantile $\hat{q}_t(\epsilon_k)$, and β_K represents the optimal parameters set of prediction model.

Although the CQR loss function addresses the concern of estimating composite quantiles simultaneously, it may also face the challenge of obtaining skewed quantile estimates due to outliers. Therefore, we combine the CQR loss function with the Huber loss function to tackle this challenge. Huber loss, with its dual approach of applying squared loss for smaller errors and linear loss for larger errors, can effectively reduce the impact of outliers.

The formula for Huber loss is as follows [28]:

$$H_\delta(d) = \begin{cases} \frac{1}{2} d^2 & \text{if } |d| \leq \delta \\ \delta \left(|d| - \frac{1}{2}\delta \right) & \text{otherwise.} \end{cases} \tag{25}$$

where δ is a tuning parameter that determines the threshold between utilizing squared loss or linear loss, which balances the trade-off between robustness to outliers and sensitivity to small prediction losses; d is the difference between the true value and the prediction.

Then, the HuberCQR loss is formulated as Equation (26), through replacing the quantile loss $\rho_{\epsilon_k}(y_t - \hat{q}_t(\epsilon_k))$ in Equation (24) with Huber loss $H_\delta(y_t - \hat{q}_t(\epsilon_k))$ of predicted quantiles $\hat{q}_t(\epsilon_k), k \in \{1, \cdots, K\}$:

$$\frac{1}{KT}\sum_{k=1}^{K}\sum_{t=1}^{T} H_\delta(y_t - \hat{q}_t(\epsilon_k)) \tag{26}$$

In addition, the HuberCQR loss in each channel needs to be gathered and averaged to obtain the overall target loss:

$$\min_{\beta} \frac{1}{NKT}\sum_{i=1}^{N}\sum_{k=1}^{K}\sum_{t=1}^{T} H_\delta(y_{i,t} - \hat{q}_{i,t}(\epsilon_k)) \tag{27}$$

Through training the PatchTST prediction model by loss function (27) and updating the model parameters $\boldsymbol{\beta}$, we achieve an optimized fit to the historical data, enabling the model to accurately predict bus voltage fluctuations at multiple quantiles in the distribution network.

4. Case Study

Numerical tests are conducted on the IEEE 33-bus radial distribution network. The topology of the distribution network is illustrated in Figure 1, where the nominal voltage and the base power are 12.66 kV and 1 MVA, respectively. The minimum and maximum voltage thresholds for each non-reference bus $V_{i,\min}$ and $V_{i,\max}$ are set to 0.95 p.u. and 1.05 p.u., respectively. Additional parameters of this network are available in [30]. To simulate dynamic loads, we aggregated hourly electricity consumption data from encrypted smart meters of users provided by the Spanish company GoiEner for each bus. This dataset [35], which was publicly released in January 2024, includes load data for various types of users, i.e., industrial, commercial, and residential. We extract the load data from 1 June 2021, 00:00 to 31 May 2022, 23:00, ensuring that the average load level matches the static load in the original network model for each bus, and the ratio of reactive power to active power is maintained.

The simulation process and the input/output data at each step are as follows. All simulations were run on a personal laptop with an Apple M2 CPU and 8 GB RAM.

Step 1: Based on the IEEE 33-bus distribution network parameters and the dynamic load data of each bus, we can compute the hourly bus voltage drop magnitude $\Xi_{i,t}$ caused by the random loads throughout the year using Equation (13).

Step 2: The calculated $\Xi_{i,t}$ from 1 June 2021 00:00 to 25 May 2022 23:00 serves as input for training the PatchTST model (with the last week of May 2022 as the test set). The hyperparameters of the PatchTST model can be found in Table 1. Additionally, the PatchTST model utilizes two loss functions, one grounded in the Gaussian Mixture Model (GMM) and the other based on the proposed HuberCQR model, for comparison purposes. GMM is a powerful parametric model that can simulate arbitrary probability distributions by combining a finite number of Gaussian components. In the simulations, the tuning parameter δ for the HuberCQR loss function is set to 0.001, and the number of mixture components for GMM loss function is set to 3. The parameters N, K and T of Equation (27) are 32, 2, and 24, respectively.

Table 1. Hyperparameters of PatchTST prediction model.

Forecast Horizon	Autoregressive Inputs Size	Patch Length	Stride of Patch	Hidden Layer Size	Number of Multi-Head	Learning Rate
168	24	8	8	64	64	0.005

Step 3: Based on the trained models under the two loss functions, with the probability of voltage exceeding both upper and lower thresholds set to less than $\epsilon = 0.1$, the voltage drop values at different quantile levels, i.e., $q_{i,t}(0.1)$ and $q_{i,t}(0.9)$, can be obtained.

Step 4: We input $q_{i,t}(0.1)$ and $q_{i,t}(0.9)$ obtained in Step 3 into the DESS CCED model after a deterministic transformation. With the DESS parameters given in Table 2 [36], we can finally output the optimal DESS scheduling scheme.

Table 2. Operational parameters of DESS.

$P^{cha}_{i,max}/P^{dis}_{i,max}$	$[SOC_{i,min}, SOC_{i,max}]$	η	E_b	λ
0.6 p.u.	[0.2, 0.9]	90%	4 p.u.	4690 $/MWh

4.1. Comparison of Prediction Accuracy

Take 26 May 2022 as the test/scheduling day, and use the trained PatchTST model to predict the hourly voltage drop range caused by random loads. The theoretical probability range is set to 80%, formed by $q_{i,t}(0.1)$ and $q_{i,t}(0.9)$.

The coverage rate (CR) is a metric that measures how well the predicted interval captures the true values. Specifically, we denote the proportion of true values that fall within the predicted interval $[q_{i,t}(0.1), q_{i,t}(0.9)]$ by CR_{actual}, and the target coverage rate by $CR_{\text{theoretical}}$, which is 80% in this case.

Figure 3 compares the Coverage Rate Deviation Ratio (CRDR) for the PatchTST model with the HuberCQR loss function and the GMM loss function. The CRDR is defined by Equation (28). A lower CRDR indicates that the actual coverage rate achieved by the model is closer to the theoretical 80% target, meaning the model is making more accurate predictions of the voltage fluctuation range:

$$CR_{\text{Diff}} = \frac{|CR_{\text{actual}} - CR_{\text{theoretical}}|}{CR_{\text{theoretical}}} \tag{28}$$

The figure shows that the CRDR for the HuberCQR-based PatchTST model is consistently lower across all non-reference buses compared to the GMM-based model. This suggests that the HuberCQR-based model performs better in prediction accuracy. This is crucial for making precise dispatch decisions for DESS, since the predicted $q_{i,t}(0.1)$ and $q_{i,t}(0.9)$ directly influence the operational limits of DESS through Equations (17) and (18).

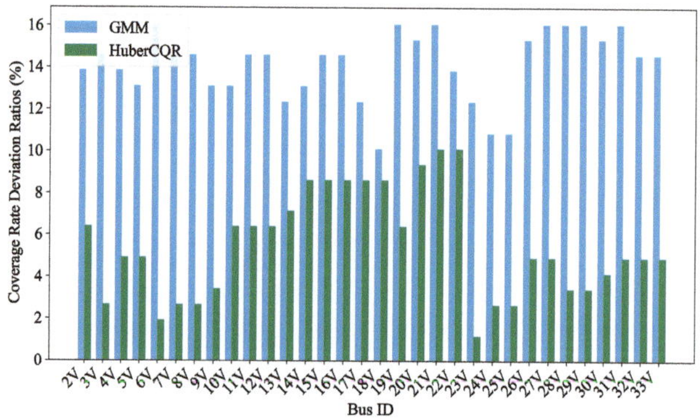

Figure 3. Coverage rate comparison under the two prediction models.

4.2. Comparison of DESS Dispatch Results

After evaluating the accuracy of the coverage rates between two probabilistic prediction models, we utilize the predictions to guide the economic scheduling of DESS and compare the cost-effectiveness of the resulting schedules. Specifically, the predicted quantiles $q_{i,t}(\epsilon)$ and $q_{i,t}(1-\epsilon)$ are substituted into the DESS CCED model, and the chance constraints for addressing the probabilities of bus voltage violations can be transformed into solvable deterministic constraints.

In the simulation, DESS integration bus is set to #18. Figure 4 shows that the more accurate estimation of bus voltage fluctuations under the HuberCQR-based PatchTST model leads to a lower DOD for the DESS, implying less degradation. The dispatch-related operating costs of the DESS further corroborate this point. Specifically, the daily dispatch cost based on the PatchTST model with the HuberCQR loss function is $1507.2 lower (a 6.23% reduction) compared to the cost based on the PatchTST model with the GMM loss function. The reduced operating cost highlights the practical benefits of the HuberCQR-based PatchTST model in optimizing DESS dispatch under uncertainty.

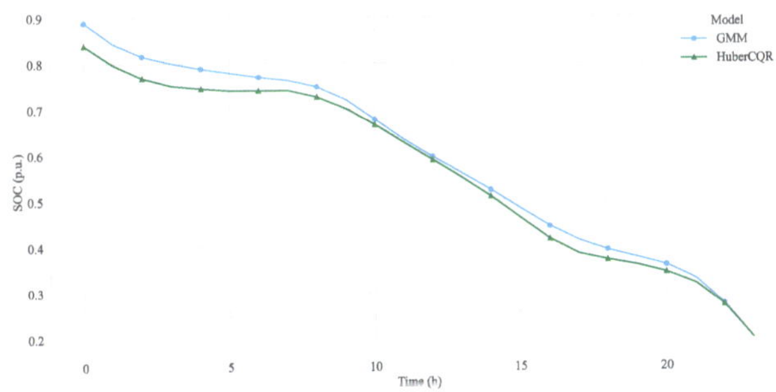

Figure 4. DESS SOC dispatch results at bus #18.

4.3. Comparison of Bus Voltage

After obtaining the DESS scheduling results based on the two prediction models, we can compare the bus voltage conditions by combining the DESS scheduling results and the actual load data to verify whether the proposed methods have handled the chance constraints properly. Figure 5 shows the voltage conditions of bus #18 in the cases without DESS and with DESS dispatched based on the two prediction models. It can be seen that without DESS, the bus voltage is below the threshold most of the time. In contrast, with dispatching DESS under the two models, the probability of bus voltage exceeding the limits is less than setting $\epsilon = 0.1$, which is actually 0.04. This result reflects that both prediction models can well predict the risk of voltage fluctuation and fully utilize the capability of DESS in voltage management. Combining the results from the Section 4.2, it can be seen that the proposed scheduling scheme based on the HuberCQR prediction results can achieve the same effectiveness of voltage management as the scheme based on the GMM model, but with lower DESS scheduling costs.

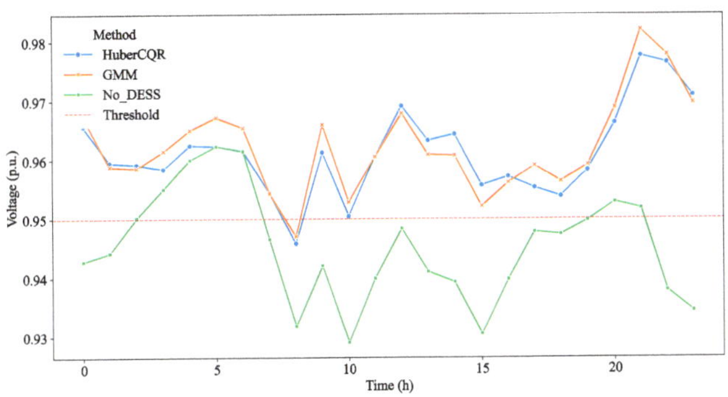

Figure 5. Voltage at bus #18.

Furthermore, we analyze the 3D surface plots of the voltage levels under the cases under the HuberCQR-based and GMM-based model at different buses and time steps. It is observed from Figure 6 that the surface plot with the HuberCQR model shows a relatively smooth surface with fewer peaks and valleys, indicating that the predicted voltages are more stable across different buses and times. In contrast, the plot of the GMM model exhibits a more rugged surface with more pronounced peaks and valleys, suggesting that this model has a greater variance in voltage prediction. This result verifies that the performance of the HuberCQR-based model is consistent with its theoretical design, and it is more robust in predicting voltage fluctuations, with stronger ability to resist outlier risks.

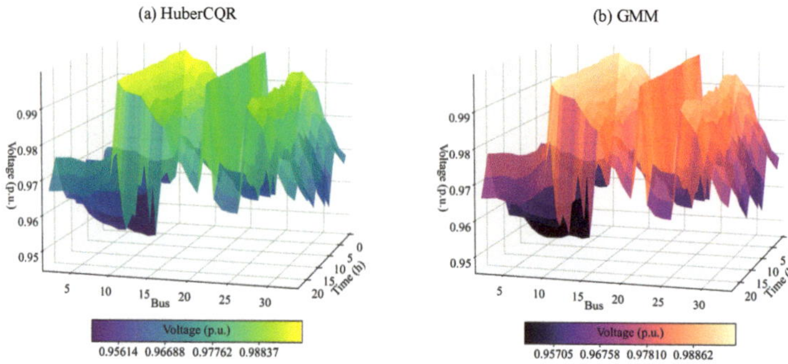

Figure 6. Three-dimensional surface plots of the voltage levels under the two prediction models.

5. Conclusions

As the types and scales of loads continue to increase, voltage issues in distribution networks become more pronounced. DESS can significantly mitigate the gradually intensifying voltage violation problems in distribution networks. Due to randomness and uncertainty of loads, the scheduling of DESS requires accurate prediction of the potential range of voltage fluctuations caused by random loads. Hence, this paper proposes a framework that combines deep learning with non-parametric probabilistic prediction method. Specifically, by utilizing a Transformer-based time series prediction model and an improved composite quantile regression technique, the DESS CCED problem considering voltage safety can be simplified into a feasible MILP problem, without the need for preset probability distributions of random variables and complex computations. Numerical experiments show that under the same voltage risk management effectiveness, the dispatching cost of DESS based on the proposed non-parametric probabilistic prediction model is lower than that based on state-of-the-art parametric models. Overall, this paper provides an efficient and economical solution for DESS dispatch considering load uncertainty and distribution network voltage safety. In the future, we hope to explore how prediction accuracy of the proposed model impacts the DESS scheduling results.

Author Contributions: Conceptualization, X.C. and T.Q.; methodology, X.C. and Y.Z.; software, X.C.; validation, X.C.; formal analysis, X.C. and Y.G.; investigation, X.C.; resources, X.C.; data curation, X.C.; writing—original draft preparation, X.C.; writing—review and editing, Y.G., Y.Z. and T.Q.; visualization, X.C.; supervision, Y.Z. and T.Q. All authors have read and agreed to the published version of the manuscript.

Funding: This research was funded by Jiangsu Provincial Key Research and Development Program (BE2020081-2), and by the Jiangsu Provincial Key Laboratory Project of Smart Grid Technology and Equipment.

Data Availability Statement: The code used in this study is available from the authors upon request.

Conflicts of Interest: The authors declare no conflicts of interest.

References

1. Murray, W.; Adonis, M.; Raji, A. Voltage Control in Future Electrical Distribution Networks. *Renew. Sustain. Energy Rev.* **2021**, *146*, 111100. [CrossRef]
2. Bendík, J.; Cenký, M.; Cintula, B.; Beláň, A.; Eleschová, Ž.; Janiga, P. Stochastic Approach for Increasing the PV Hosting Capacity of a Low-Voltage Distribution Network. *Processes* **2023**, *11*, 9. [CrossRef]
3. Jabir, H.J.; Teh, J.; Ishak, D.; Abunima, H. Impacts of Demand-Side Management on Electrical Power Systems: A Review. *Energies* **2018**, *11*, 1050. [CrossRef]
4. Abou El-Ela, A.A.; El-Sehiemy, R.A.; Salah Ali, E.; Kinawy, A.M. Minimisation of Voltage Fluctuation Resulted from Renewable Energy Sources Uncertainty in Distribution Systems. *IET Gener. Transm. Distrib.* **2019**, *13*, 2339–2351. [CrossRef]
5. Zhang, D.; Li, J.; Hui, D. Coordinated Control for Voltage Regulation of Distribution Network Voltage Regulation by Distributed Energy Storage Systems. *Prot. Control Mod. Power Syst.* **2018**, *3*, 1–8. [CrossRef]
6. Li, X.; Ma, R.; Gan, W.; Yan, S. Optimal Dispatch for Battery Energy Storage Station in Distribution Network Considering Voltage Distribution Improvement and Peak Load Shifting. *J. Mod. Power Syst. Clean Energy* **2020**, *10*, 131–139. [CrossRef]
7. Han, R.; Hu, Q.; Cui, H.; Chen, T.; Quan, X.; Wu, Z. An optimal bidding and scheduling method for load service entities considering demand response uncertainty. *Appl. Energy* **2022**, *328*, 120167. [CrossRef]
8. Kheirkhah, A.R.; Meschini Almeida, C.F.; Kagan, N.; Leite, J.B. Optimal Probabilistic Allocation of Photovoltaic Distributed Generation: Proposing a Scenario-Based Stochastic Programming Model. *Energies* **2023**, *16*, 7261. [CrossRef]
9. Ramadan, A.; Ebeed, M.; Kamel, S.; Abdelaziz, A.Y.; Haes Alhelou, H. Scenario-Based Stochastic Framework for Optimal Planning of Distribution Systems Including Renewable-Based DG Units. *Sustainability* **2021**, *13*, 3566. [CrossRef]
10. Jeddi, B.; Vahidinasab, V.; Ramezanpour, P.; Aghaei, J.; Shafie-khah, M.; Catalão, J.P. Robust Optimization Framework for Dynamic Distributed Energy Resources Planning in Distribution Networks. *Int. J. Electr. Power Energy Syst.* **2019**, *110*, 419–433. [CrossRef]
11. Sun, Q.; Chen, Q. Fully Decentralized Robust Modelling and Optimization of Radial Distribution Networks Considering Uncertainties. *IEEE Trans. Smart Grid* **2021**, *13*, 1012–1022. [CrossRef]
12. Cao, Y.; Tan, Y.; Li, C.; Rehtanz, C. Chance-Constrained Optimization-Based Unbalanced Optimal Power Flow for Radial Distribution Networks. *IEEE Trans. Power Deliv.* **2013**, *28*, 1855–1864.
13. Nazir, F.U.; Pal, B.C.; Jabr, R.A. A Two-Stage Chance Constrained Volt/Var Control Scheme for Active Distribution Networks with Nodal Power Uncertainties. *IEEE Trans. Power Syst.* **2018**, *34*, 314–325. [CrossRef]
14. Zhang, J.; Cheng, C.; Yu, S.; Su, H. Chance-Constrained Co-Optimization for Day-Ahead Generation and Reserve Scheduling of Cascade Hydropower—Variable Renewable Energy Hybrid Systems. *Appl. Energy* **2022**, *324*, 119732. [CrossRef]
15. Zhang, Z.S.; Sun, Y.Z.; Gao, D.W.; Lin, J.; Cheng, L. A Versatile Probability Distribution Model for Wind Power Forecast Errors and Its Application in Economic Dispatch. *IEEE Trans. Power Syst.* **2013**, *28*, 3114–3125. [CrossRef]
16. Yu, J.; Li, Z.; Guo, Y.; Sun, H. Decentralized Chance-Constrained Economic Dispatch for Integrated Transmission-District Energy Systems. *IEEE Trans. Smart Grid* **2019**, *10*, 6724–6734. [CrossRef]
17. Wan, C.; Lin, J.; Wang, J.; Song, Y.; Dong, Z.Y. Direct Quantile Regression for Nonparametric Probabilistic Forecasting of Wind Power Generation. *IEEE Trans. Power Syst.* **2016**, *32*, 2767–2778. [CrossRef]
18. Akhavan-Hejazi, H.; Mohsenian-Rad, H. Energy Storage Planning in Active Distribution Grids: A Chance-Constrained Optimization with Non-Parametric Probability Functions. *IEEE Trans. Smart Grid* **2016**, *9*, 1972–1985.
19. Qian, T.; Ming, W.; Shao, C.; Hu, Q.; Wang, X.; Wu, J.; Wu, Z. An Edge Intelligence-Based Framework for Online Scheduling of Soft Open Points With Energy Storage. *IEEE Trans. Smart Grid* 2023, *Early Access*. [CrossRef]
20. Qian, T.; Shao, C.; Wang, X.; Shahidehpour, M. Deep reinforcement learning for EV charging navigation by coordinating smart grid and intelligent transportation system. *IEEE Trans. Smart Grid* **2019**, *11*, 1714–1723. [CrossRef]
21. Qian, T.; Shao, C.; Shi, D.; Wang, X.; Wang, X. Automatically Improved VCG Mechanism for Local Energy Markets via Deep Learning. *IEEE Trans. Smart Grid* **2021**, *13*, 1261–1272. [CrossRef]
22. Shi, H.; Xu, M.; Li, R. Deep Learning for Household Load Forecasting-A Novel Pooling Deep RNN. *IEEE Trans. Smart Grid* **2017**, *9*, 5271–5280. [CrossRef]
23. Zhu, J.; Yang, Z.; Mourshed, M.; Guo, Y.; Zhou, Y.; Chang, Y.; Wei, Y.; Feng, S. Electric Vehicle Charging Load Forecasting: A Comparative Study of Deep Learning Approaches. *Energies* **2019**, *12*, 2692. [CrossRef]
24. Li, Z.; Li, Y.; Liu, Y.; Wang, P.; Lu, R.; Gooi, H.B. Deep Learning Based Densely Connected Network for Load Forecasting. *IEEE Trans. Power Syst.* **2020**, *36*, 2829–2840. [CrossRef]
25. Zhang, Y.; Qian, W.; Ye, Y.; Li, Y.; Tang, Y.; Long, Y.; Duan, M. A Novel Non-Intrusive Load Monitoring Method Based on ResNet-Seq2Seq Networks for Energy Disaggregation of Distributed Energy Resources Integrated with Residential Houses. *Appl. Energy* **2023**, *349*, 121703. [CrossRef]
26. Nie, Y.; Nguyen, N.H.; Sinthong, P.; Kalagnanam, J. A Time Series Is Worth 64 Words: Long-Term Forecasting with Transformers. *arXiv* **2022**, arXiv:2211.14730.
27. Meyer G.P. An alternative probabilistic interpretation of the Huber loss. In Proceedings of the IEEE/CVF Conference on Computer Vision and Pattern Recognition (CVPR), Online, 19–25 June 2021; pp. 5261–5269.
28. Huber, P.J. Robust estimation of a location parameter. In *Breakthroughs in Statistics: Methodology and Distribution*; Springer: New York, NY, USA, 1992; pp. 492–518.

29. Zou, H.; Yuan, M. Composite Quantile Regression and the Oracle Model Selection Theory. *Ann. Statist.* **2008**, *36*, 1108–1126. [CrossRef]
30. Kashem, M.A.; Ganapathy, V.; Jasmon, G.B.; Buhari, M.I. A Novel Method for Loss Minimization in Distribution Networks. In Proceedings of the DRPT2000. International Conference on Electric Utility Deregulation and Restructuring and Power Technologies, London, UK, 4–7 April 2000; pp. 251–256.
31. Baran, M.E.; Wu, F.F. Network Reconfiguration in Distribution Systems for Loss Reduction and Load Balancing. *IEEE Trans. Power Deliv.* **1989**, *4*, 1401–1407. [CrossRef]
32. Farivar, M.; Chen, L.; Low, S. Equilibrium and dynamics of local voltage control in distribution systems. In Proceedings of the 52nd IEEE Conference on Decision and Control, Firenze, Italy, 10–13 December 2013; pp. 4329–4334.
33. Zheng, Y.; Zhao, J.; Song, Y.; Luo, F.; Meng, K.; Qiu, J.; Hill, D.J. Optimal Operation of Battery Energy Storage System Considering Distribution System Uncertainty. *IEEE Trans. Sustain. Energy* **2017**, *9*, 1051–1060. [CrossRef]
34. Koenker, R.; Bassett, G., Jr. Regression Quantiles. *Econom. J. Econ. Soc.* **1978**, *46*, 33–50. [CrossRef]
35. Quesada, C.; Astigarraga, L.; Merveille, C.; Borges, C.E. An electricity smart meter dataset of Spanish households: Insights into consumption patterns. *Sci Data* **2024**, *11*, 59. [CrossRef] [PubMed]
36. Liu, G.; Sun, W.; Hong, H.; Shi, G. Coordinated Configuration of SOPs and DESSs in an Active Distribution Network Considering Social Welfare Maximization. *Sustainability* **2024**, *16*, 2247. [CrossRef]

processes
MDPI

Article

Maximum Power Point Tracking Constraint Conditions and Two Control Methods for Isolated Photovoltaic Systems

Jingxun Fan [1,2], Shaowu Li [1,*], Sanjun Liu [1,*], Xiaoqing Deng [1,3] and Xianping Zhu [1,4]

1 College of Intelligent Systems Science and Engineering, Hubei Minzu University, Enshi 445000, China; jingxun_fan@163.com (J.F.)
2 Jiangmen Pengjiang Power Supply Bureau of Guangdong Power Grid Co., Ltd., Jiangmen 529000, China
3 Hubei Chuangsinuo Electrical Technology Corp., Enshi 445000, China
4 Guilin Power Supply Bureau of Guangxi Power Grid Co., Ltd., Guilin 541000, China
* Correspondence: xidu_surfer@163.com (S.L.); liusanjinbox1@126.com (S.L.)

Abstract: A maximum power point (MPP) always exists in photovoltaic (PV) cells, but a mismatch between PV system circuit parameters, weather conditions and system structure leads to the possibility that the MPP may not be tracked successfully. In addition, the introduction of an isolation transformer into a basic PV system allows for moderate values of the converter duty cycle and electrical isolation. However, there is no comprehensive research on MPPT (maximum power point tracking) constraint conditions for different isolated PV systems, which seriously hinders the application of isolated PV systems and the development of a related linear control theory. Therefore, in this paper, the overall mathematical models of different isolated PV systems are first established based on the PV cell engineering model and the MPP linear model, and then, two sets of constraint conditions are found for the successful realization of MPPT. These MPPT constraint conditions (MCCs) describe in detail the direct mathematical relationships between PV cell parameters, weather conditions and circuit parameters. Finally, based on the MPP linear model and MCCs, two new MPPT methods are designed for isolated PV systems. Considering the MCCs proposed in this paper, a suitable range of load and transformer ratios can be estimated from the measured data of irradiance and temperature in a certain area, and the range of MPPs existing in PV systems with different structures can be estimated, which is a good guide for circuit design, theoretical derivation and product selection for PV systems. Meanwhile, comparative experiments confirm the rapidity and accuracy of the two proposed MPPT methods, with the MPPT time improving from 0.23 s to 0.03 s, and they have the advantages of a simple program, small computational volume and low hardware cost.

Keywords: isolated PV system; MPPT constraint conditions; linear cell model

Citation: Fan, J.; Li, S.; Liu, S.; Deng, X.; Zhu, X. Maximum Power Point Tracking Constraint Conditions and Two Control Methods for Isolated Photovoltaic Systems. *Processes* **2023**, *11*, 3245. https://doi.org/10.3390/pr11113245

Academic Editors: Chenyu Wu, Zhongkai Yi and Chenhui Lin

Received: 30 September 2023
Revised: 27 October 2023
Accepted: 8 November 2023
Published: 17 November 2023

1. Introduction

To carry out a theoretical analysis and practical verification of a PV system, an accurate model of the PV cell should be established first. Nowadays, a large number of studies on PV systems and PV cells are carried out, and they have led to a lot of breakthroughs and innovations in mathematical and circuit model optimization, as well as MPPT and parameter extraction methods for PV cells. However, the model used cannot be completely compatible with the required accuracy, the complexity of the calculations and the environmental conditions [1]. There are nine commonly used circuit models and mathematical models of PV cells categorized in Ref. [2], which can accurately reflect the output characteristics of PV cells but are not convenient for engineering applications, so simplified engineering models of PV cells have been widely investigated [3]. Many scholars have investigated how to model PV cells using four important parameters (I_{sc}, V_{oc}, I_m and V_m) provided by manufacturers and, based on the derivation of the circuit model, to simplify the modeling process, which is called engineering modeling. Under standard test conditions (STC; solar irradiance S is 1000 W/m^2, and PV cell temperature T is 25 °C), the PV cell engineering model is obtained using Equations (1)–(3), where I, V, I_{sc}, V_{oc}, I_m and V_m represent the output current, voltage of the PV cell, short-circuit current, open-circuit voltage, MPP current and voltage of the PV cell at STC, respectively [4].

$$I = I_{sc}\left[1 - C_1\left(e^{\frac{V}{C_2 V_{oc}}} - 1\right)\right] \tag{1}$$

$$C_1 = \left(1 - \frac{I_m}{I_{sc}}\right)e^{-\frac{V_m}{C_2 V_{oc}}} \tag{2}$$

$$C_2 = \frac{\frac{V_m}{V_{oc}} - 1}{\ln(1 - \frac{I_m}{I_{sc}})} \tag{3}$$

However, when there are obstacles such as tall buildings and trees, the illumination of PV modules is no longer uniform, resulting in partial shadow problems in which the power curve has multiple peaks. So, it is necessary to establish a PV model under partial shadow conditions and to simulate and analyze its output characteristics [5,6]. This is a steady-state model of PV cells, but MPPT is a dynamic optimization process, so the dynamic characteristics of PV cells have also been studied in a number of ways [7]. All of the above models are nonlinear models of PV cells, which require complex iterations and calculations to extract parameters and conduct studies, so scholars have proposed some linearized models, such as segmented linear models, which replace the nonlinear PV relationship with multi-segmented linear equations through segmented linearization [8]. In [9], the authors proposed a new segmented linear shunt branch model that approximates the nonlinear *I-V* curve of a PV cell via an equivalent circuit. The segmented linearized model simplifies the workload in the nonlinear PV cell model and obtains comparable accuracy under certain conditions, but the number of segments must be increased in the segmented linear model if higher accuracy is required, which undoubtedly increases the computational complexity. The authors of [10] derived a linearized model that relates changes in the inputs to the system, such as irradiance and temperature, to its outputs, such as the array current and power. The authors of [11] derived a set of nonlinear state-space equations based on the average switching technique, which was implemented using MATLAB2016b. The authors of [12] linearized the voltage–current characteristics of PV cells at the MPP in order to completely remove the obstacle of nonlinear PV cells to the overall linearization of the PV system by proposing two equivalent linear models, the Thevenin equivalent model and the Norton equivalent model, as shown in Figure 1. In contrast, the MPP linear model can better overcome these problems in the segmented linear model. On this basis, it is feasible and reasonable to linearize the PV system as a whole, and the PV system can be conveniently studied using the traditional linear theory or law.

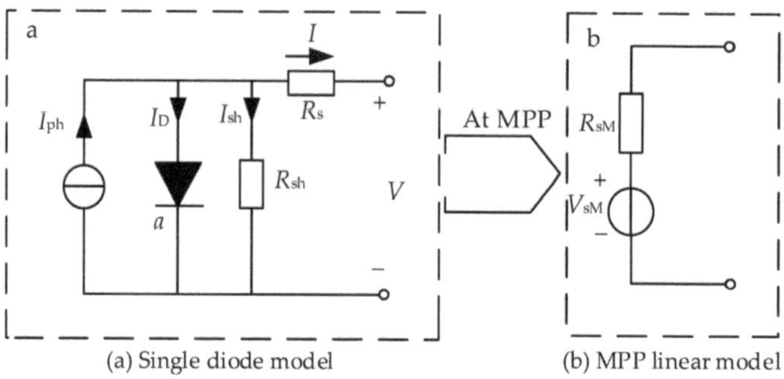

Figure 1. Relationship between single-diode model and MPP linear model.

The DC/DC converters in PV systems are categorized into non-isolated and isolated DC/DC converters. Non-isolated DC/DC converters, such as buck, boost, buck-boost and Sepic converters, are widely used as the MPPT control circuits of PV systems. Isolated DC/DC converters usually include forward, flyback, push–pull, half-bridge and full-bridge converters. The introduction of an isolation transformer into a basic non-isolated DC/DC converter can realize electrical isolation between the converter's input power supply and load. Meanwhile, it can improve the safety and reliability of converter operation and electromagnetic compatibility. In addition, it can make the duty cycle of the DC/DC converter change near a moderate value. Usually, in this case, a high boosted voltage can be achieved by using a high-transformation-ratio transformer and a voltage multiplier [13]. The analysis shown in Reference [14] verifies that isolation not only ensures safety but also increases the MPPT capability. Meanwhile, it shows that isolated converters have the highest MPPT capability without considering the hardware implementation.

At present, MPPT methods can be classified into five categories: (1) classical methods, such as perturbation observation, constant voltage and conductance increment methods [15]; (2) intelligent methods, such as artificial neural networks (ANNs), fuzzy logic controllers (FLCs) and sliding-mode control (SMC) [16,17]; (3) optimization methods, such as cuckoo search (CS), the particle swarm algorithm (PSO), the gray wolf algorithm (GWO), the ant colony algorithm (ACO) and the artificial bee colony algorithm (ABC) [18,19]; (4) hybrid methods, such as fuzzy particle swarm optimization (FPSO) and the adaptive neuro-fuzzy inference system (ANFIS) [20]; (5) other methods, such as the variable-weather parameter (VWP) method [21]. Under specific environmental conditions and requirements, good performance can be obtained with all five of the above-mentioned MPPT methods. However, the nonlinear model of the PV cell is one of the fundamental reasons why the linear control theory cannot be widely applied in the MPPT control of PV systems at present. And since the MPP must always exist in the process of use, it is easy to cause errors if its constraints are not analyzed. In order to solve this problem, some expressions have been proposed in Reference [22] to ensure the existence of the MPP in PV systems with buck, boost, buck/boost and other non-isolated DC/DC converters.

Therefore, the research objective of this paper is as follows: to find the relationship between the circuit parameters and the control signals of an isolated PV system by directly utilizing the weather conditions so as to find the range of circuit parameters for which it is capable of successful MPPT control and, accordingly, to propose two new MPPT methods.

The innovations and contributions of this work are as follows:

(1) The mathematical models of isolated PV systems are established, and the mathematical relationships between the output power of the PV systems and the weather conditions are found.
(2) The MCCs of isolated PV systems are found based on the engineering model and the MPP linear model. The relationships between MCCs and the weather conditions, circuit parameters and system structure are obtained.
(3) The practicality of the MPPT control algorithm can be enhanced. The problem of MPPT failure can be avoided by fully considering the MCCs in the design and improvement of the MPPT algorithm. Therefore, two MPPT methods, which are applicable to different PV system structures, are proposed to improve the stability, applicability and rapidity of MPPT control.

The section arrangement of this paper is as follows: Two MPPT constraint conditions and two new MPPT methods are presented in Section 2. Some simulation experiments are presented in Section 3. Finally, a discussion and some conclusions are given in Sections 4 and 5, respectively.

2. Materials and Methods

2.1. Integrative Model of Isolated PV Systems

The structure of the isolated PV system is shown in Figure 2. I and V denote the output current and voltage of the PV cell, respectively. I_o and V_o denote the output current and output voltage of the isolated DC/DC converter, respectively. R_i and R_L denote the equivalent resistances after the PV cell and after the isolated DC/DC converter, respectively.

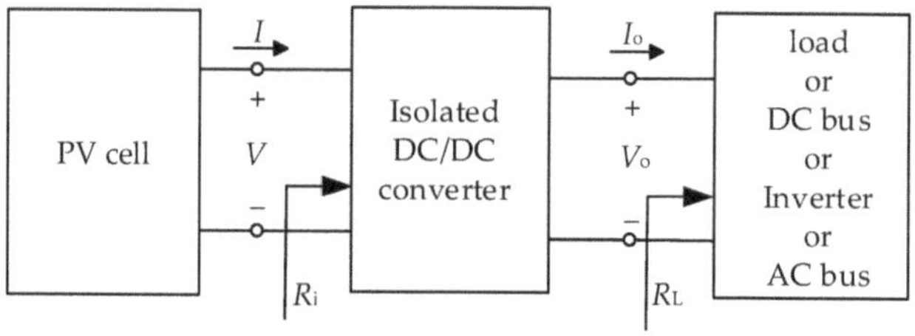

Figure 2. Isolated PV system structure.

The basic circuits of isolated DC/DC converters include the forward converter, flyback converter, half-bridge converter, full-bridge converter and push–pull converter. They are associated with the PV cell to produce the PV-Forward system, PV-Flyback system, PV-Half-bridge system, PV-Full-bridge system and PV-push–pull system, respectively. The isolated DC/DC converter is generally connected to a resistor, DC bus, inverter or AC bus (shown in Figure 3). The different system structures also lead to differences in the mathematical model and MPPT method.

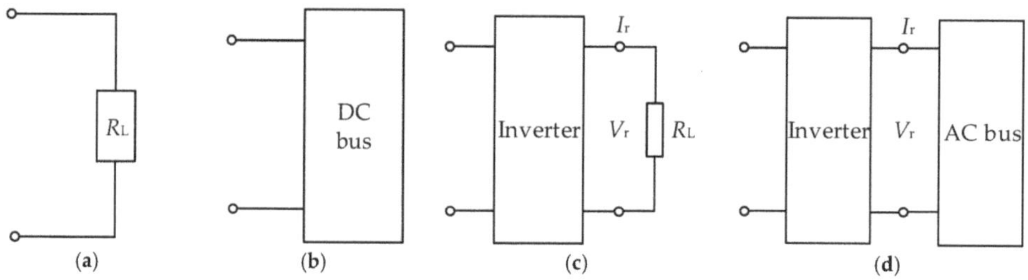

Figure 3. Four types of output. (**a**) Load; (**b**) DC bus; (**c**) inverter; (**d**) AC bus.

In order to derive a theoretical mathematical model, two assumptions need to be made for isolated PV systems:

(1) All circuit components are ideal;
(2) The isolated DC/DC converter operates in the continuous-current mode (CCM).

Firstly, according to Figure 2, it can be obtained by the power balance relationship:

$$VI = V_o I_o = P_o \tag{4}$$

$$R_i = \frac{V}{I} \tag{5}$$

$$R_L = \frac{V_o}{I_o} \tag{6}$$

P_o denotes the output power of the PV system.

The input-and-output-voltage relationships of forward, flyback, half-bridge, full-bridge and push–pull converters can be expressed by Equations (7)–(11), respectively [23]. D denotes the duty cycle of the PWM wave for the isolated DC/DC converter, and the isolation transformer ratio n is equal to N_1/N_2.

$$V_o = \frac{DV}{n} \tag{7}$$

$$V_o = \frac{DV}{n(1-D)} \tag{8}$$

$$V_o = \frac{DV}{n} \tag{9}$$

$$V_o = \frac{2DV}{n} \tag{10}$$

$$V_o = \frac{DV}{n} \tag{11}$$

It can be seen that Equations (7), (9) and (11) are the same, which means that the input–output-voltage relationships are the same for forward, half-bridge and push–pull converters.

According to Figure 2, Equation (12) is satisfied.

$$P_o = \frac{V_o^2}{R_L} \tag{12}$$

The mathematical model of the PV-Forward system can be obtained by combining Equations (1), (4), (7) and (12).

$$P_o = \frac{n^2 R_L I_{sc}^2}{D^2}\left[1 - C_1\left(e^{\frac{n\sqrt{P_o R_L}}{C_2 D V_{oc}}} - 1\right)\right]^2 \tag{13}$$

Since the forward, half-bridge and push–pull converters have the same input–output-voltage relationships, the mathematical models of the PV-Forward, PV-Half-bridge and PV-Push–pull systems are also the same, all of which are expressed in Equation (13) and will not be repeated below.

Similarly, the mathematical models of the PV-Flyback and PV-Full-bridge systems can also be obtained.

$$P_o = \frac{n^2 R_L I_{sc}^2 (1-D)^2}{D^2}\left[1 - C_1\left(e^{\frac{n(1-D)\sqrt{P_o R_L}}{C_2 D V_{oc}}} - 1\right)\right]^2 \tag{14}$$

$$P_o = \frac{n^2 R_L I_{sc}^2}{2D^2}\left[1 - C_1\left(e^{\frac{n\sqrt{P_o R_L}}{2C_2 D V_{oc}}} - 1\right)\right]^2 \tag{15}$$

For the DC bus, Equation (16) is satisfied.

$$V_o = V_{Dbus} \tag{16}$$

The mathematical model of the PV-Forward-Dbus system can be obtained by combining Equations (1), (4), (7) and (16).

$$P_o = \frac{n V_{Dbus} I_{sc}}{D}\left[1 - C_1\left(e^{\frac{n V_{Dbus}}{C_2 D V_{oc}}} - 1\right)\right] \tag{17}$$

Similarly, the mathematical models of the PV-Flyback-Dbus and PV-Full-bridge-Dbus systems can also be obtained.

$$P_o = \frac{n V_{Dbus} I_{sc} (1-D)}{D}\left[1 - C_1\left(e^{\frac{n V_{Dbus}(1-D)}{C_2 D V_{oc}}} - 1\right)\right] \tag{18}$$

$$P_o = \frac{n V_{Dbus} I_{sc}}{2D}\left[1 - C_1\left(e^{\frac{n V_{Dbus}}{2C_2 D V_{oc}}} - 1\right)\right] \tag{19}$$

The mathematical models of the inverter (SPWM control) and AC load can be represented by Equations (20) and (21), respectively. M denotes the SPWM wave modulation ratio. V_r and I_r denote the RMS values of the output AC voltage and AC current for the inverter, respectively.

$$V_r = \frac{MV_o}{\sqrt{2}} \tag{20}$$

$$R_L = \frac{V_r}{I_r} \tag{21}$$

The mathematical model of the PV-Forward-INV system can be obtained by combining Equations (1), (4), (7), (20) and (21).

$$P_o = \frac{2n^2 R_L I_{sc}^2}{D^2 M^2}\left[1 - C_1\left(e^{\frac{n\sqrt{2P_o R_L}}{C_2 D M V_{oc}}} - 1\right)\right]^2 \tag{22}$$

Similarly, the mathematical models of the PV-Flyback-INV and PV-Full-bridge-INV systems can also be obtained.

$$P_o = \frac{2n^2 R_L I_{sc}^2 (1-D)^2}{D^2 M^2}\left[1 - C_1\left(e^{\frac{n(1-D)\sqrt{2P_o R_L}}{C_2 D M V_{oc}}} - 1\right)\right]^2 \tag{23}$$

$$P_o = \frac{n^2 R_L I_{sc}^2}{2D^2 M^2}\left[1 - C_1\left(e^{\frac{n\sqrt{2P_o R_L}}{2C_2 D M V_{oc}}} - 1\right)\right]^2 \tag{24}$$

For the AC bus, Equation (25) is satisfied.

$$V_r = V_{Abus} \tag{25}$$

The mathematical model of the PV-Forward-INV-Abus system can be obtained by combining Equations (1), (10), (13) and (25).

$$P_o = \frac{\sqrt{2} n V_{Abus} I_{sc}}{DM}\left[1 - C_1\left(e^{\frac{\sqrt{2} n V_{Abus}}{C_2 D M V_{oc}}} - 1\right)\right] \tag{26}$$

Similarly, the mathematical models of the PV-Flyback-INV-Abus and PV-Full-bridge-INV-Abus systems can also be obtained.

$$P_o = \frac{\sqrt{2} n V_{Abus} I_{sc}(1-D)}{DM}\left[1 - C_1\left(e^{\frac{\sqrt{2} n V_{Abus}(1-D)}{C_2 D M V_{oc}}} - 1\right)\right] \tag{27}$$

$$P_o = \frac{\sqrt{2} n V_{Abus} I_{sc}}{2DM}\left[1 - C_1\left(e^{\frac{\sqrt{2} n V_{Abus}}{2C_2 D M V_{oc}}} - 1\right)\right] \tag{28}$$

Equations (13)–(15), (17)–(19), (22)–(24) and (26)–(28) are the theoretical basis for the MCCs of PV systems with these five isolated DC/DC converters connected to the load, DC bus, inverter and AC bus, respectively.

It can be concluded that P_{omax} appears in the slope of the curve at 0. Therefore, in order to find the MCCs of PV systems with different structures, their mathematical models are analyzed by substituting each of them into Equation (29).

$$\frac{dP_o}{dD} = 0 \tag{29}$$

For the PV-Forward, PV-Flyback, PV-Full-bridge and PV-Forward-Dbus systems, substituting Equations (13)–(15) and (17) into Equation (29), respectively, give Equations (30)–(33), where the parameter C_3 is represented by Equation (34).

$$D_{\max} = \frac{n\sqrt{P_{\text{omax}}R_{\text{L}}}}{C_3} \tag{30}$$

$$D_{\max} = \frac{\sqrt{P_{\text{omax}}R_{\text{L}}}}{C_3/n + \sqrt{P_{\text{omax}}R_{\text{L}}}} = 1 - \frac{C_3/n}{C_3/n + \sqrt{P_{\text{omax}}R_{\text{L}}}} \tag{31}$$

$$D_{\max} = \frac{n\sqrt{P_{\text{omax}}R_{\text{L}}}}{2C_3} \tag{32}$$

$$V = C_3 \tag{33}$$

$$C_3 = C_2 V_{\text{oc}}[\text{lambertw}(\text{e} \times \frac{1+C_1}{C_1}) - 1] \tag{34}$$

According to Equation (34), it can be concluded that the value of C_3 is only related to the parameters of the PV cell itself (S and T). The simulation experiments revealed that P_{omax} is only affected by S and T and is independent of R_{L} and n. Therefore, only the values of C_3 and P_{omax} under different weather conditions are required to derive the relationship between $D_{\max}$ and R_{L}, n. This leads to the MPPT control of isolated PV systems to improve the efficiency. The C_3-S, C_3-T, P_{omax}-S and P_{omax}-T curves under different weather conditions were plotted using MATLAB, and by applying the curve-fitting method, Equations (35) and (36) can be obtained.

$$C_3 = 0.0057 \times S - 0.086 \times T + 26.15 \tag{35}$$

$$P_{\text{omax}} = \begin{cases} -5.5 \times 10^{-9} \times S^3 + 5.3 \times 10^{-5} \times S^2 + 0.17 \times S - 0.09 \times T - 1.45 & 0 \le T \le 40 \\ -5.5 \times 10^{-9} \times S^3 + 5.3 \times 10^{-5} \times S^2 + 0.17 \times S - 2.7 & -20 \le T < 0 \end{cases} \tag{36}$$

According to Equations (35) and (36), C_3 and P_{omax} can be easily derived from the weather conditions. Meanwhile, in order to find the MCCs and improve the MPPT methodology of isolated PV systems, $D_{\max}$ can also be derived by combining the circuit parameters R_{L} and n.

Figure 4 shows the equivalent model of the isolated PV system at the MPP [12], where R_{iMPP}, V_{MPP} and I_{MPP} represent the values of R_{i}, V and I at the MPP in Figure 2, respectively.

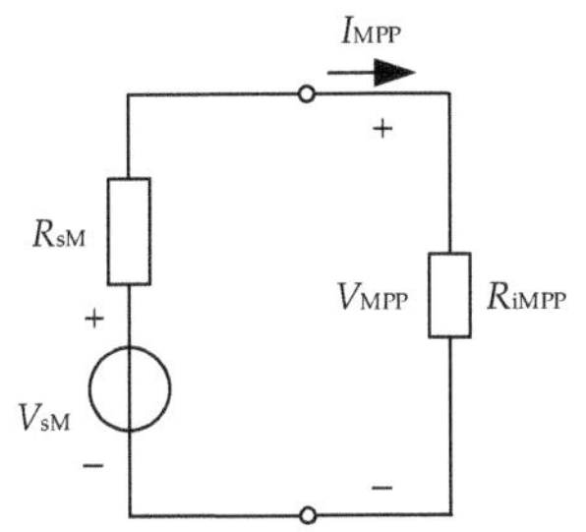

Figure 4. Isolated PV system with MPP linear model.

At the MPP, Equations (37) and (38) can be given by the circuit theorem [24].

$$R_{\text{iMPP}} = \frac{V_{\text{MPP}}}{I_{\text{MPP}}} \tag{37}$$

$$P_{\text{omax}} = V_{\text{MPP}} \times I_{\text{MPP}} \tag{38}$$

Equations (33), (37) and (38) are combined to obtain Equation (39).

$$R_{\text{iMPP}} = \frac{C_3^2}{P_{\text{omax}}} \tag{39}$$

According to the maximum power transfer theorem [24], the isolated PV system can operate at the MPP when Equation (40) is satisfied.

$$R_{\text{iMPP}} = R_{\text{sM}} \tag{40}$$

Meanwhile, according to the circuit theorem [24], Equation (41) is satisfied.

$$V_{\text{sM}} = 2C_3 \tag{41}$$

Using Equations (35), (36), (39) and (41), Equations (42) and (43) can be obtained.

$$R_{\text{sM}}(S,T) = \frac{[C_3(S,T)]^2}{P_{\text{omax}}(S,T)} \tag{42}$$

$$V_{\text{sM}}(S,T) = 2C_3(S,T) \tag{43}$$

According to Equations (42) and (43), the MPP linear model of the PV cell can be built using MATLAB/Simulink. When the weather conditions change, R_{sM} is involved in the design of MPPT as the output signal of the model.

2.2. MCCs Based on the Engineering Model

The relationship between circuit parameters, weather conditions and control parameters has been derived in Section 2.1 when the output of the isolated DC/DC converter is a load resistor. This section continues to derive the MCCs for isolated PV systems with different topologies and outputs on the basis of the engineering cell model.

The circuit topologies of forward and flyback converters determine their D to satisfy Equation (44), those of half-bridge and push–pull converters determine their D to satisfy Equation (45), and that of the full-bridge converter determines its D to satisfy Equation (46) [23]. These three formulas are also the basis of the analysis of MCCs carried out in a later section. D_{max} represents D at the MPP.

$$0 < D_{\text{max}} < 1 \tag{44}$$

$$0 < D_{\text{max}} < 0.5 \tag{45}$$

$$0 < D_{\text{max}} \leq 0.5 \tag{46}$$

Substituting Equation (30) into Equation (44), it can be seen that Equation (47) is satisfied. This is the R_{L} range in which the PV-Forward system can successfully track the MPP.

$$0 < R_{\text{L}} < \frac{C_3^2}{n^2 P_{\text{omax}}} \tag{47}$$

If the transformer ratio n is the object of study, Equation (47) can be replaced by Equation (48).

$$0 < n < \frac{C_3}{\sqrt{P_{\text{omax}} R_{\text{L}}}} \tag{48}$$

Similarly, the MCCs in the ideal case using the different PV systems are displayed in Table 1. These expressions are the prerequisites of successful MPPT control for isolated PV systems in the ideal case.

Table 1. Theoretical expressions of MCCs.

PV System	Range of the Output	Range of n
PV-Forward	$0 < R_L < \frac{C_3^2}{n^2 P_{omax}}$	$0 < n < \frac{C_3}{\sqrt{P_{omax} R_L}}$
PV-Flyback	$0 < R_L$	$0 < n$
PV-Half-bridge	$0 < R_L < \frac{C_3^2}{4n^2 P_{omax}}$	$0 < n < \frac{C_3}{2\sqrt{P_{omax} R_L}}$
PV-Full-bridge	$0 < R_L \le \frac{C_3^2}{n^2 P_{omax}}$	$0 < n \le \frac{C_3}{\sqrt{P_{omax} R_L}}$
PV-Forward-Dbus	$0 < V_{Dbus} < \frac{C_3}{n}$	$0 < n < \frac{C_3}{V_{Dbus}}$
PV-Flyback-Dbus	$0 < V_{Dbus}$	$0 < n$
PV-Half-bridge-Dbus	$0 < V_{Dbus} < \frac{C_3}{2n}$	$0 < n < \frac{C_3}{2V_{Dbus}}$
PV-Full-bridge-Dbus	$0 < V_{Dbus} \le \frac{C_3}{n}$	$0 < n \le \frac{C_3}{V_{Dbus}}$
PV-Forward-INV	$0 < R_L < \frac{M^2 C_3^2}{2n^2 P_{omax}}$	$0 < n < \frac{MC_3}{\sqrt{2P_{omax} R_L}}$
PV-Flyback-INV	$0 < R_L$	$0 < n$
PV-Half-bridge-INV	$0 < R_L < \frac{M^2 C_3^2}{8n^2 P_{omax}}$	$0 < n < \frac{MC_3}{2\sqrt{2P_{omax} R_L}}$
PV-Full-bridge-INV	$0 < R_L \le \frac{M^2 C_3^2}{2n^2 P_{omax}}$	$0 < n \le \frac{MC_3}{\sqrt{2P_{omax} R_L}}$
PV-Forward-INV-Abus	$0 < V_{Abus} < \frac{C_3 M}{\sqrt{2}n}$	$0 < n < \frac{C_3 M}{\sqrt{2}V_{Abus}}$
PV-Flyback-INV-Abus	$0 < V_{Abus}$	$0 < n$
PV-Half-bridge-INV-Abus	$0 < V_{Abus} < \frac{C_3 M}{2\sqrt{2}n}$	$0 < n < \frac{C_3 M}{2\sqrt{2}V_{Abus}}$
PV-Full-bridge-INV-Abus	$0 < V_{Abus} \le \frac{C_3 M}{\sqrt{2}n}$	$0 < n \le \frac{C_3 M}{\sqrt{2}V_{Abus}}$

From the practical application point of view, the isolated PV system is a non-ideal circuit, and the expressions in Table 1 need to be improved. The duty cycle of the isolated DC/DC converter cannot be too small or too large due to the losses of the switching devices and the isolation transformer itself, the limitations on the switching device's opening and closing times and the through-current withstand voltage, the transmission delay of the controller and the PWM sampling delay. Therefore, in order to find the MCCs in practical applications, it is assumed that the minimum D of the forward and flyback converters is D_{L1}, while their maximum D is D_{U1}, and the minimum D of the half-bridge, full-bridge and push–pull converters is D_{L2}, while their maximum D is D_{U2}. At this point, the duty cycle ranges of the forward and flyback converters can be expressed by Equation (49), and the half-bridge, full-bridge and push–pull converter duty cycle ranges can be expressed by Equation (50).

$$D_{L1} \le D_{max} \le D_{U1} \tag{49}$$

$$D_{L2} \le D_{max} \le D_{U2} \tag{50}$$

Substituting Equation (30) into Equation (49), it can be seen that Equation (51) can be obtained. This is the R_L range in which the PV-Forward system can successfully track the MPP in practical applications.

$$\frac{D_{L1}^2 C_3^2}{n^2 P_{omax}} \le R_L \le \frac{D_{U1}^2 C_3^2}{n^2 P_{omax}} \tag{51}$$

If the transformer ratio n is the object of study, Equation (51) can be replaced by Equation (52).

$$\frac{D_{L1} C_3}{\sqrt{P_{omax} R_L}} \le n \le \frac{D_{U1} C_3}{\sqrt{P_{omax} R_L}} \tag{52}$$

Similarly, the MCCs of various isolated PV systems can be derived when the D limitation in a practical situation is considered, as shown in Table 2. These expressions are the prerequisites of successful MPPT control for isolated PV systems in practical applications.

Table 2. Practical expressions of MCCs.

PV System	Range of the Output	Range of n
PV-Forward	$\frac{D_{L1}^2 C_3^2}{n^2 P_{omax}} \le R_L \le \frac{D_{U1}^2 C_3^2}{n^2 P_{omax}}$	$\frac{D_{L1} C_3}{\sqrt{P_{omax} R_L}} \le n \le \frac{D_{U1} C_3}{\sqrt{P_{omax} R_L}}$
PV-Flyback	$\frac{D_{L1}^2 C_3^2}{n^2 (1-D_{L1})^2 P_{omax}} \le R_L \le \frac{D_{U1}^2 C_3^2}{n^2 (1-D_{U1})^2 P_{omax}}$	$\frac{D_{L1} C_3}{(1-D_{L1})\sqrt{P_{omax} R_L}} \le n \le \frac{D_{U1} C_3}{(1-D_{U1})\sqrt{P_{omax} R_L}}$
PV-Half-bridge	$\frac{D_{L2}^2 C_3^2}{n^2 P_{omax}} \le R_L \le \frac{D_{U2}^2 C_3^2}{n^2 P_{omax}}$	$\frac{D_{L2} C_3}{\sqrt{P_{omax} R_L}} \le n \le \frac{D_{U2} C_3}{\sqrt{P_{omax} R_L}}$
PV-Full-bridge	$\frac{4D_{L2}^2 C_3^2}{n^2 P_{omax}} \le R_L \le \frac{4D_{U2}^2 C_3^2}{n^2 P_{omax}}$	$\frac{2D_{L2} C_3}{\sqrt{P_{omax} R_L}} \le n \le \frac{2D_{U2} C_3}{\sqrt{P_{omax} R_L}}$
PV-Forward-Dbus	$\frac{C_3 D_{L1}}{n} \le V_{Dbus} \le \frac{C_3 D_{U1}}{n}$	$\frac{C_3 D_{L1}}{V_{Dbus}} \le n \le \frac{C_3 D_{U1}}{V_{Dbus}}$
PV-Flyback-Dbus	$\frac{C_3 D_{L1}}{n(1-D_{L1})} < V_{Dbus} \le \frac{C_3 D_{U1}}{n(1-D_{U1})}$	$\frac{C_3 D_{L1}}{V_{Dbus}(1-D_{L1})} < n \le \frac{C_3 D_{U1}}{V_{Dbus}(1-D_{U1})}$
PV-Half-bridge-Dbus	$\frac{C_3 D_{L2}}{n} \le V_{Dbus} \le \frac{C_3 D_{U2}}{n}$	$\frac{C_3 D_{L2}}{V_{Dbus}} \le n \le \frac{C_3 D_{U2}}{V_{Dbus}}$
PV-Full-bridge-Dbus	$\frac{2C_3 D_{L2}}{n} \le V_{Dbus} \le \frac{2C_3 D_{U2}}{n}$	$\frac{2C_3 D_{L2}}{V_{Dbus}} \le n \le \frac{2C_3 D_{U2}}{V_{Dbus}}$
PV-Forward-INV	$\frac{M^2 C_3^2 D_{L1}^2}{2n^2 P_{omax}} \le R_L \le \frac{M^2 C_3^2 D_{U1}^2}{2n^2 P_{omax}}$	$\frac{M D_{L1} C_3}{\sqrt{2 P_{omax} R_L}} \le n \le \frac{M D_{U1} C_3}{\sqrt{2 P_{omax} R_L}}$
PV-Flyback-INV	$\frac{M^2 C_3^2 D_{L1}^2}{2n^2 P_{omax}(1-D_{L1})^2} \le R_L \le \frac{M^2 C_3^2 D_{U1}^2}{2n^2 P_{omax}(1-D_{U1})^2}$	$\frac{M D_{L1} C_3}{(1-D_{L1})\sqrt{2 P_{omax} R_L}} \le n \le \frac{M D_{U1} C_3}{(1-D_{U1})\sqrt{2 P_{omax} R_L}}$
PV-Half-bridge-INV	$\frac{M^2 C_3^2 D_{L2}^2}{2n^2 P_{omax}} \le R_L \le \frac{M^2 C_3^2 D_{U2}^2}{2n^2 P_{omax}}$	$\frac{M D_{L2} C_3}{\sqrt{2 P_{omax} R_L}} \le n \le \frac{M D_{U2} C_3}{\sqrt{2 P_{omax} R_L}}$
PV-Full-bridge-INV	$\frac{2M^2 C_3^2 D_{L2}^2}{n^2 P_{omax}} \le R_L \le \frac{2M^2 C_3^2 D_{U2}^2}{n^2 P_{omax}}$	$\frac{\sqrt{2} M D_{L2} C_3}{\sqrt{P_{omax} R_L}} \le n \le \frac{\sqrt{2} M D_{U2} C_3}{\sqrt{P_{omax} R_L}}$
PV-Forward-INV-Abus	$\frac{C_3 M D_{L1}}{\sqrt{2} n} \le V_{Abus} \le \frac{C_3 M D_{U1}}{\sqrt{2} n}$	$\frac{C_3 M D_{L1}}{\sqrt{2} V_{Abus}} \le n \le \frac{C_3 M D_{U1}}{\sqrt{2} V_{Abus}}$
PV-Flyback-INV-Abus	$\frac{C_3 M D_{L1}}{\sqrt{2} n(1-D_{L1})} \le V_{Abus} \le \frac{C_3 M D_{U1}}{\sqrt{2} n(1-D_{U1})}$	$\frac{C_3 M D_{L1}}{\sqrt{2} V_{Abus}(1-D_{L1})} \le n \le \frac{C_3 M D_{U1}}{\sqrt{2} V_{Abus}(1-D_{U1})}$
PV-Half-bridge-INV-Abus	$\frac{C_3 M D_{L2}}{\sqrt{2} n} \le V_{Abus} \le \frac{C_3 M D_{U2}}{\sqrt{2} n}$	$\frac{C_3 M D_{L2}}{\sqrt{2} V_{Abus}} \le n \le \frac{C_3 M D_{U2}}{\sqrt{2} V_{Abus}}$
PV-Full-bridge-INV-Abus	$\frac{\sqrt{2} C_3 M D_{L2}}{n} \le V_{Abus} \le \frac{\sqrt{2} C_3 M D_{U2}}{n}$	$\frac{\sqrt{2} C_3 M D_{L2}}{V_{Abus}} \le n \le \frac{\sqrt{2} C_3 M D_{U2}}{V_{Abus}}$

2.3. *MCCs Based on the MPP Linear Model*

2.3.1. Expression of MCCs

The analysis in Section 2.2 has produced the ranges of circuit parameters for twenty isolated PV systems capable of MPPT control based on the engineering model. This section continues with an in-depth study of these circuit parameter ranges based on the MPP linear model. After the engineering model has been linearized by using the methodology in Section 2.1, the isolated PV system structure can be replaced by the system shown in Figure 5. The flyback converter is selected as an example, where V_{sM} and R_{sM} are quantities that vary with the weather conditions (S and T).

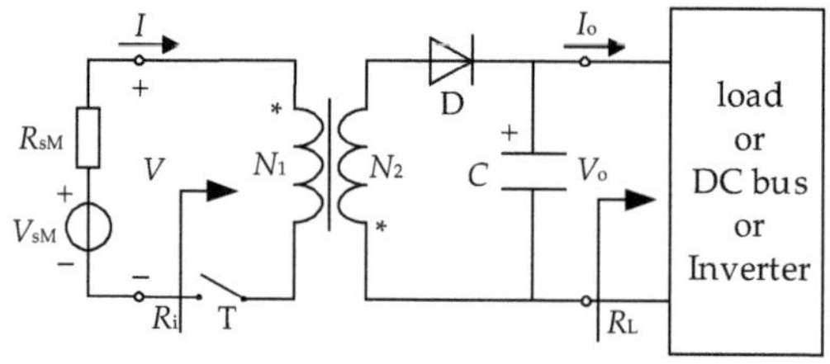

Figure 5. Isolated PV system based on MPP linear model. (* indicates the eponymous end of the induced electromotive force of the winding).

In order to find the MCCs in the ideal case, according to the maximum power transfer theorem, it can be seen that Equations (53) and (54) are satisfied when the PV system is operating at the MPP.

$$R_{\mathrm{i}} = R_{\mathrm{sM}} \tag{53}$$

$$V_{\mathrm{sM}} = 2V \tag{54}$$

The R_{i} of the PV-Forward system can be expressed by Equation (55), and R_{i} will vary with the different output devices and the transformations of isolated DC/DC converters.

$$R_{\mathrm{i}} = \frac{n^2 R_{\mathrm{L}}}{D^2} \tag{55}$$

The R_{i} of the PV-Forward-INV system can be expressed by Equation (56).

$$R_{\mathrm{i}} = \frac{2n^2 R_{\mathrm{L}}}{M^2 D^2} \tag{56}$$

Equation (56) reveals the mathematical relationship between the circuit parameters (R_{i}, R_{L} and n) and the control signals (D and M). On the basis of these expressions, the MCCs can be found.

When the output of the PV cell is connected to a resistor, Equation (55) is substituted into Equation (53), and then Equation (57) can be obtained.

$$D_{\mathrm{max}} = n\sqrt{\frac{R_{\mathrm{L}}}{R_{\mathrm{sM}}}} \tag{57}$$

Substituting Equation (56) into Equation (53), it can be seen that Equation (58) is satisfied. This is the R_{L} range in which the PV-Forward system can successfully track the MPP.

$$0 < R_{\mathrm{L}} < \frac{R_{\mathrm{sM}}}{n^2} \tag{58}$$

If the transformer ratio n is the object of study, Equation (58) can be replaced by Equation (59).

$$0 < n < \sqrt{\frac{R_{\mathrm{sM}}}{R_{\mathrm{L}}}} \tag{59}$$

Similarly, the MCCs of the different PV systems in the ideal case are displayed in Table 3. These expressions are the prerequisites of successful MPPT control for isolated PV systems in the ideal case.

Table 3 shows that under ideal conditions, an R_{L} or n value always exists in the PV-Flyback system to match the conditions for the use of the MPP linear model. Also, Table 3 shows that under ideal conditions, a V_{Dbus} or n value always exists in the PV-Flyback-Dbus system to match the use of the linear model. In contrast, for other PV systems, some constraints always exist. In addition, the use of inverters in isolated PV systems also affects the ranges of R_{L} and n. For the PV-Forward-INV, PV-Half-bridge-INV and PV-Full-bridge-INV systems, the presence of inverters narrows the ranges of R_{L} and n. Obviously, the expressions shown in Table 3 are the theoretical expressions of the MCCs, which can be used as the basis for designing the MPPT control process and proposing the MPPT control strategy under ideal conditions.

Table 3. Theoretical expressions of MCCs.

PV System	Range of the Output	Range of n
PV-Forward	$0 < R_L < \frac{R_{sM}}{n^2}$	$0 < n < \sqrt{\frac{R_{sM}}{R_L}}$
PV-Flyback	$0 < R_L$	$0 < n$
PV-Half-bridge	$0 < R_L < \frac{R_{sM}}{4n^2}$	$0 < n < \frac{1}{2}\sqrt{\frac{R_{sM}}{R_L}}$
PV-Full-bridge	$0 < R_L \leq \frac{R_{sM}}{n^2}$	$0 < n \leq \sqrt{\frac{R_{sM}}{R_L}}$
PV-Forward-Dbus	$0 < V_{Dbus} < \frac{V_{sM}}{2n}$	$0 < n < \frac{V_{sM}}{2V_{Dbus}}$
PV-Flyback-Dbus	$0 < V_{Dbus}$	$0 < n$
PV-Half-bridge-Dbus	$0 < V_{Dbus} \leq \frac{V_{sM}}{4n}$	$0 < n < \frac{V_{sM}}{4V_{Dbus}}$
PV-Full-bridge-Dbus	$0 < V_{Dbus} \leq \frac{V_{sM}}{2n}$	$0 < n \leq \frac{V_{sM}}{2V_{Dbus}}$
PV-Forward-INV	$0 < R_L < \frac{M^2 R_{sM}}{2n^2}$	$0 < n < M\sqrt{\frac{R_{sM}}{2R_L}}$
PV-Flyback-INV	$0 < R_L$	$0 < n$
PV-Half-bridge-INV	$0 < R_L < \frac{M^2 R_{sM}}{8n^2}$	$0 < n < \frac{M}{2}\sqrt{\frac{R_{sM}}{2R_L}}$
PV-Full-bridge-INV	$0 < R_L \leq \frac{M^2 R_{sM}}{2n^2}$	$0 < n \leq M\sqrt{\frac{R_{sM}}{2R_L}}$

From the practical application point of view, the isolated PV system is a non-ideal circuit, and the expressions in Table 3 need to be improved. The duty cycle of the isolated DC/DC converter cannot be too small or too large. Therefore, in order to find the range of circuit parameters in practical applications, the duty cycle ranges of the forward, flyback, half-bridge, full-bridge and push–pull converters are expressed by Equations (49) and (50).

Substituting Equation (57) into Equation (49), it can be seen that Equation (60) is satisfied. This is the R_L range in which MPPT control can be successfully realized in practical applications for the PV-Forward system.

$$\frac{D_{L1}^2 R_{sM}}{n^2} \leq R_L \leq \frac{D_{U1}^2 R_{sM}}{n^2} \tag{60}$$

If the transformer ratio n is the object of study, Equation (60) can be replaced by Equation (61).

$$D_{L1}\sqrt{\frac{R_{sM}}{R_L}} \leq n \leq D_{U1}\sqrt{\frac{R_{sM}}{R_L}} \tag{61}$$

Similarly, the ranges of circuit parameters in which various isolated PV systems are capable of successfully realizing MPPT are shown in Table 4, when considering the limited range of D in practical situations. These expressions are the prerequisites of successful MPPT control for isolated PV systems in practical situations.

Table 4 shows significantly smaller ranges for R_L and V_{Dbus} when compared with those in Table 3. Unlike the ideal case, the PV-Flyback, PV-Flyback-Dbus and PV-Flyback-INV systems have certain constraints in practical applications. Obviously, the expressions in Table 4 provide a theoretical basis for isolated PV systems on the basis of the MPP linear model in practical applications.

2.3.2. Range of MCCs

The ranges of V_{sM} and R_{sM} have been derived for changing weather conditions. Therefore, the extreme values of MCCs for practical applications are shown in Table 5. It can be seen that the maximum range of R_L (or V_{Dbus}) is necessary for each PV system to be modeled with the MPP linear cell. By contrast, the minimum range of R_L (or V_{Dbus}) is a sufficient condition for every PV system to use the MPP linear model. Similarly, the maximum and minimum ranges of the variable ratio n can be derived analogously.

Table 4. Practical expressions of MCCs.

PV System	Range of the Output	Range of n
PV-Forward	$\frac{D_{L1}^2 R_{sM}}{n^2} \leq R_L \leq \frac{D_{U1}^2 R_{sM}}{n^2}$	$D_{L1}\sqrt{\frac{R_{sM}}{R_L}} \leq n \leq D_{U1}\sqrt{\frac{R_{sM}}{R_L}}$
PV-Flyback	$\frac{D_{L1}^2 R_{sM}}{n^2(1-D_{L1})^2} \leq R_L \leq \frac{D_{U1}^2 R_{sM}}{n^2(1-D_{U1})^2}$	$\frac{D_{L1}}{(1-D_{L1})}\sqrt{\frac{R_{sM}}{R_L}} \leq n \leq \frac{D_{U1}}{(1-D_{U1})}\sqrt{\frac{R_{sM}}{R_L}}$
PV-Half-bridge	$\frac{D_{L2}^2 R_{sM}}{n^2} \leq R_L \leq \frac{D_{U2}^2 R_{sM}}{n^2}$	$D_{L2}\sqrt{\frac{R_{sM}}{R_L}} \leq n \leq D_{U2}\sqrt{\frac{R_{sM}}{R_L}}$
PV-Full-bridge	$\frac{4D_{L2}^2 R_{sM}}{n^2} \leq R_L \leq \frac{4D_{U2}^2 R_{sM}}{n^2}$	$2D_{L2}\sqrt{\frac{R_{sM}}{R_L}} \leq n \leq 2D_{U2}\sqrt{\frac{R_{sM}}{R_L}}$
PV-Forward-Dbus	$\frac{D_{L1} V_{sM}}{2n} \leq V_{Dbus} \leq \frac{D_{U1} V_{sM}}{2n}$	$\frac{D_{L1} V_{sM}}{2V_{Dbus}} \leq n \leq \frac{D_{U1} V_{sM}}{2V_{Dbus}}$
PV-Flyback-Dbus	$\frac{D_{L1} V_{sM}}{2n(1-D_{L1})} < V_{Dbus} \leq \frac{D_{U1} V_{sM}}{2n(1-D_{U1})}$	$\frac{D_{L1} V_{sM}}{2V_{Dbus}(1-D_{L1})} < n \leq \frac{D_{U1} V_{sM}}{2V_{Dbus}(1-D_{U1})}$
PV-Half-bridge-Dbus	$\frac{D_{L2} V_{sM}}{2n} \leq V_{Dbus} \leq \frac{D_{U2} V_{sM}}{2n}$	$\frac{D_{L2} V_{sM}}{2V_{Dbus}} \leq n \leq \frac{D_{U2} V_{sM}}{2V_{Dbus}}$
PV-Full-bridge-Dbus	$\frac{D_{L2} V_{sM}}{n} \leq V_{Dbus} \leq \frac{D_{U2} V_{sM}}{n}$	$\frac{D_{L2} V_{sM}}{V_{Dbus}} \leq n \leq \frac{D_{U2} V_{sM}}{V_{Dbus}}$
PV-Forward-INV	$\frac{D_{L1}^2 M^2 R_{sM}}{2n^2} \leq R_L \leq \frac{D_{U1}^2 M^2 R_{sM}}{2n^2}$	$MD_{L1}\sqrt{\frac{R_{sM}}{2R_L}} \leq n \leq MD_{U1}\sqrt{\frac{R_{sM}}{2R_L}}$
PV-Flyback-INV	$\frac{D_{L1}^2 M^2 R_{sM}}{2n^2(1-D_{L1})^2} \leq R_L \leq \frac{D_{U1}^2 M^2 R_{sM}}{2n^2(1-D_{U1})^2}$	$\frac{MD_{L1}}{(1-D_{L1})}\sqrt{\frac{R_{sM}}{2R_L}} \leq n \leq \frac{MD_{U1}}{(1-D_{U1})}\sqrt{\frac{R_{sM}}{2R_L}}$
PV-Half-bridge-INV	$\frac{D_{L2}^2 M^2 R_{sM}}{2n^2} \leq R_L \leq \frac{D_{U2}^2 M^2 R_{sM}}{2n^2}$	$MD_{L2}\sqrt{\frac{R_{sM}}{2R_L}} \leq n \leq MD_{U2}\sqrt{\frac{R_{sM}}{2R_L}}$
PV-Full-bridge-INV	$\frac{2D_{L2}^2 M^2 R_{sM}}{n^2} \leq R_L \leq \frac{2D_{U2}^2 M^2 R_{sM}}{n^2}$	$MD_{L2}\sqrt{\frac{2R_{sM}}{R_L}} \leq n \leq MD_{U2}\sqrt{\frac{2R_{sM}}{R_L}}$

Table 5. Ranges of MCCs.

PV System	Maximum Range	Minimum Range
PV-Forward	$\frac{D_{L1}^2 R_{sMmin}}{n^2} \leq R_L \leq \frac{D_{U1}^2 R_{sMmax}}{n^2}$	$\frac{D_{L1}^2 R_{sMmax}}{n^2} \leq R_L \leq \frac{D_{U1}^2 R_{sMmin}}{n^2}$
PV-Flyback	$\frac{D_{L1}^2 R_{sMmin}}{n^2(1-D_{L1})^2} \leq R_L \leq \frac{D_{U1}^2 R_{sMmax}}{n^2(1-D_{U1})^2}$	$\frac{D_{L1}^2 R_{sMmax}}{n^2(1-D_{L1})^2} \leq R_L \leq \frac{D_{U1}^2 R_{sMmin}}{n^2(1-D_{U1})^2}$
PV-Half-bridge	$\frac{D_{L2}^2 R_{sMmin}}{n^2} \leq R_L \leq \frac{D_{U2}^2 R_{sMmax}}{n^2}$	$\frac{D_{L2}^2 R_{sMmax}}{n^2} \leq R_L \leq \frac{D_{U2}^2 R_{sMmin}}{n^2}$
PV-Full-bridge	$\frac{4D_{L2}^2 R_{sMmin}}{n^2} \leq R_L \leq \frac{4D_{U2}^2 R_{sMmax}}{n^2}$	$\frac{4D_{L2}^2 R_{sMmax}}{n^2} \leq R_L \leq \frac{4D_{U2}^2 R_{sMmin}}{n^2}$
PV-Forward-Dbus	$\frac{D_{L1} V_{sMmin}}{2n} \leq V_{Dbus} \leq \frac{D_{U1} V_{sMmax}}{2n}$	$\frac{D_{L1} V_{sMmax}}{2n} \leq V_{Dbus} \leq \frac{D_{U1} V_{sMmin}}{2n}$
PV-Flyback-Dbus	$\frac{D_{L1} V_{sMmin}}{2n(1-D_{L1})} < V_{Dbus} \leq \frac{D_{U1} V_{sMmax}}{2n(1-D_{U1})}$	$\frac{D_{L1} V_{sMmax}}{2n(1-D_{L1})} < V_{Dbus} \leq \frac{D_{U1} V_{sMmin}}{2n(1-D_{U1})}$
PV-Half-bridge-Dbus	$\frac{D_{L2} V_{sMmin}}{2n} \leq V_{Dbus} \leq \frac{D_{U2} V_{sMmax}}{2n}$	$\frac{D_{L2} V_{sMmax}}{2n} \leq V_{Dbus} \leq \frac{D_{U2} V_{sMmin}}{2n}$
PV-Full-bridge-Dbus	$\frac{D_{L2} V_{sMmin}}{n} \leq V_{Dbus} \leq \frac{D_{U2} V_{sMmax}}{n}$	$\frac{D_{L2} V_{sMmax}}{n} \leq V_{Dbus} \leq \frac{D_{U2} V_{sMmin}}{n}$
PV-Forward-INV	$\frac{D_{L1}^2 M^2 R_{sMmin}}{2n^2} \leq R_L \leq \frac{D_{U1}^2 M^2 R_{sMmax}}{2n^2}$	$\frac{D_{L1}^2 M^2 R_{sMmax}}{2n^2} \leq R_L \leq \frac{D_{U1}^2 M^2 R_{sMmin}}{2n^2}$
PV-Flyback-INV	$\frac{D_{L1}^2 M^2 R_{sMmin}}{2n^2(1-D_{L1})^2} \leq R_L \leq \frac{D_{U1}^2 M^2 R_{sMmax}}{2n^2(1-D_{U1})^2}$	$\frac{D_{L1}^2 M^2 R_{sMmax}}{2n^2(1-D_{L1})^2} \leq R_L \leq \frac{D_{U1}^2 M^2 R_{sMmin}}{2n^2(1-D_{U1})^2}$
PV-Half-bridge-INV	$\frac{D_{L2}^2 M^2 R_{sMmin}}{2n^2} \leq R_L \leq \frac{D_{U2}^2 M^2 R_{sMmax}}{2n^2}$	$\frac{D_{L2}^2 M^2 R_{sMmax}}{2n^2} \leq R_L \leq \frac{D_{U2}^2 M^2 R_{sMmin}}{2n^2}$
PV-Full-bridge-INV	$\frac{2D_{L2}^2 M^2 R_{sMmin}}{n^2} \leq R_L \leq \frac{2D_{U2}^2 M^2 R_{sMmax}}{n^2}$	$\frac{2D_{L2}^2 M^2 R_{sMmax}}{n^2} \leq R_L \leq \frac{2D_{U2}^2 M^2 R_{sMmin}}{n^2}$

In practical applications, Tables 3–5 are a good guide for the circuit design, theoretical derivation and product selection of isolated PV systems. On the one hand, it is complicated to adjust the output under changing weather conditions. In order to realize MPPT control, they can be used to select the types of isolated DC/DC converters and circuit components. On the other hand, they can also be used as a basis for the study of MPPT methods. Meanwhile, they can be used to estimate the MPPT effect based on the recorded historical data of S and T in the application area. In addition, the results shown in Table 5 can provide a theoretical basis when the overall linearized model of the isolated PV system is investigated.

2.4. Two New MPPT Methods Based on MPP Linear Modeling

Two new MPPT methods based on the MPP linear model are proposed. Here, the PV-Flyback and PV-Flyback-Dbus systems are used as examples.

2.4.1. MPPT Method for PV Systems with Resistive Output (RMPPT)

Substituting Equations (4)–(6), (8) and (42) into (53), Equation (62) is satisfied. It relates D_{max} to the weather conditions (S and T) and the circuit parameters (R_L and n) when the PV-Flyback system operates at the MPP.

$$D_{max} = \frac{n\sqrt{R_L}}{\sqrt{R_{sM}(S,T)} + n\sqrt{R_L}} \tag{62}$$

According to Equation (62), it can be seen that RMPPT can be used when R_L and n are measured or known. Equation (62) is the theoretical basis of RMPPT, which can be described as follows: by measuring or knowing S and T as well as R_L and n, the duty cycle D_{max} at the MPP for the isolated PV system can be calculated, and the microcontroller or chip can realize MPPT control by controlling $D = D_{max}$.

The structure of the isolated PV system using RMPPT is shown in Figure 6. As can be seen in Figure 6, the D_{max} value of the PV system when it is located at the MPP attachment can be simply calculated by using a microcontroller or chip to measure or know the weather parameters (S and T) and the circuit parameters (n, V_o and I_o), calculating the load resistor R_L and then substituting these parameters into Equation (62). When the input is a PV array, the cost of the sensor can be reduced by sharing the irradiance sensor if S is uniform in a certain area. Also, the cost of voltage sampling and current sampling can be reduced if R_L is essentially the same for each PV system. It can be seen that the implementation of RMPPT requires only a simple process with low computational complexity, which can greatly reduce the hardware cost and program design of an isolated PV system.

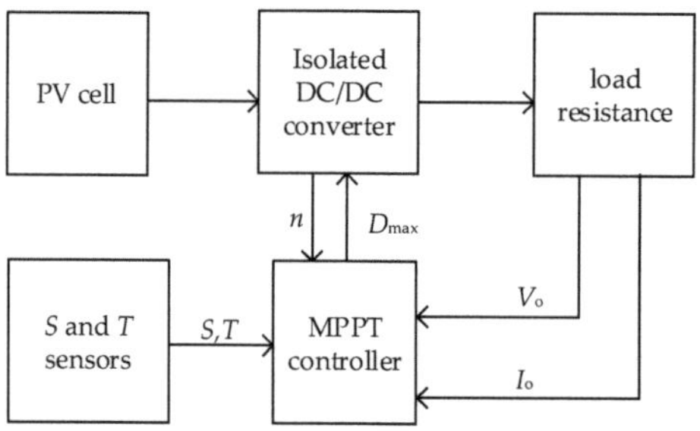

Figure 6. Isolated PV system structure of RMPPT.

2.4.2. MPPT Method with Output as DC Bus (BMPPT)

Substituting Equations (8), (16), (43) into (54), Equation (63) can be obtained. It relates D_{max} to the weather conditions (S and T) and the circuit parameters (V_{Dbus} and n) when the PV-Flyback-Dbus system operates at MPP.

$$D_{max} = \frac{2nV_{Dbus}}{V_{sM}(S,T) + 2nV_{Dbus}} \tag{63}$$

According to Equation (63), it can be seen that when V_{Dbus} and n can be measured or known, BMPPT can be used. Equation (63) is the theoretical basis of BMPPT, which can be described as follows: From the measured or known S and T, as well as V_{Dbus} and n, the duty cycle at the MPP D_{max} of the isolated PV system can be calculated. Then, the microcontroller or chip makes the duty cycle of the PWM wave equal to D_{max}, thereby achieving MPPT control. In contrast to RMPPT, BMPPT need not collect the output current. Eliminating the current-sampling device from the hardware design reduces the design difficulty and cost of the PV system and also reduces the current-sampling program designed for the software. When the output is a DC bus, BMPPT has an obvious advantage.

The structure of the isolated PV system using BMPPT is shown in Figure 7. As can be seen in Figure 7, the value of D_{max} for a PV system located at the MPP attachment can be simply calculated by using a microcontroller or chip to measure or know the weather conditions (S and T) and the circuit parameters (n and V_{Dbus}) and then substituting these parameters into Equation (63). Similarly, in the case of multiple PV arrays at the input, the cost of the sensors can be reduced by sharing irradiance sensors if S is uniform in a certain area. At the same time, multiple PV cells simplify the design of voltage-sampling circuits and reduce hardware and software costs by sharing a common set of DC buses.

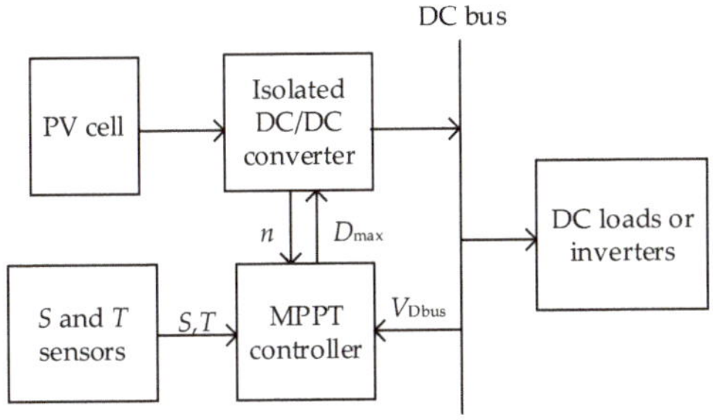

Figure 7. Isolated PV system structure of BMPPT.

3. Results

3.1. Simulation Verification of MCCs Based on the Engineering Model

In Table 2, it can be seen that the MCCs for PV systems with forward, half-bridge, full-bridge and push–pull converters are similar, as are the MCCs for PV systems with and without inverters. In this section, only the PV-Flyback, PV-Full-bridge, PV-Flyback-Dbus and PV-Full-bridge-Dbus systems are verified, and other PV systems with different structures can be verified analogously. In order to verify the accuracy of Table 2, some simulation experiments were carried out for PV systems with a flyback converter and full-bridge converter at STC with $n = 1/10$ or $R_L = 5\ \Omega$ or $V_{Dbus} = 500$ V for three cases, respectively. The experimental results are shown in Figure 8. The four factory parameter settings of this PV cell model are the same as in the first PV cell (1Soltech 1STH-215-P) of the PV array module in MATLAB/Simulink, which are $I_{sc} = 7.84$ A, $V_{oc} = 36.3$ V, $I_m = 7.35$ A and $V_m = 29$ V, respectively.

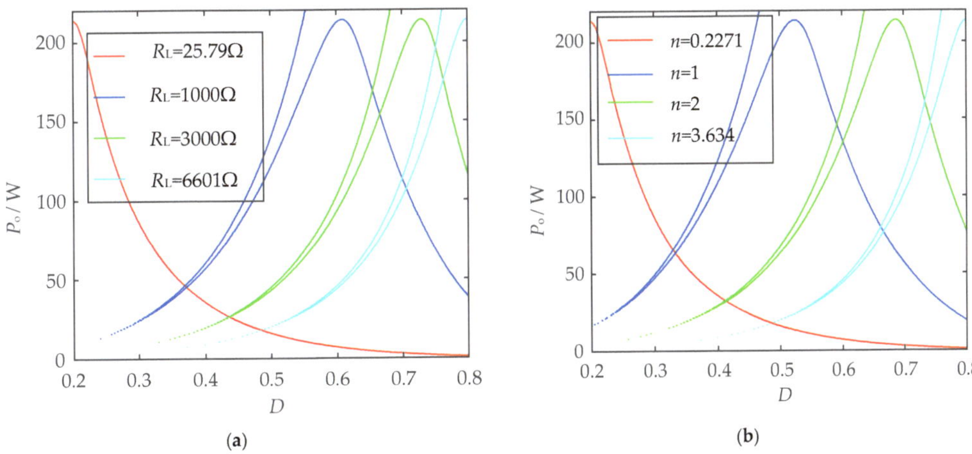

Figure 8. *Cont.*

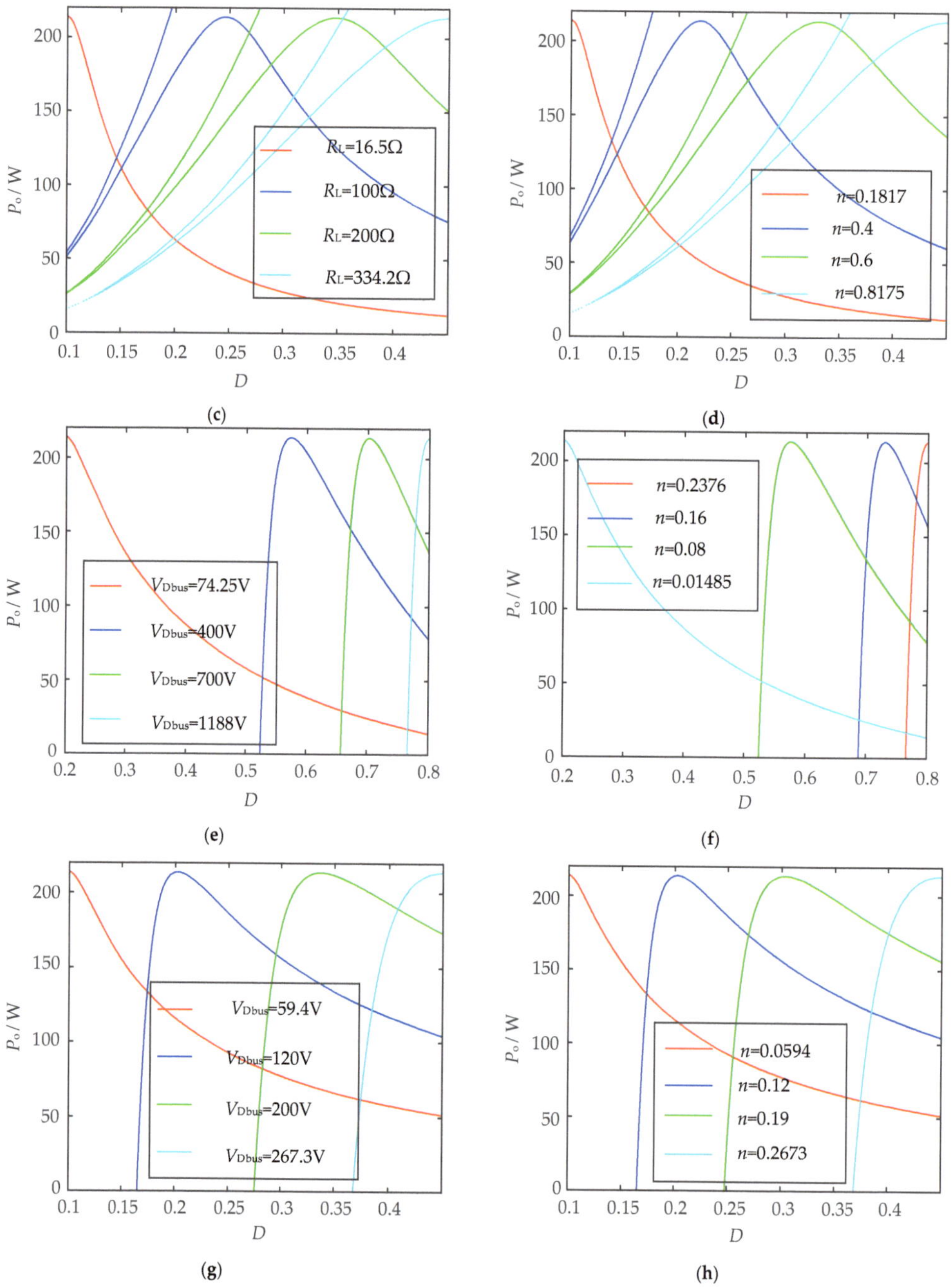

Figure 8. P_o-D curves of the different outputs and n. (**a**) P_o-D curves of PV-Flyback system for different R_L; (**b**) P_o-D curves of PV-Flyback system for different n; (**c**) P_o-D curves of PV-Full-bridge system for different R_L; (**d**) P_o-D curves of PV-Full-bridge system for different n; (**e**) P_o-D curves of PV-Flyback-Dbus for different V_{Dbus}; (**f**) P_o-D curves of PV-Flyback-Dbus for different n; (**g**) P_o-D curves of PV-Full-bridge-Dbus for different V_{Dbus}; (**h**) P_o-D curves of PV-Full-bridge-Dbus for different n.

Assume that D_{L1}, D_{U1}, D_{L2} and D_{U2} are taken as 0.2, 0.8, 0.1 and 0.45, respectively, that $R_L = 5\ \Omega$ or $V_{Dbus} = 500$ V for the study of the range of n, and that $n = 0.1$ for the study of the range of R_L or V_{Dbus}. The calculated maximum and minimum values of the circuit parameter range for a PV system with a forward converter and a full-bridge converter capable of successful MPPT are shown in Table 6, where R_{Lmax} and R_{Lmin} denote the maximum and minimum values of R_L, respectively, n_{max} and n_{min} denote the maximum and minimum values of n, respectively, and V_{Dmax} and V_{Dmin} denote the maximum and minimum values of $V_{Dbus,}$ respectively. These data are compared with Figure 8 to analyze the reasonableness and accuracy of the MCCs.

Table 6. The extreme values of MCCs.

PV System	R_{Lmin} or V_{Dmin}	R_{Lmax} or V_{Dmax}	n_{min}	n_{max}
PV-Flyback	25.79 Ω	6601 Ω	0.2271	3.634
PV-Full-bridge	16.5 Ω	334.2 Ω	0.1817	0.8175
PV-Flyback-Dbus	74.25 V	1188 V	0.01485	0.2376
PV-Full-bridge-Dbus	59.4 V	267.3 V	0.0594	0.2673

According to Figure 8a,b,e,f, for the PV-Flyback and PV-Flyback-Dbus systems, when n is certain and $R_L = R_{Lmin}$ or $V_{Dbus} = V_{Dmin}$ is satisfied, the MPP is reached exactly at $D = D_{L1}$. When $R_L = R_{Lmax}$ or $V_{Dbus} = V_{Dmax}$ is satisfied, the MPP is reached exactly at $D = D_{U1}$. When R_L or V_{Dbus} is certain and $n = n_{min}$ is satisfied, the MPP is reached exactly at $D = D_{L1}$. When $n = n_{max}$, the MPP is reached exactly at $D = D_{U1}$.

In Figure 8c,d,g,h, it can be seen that, for the PV-Full-bridge and PV-Full-bridge-Dbus systems, the MPP is reached exactly at $D = D_{L2}$ when n is certain and $R_L = R_{Lmin}$ or $V_{Dbus} = V_{Dmin}$ is satisfied. When $R_L = R_{Lmax}$ or $V_{Dbus} = V_{Dmax}$ is satisfied, the MPP is reached exactly at $D = D_{U2}$. When R_L or V_{Dbus} is certain and $n = n_{min}$ is satisfied, the MPP is reached exactly at $D = D_{L2}$. When $n = n_{max}$ is satisfied, the MPP is reached exactly at $D = D_{U2}$.

In Table 6, it can be seen that the range of MCCs for the PV-Flyback and PV-Flyback-Dbus systems is much larger than that of the PV-Full-bridge and PV-Full-bridge-Dbus systems. However, in the small-load and low-variable-ratio segments, the range of MCCs for the PV-Full-bridge and PV-Full-bridge-Dbus systems is slightly larger than that for the PV-Flyback and PV-Flyback-Dbus PV systems.

In conclusion, according to Figure 8, the MCCs shown in Table 2 are accurate in practical applications when the duty cycle constraints of isolated DC/DC converters are considered.

Obviously, the MCCs of PV systems are influenced by the changing irradiance and temperature. Therefore, in the research and application of PV systems, we can judge the effect of MPPT control and estimate the range of its circuit parameters according to local historical meteorological data.

3.2. Simulation Verification of MCCs Based on MPP Linear Model

3.2.1. Accuracy Verification of MCCs

Table 7 shows the four weather conditions of the PV system, and simulation experiments were conducted for the MCCs. Meanwhile, the results in Tables 3–5 and other weather conditions can be verified analogously.

Table 7. Simulated weather parameters of PV system.

Weather Conditions	(a)	(b)	(c)	(d)
S (W/m^2)	1300	850	550	350
T (°C)	40	25	20	15

When the output of the PV system is resistive, D_{L1}, D_{U1}, D_{L2} and D_{U2} are taken as 0.2, 0.8, 0.1 and 0.45, respectively, and R_L is equal to 0.5 Ω. The simulation results are shown in Figure 9. Figure 9 compares the curves of D_{max} variation with n for the PV-Forward, PV-Flyback, PV-Half-bridge and PV-Full-bridge systems under four weather conditions. Meanwhile, the MCCs in Table 4 are calculated, and the results are shown in Table 8. They can verify the accuracy of the simulation results in Figure 9 and Table 4.

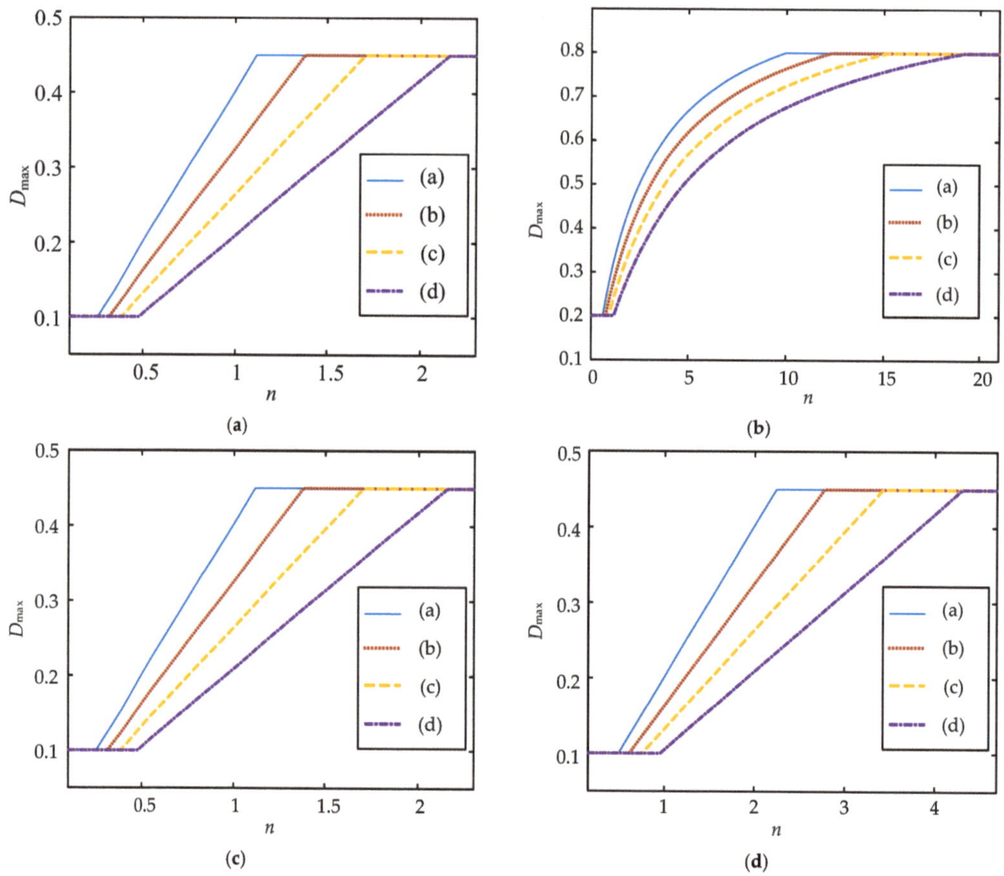

Figure 9. P_o-D curves of different PV systems. (**a**) D_{max}-n curves of PV-Forward system; (**b**) D_{max}-n curves of PV-Flyback system; (**c**) D_{max}-n curves of PV-Half-bridge system; (**d**) D_{max}-n curves of PV-Full-bridge system.

Table 8. Calculated values of MCCs.

Weather Conditions	(a)	(b)	(c)	(d)
PV-Forward	0.497 1.989	0.615 2.462	0.760 3.038	0.957 3.829
PV-Flyback	0.622 9.947	0.769 12.31	0.949 15.19	1.197 19.15
PV-Half-bridge	0.249 1.119	0.308 1.385	0.380 1.709	0.479 2.154
PV-Full-bridge	0.497 2.238	0.615 2.770	0.760 3.418	0.957 4.308

In Figure 9a, it can be seen that, for the PV-Forward system, D_{max} remains at 0.2 when $n < D_{L1}\sqrt{R_{sM}}/\sqrt{R_L}$ and 0.8 when $n > D_{U1}\sqrt{R_{sM}}/\sqrt{R_L}$, which implies that the MPP does not exist outside the range of n, and the MPP linear model cannot be used. In Figure 9b, it can be seen that, for the PV-Flyback system, D_{max} stays at 0.2 when $n < D_{L1}\sqrt{R_{sM}}/[\sqrt{R_L}(1 - D_{L1})]$, while when $n > D_{U1}\sqrt{R_{sM}}/[\sqrt{R_L}(1 - D_{U1})]$, Dmax stays at 0.8, which means that the MPP does not exist outside the range of n, and the MPP linear model cannot be used. In Figure 9c, it can be seen that, for the PV-Half-bridge system, D_{min} stays at 0.1 when $n < D_{L2}\sqrt{R_{sM}}/\sqrt{R_L}$, while D_{max} stays at 0.45 when $n > D_{U2}\sqrt{R_{sM}}/\sqrt{R_L}$, which implies that the MPP does not exist outside of the range of n, and the MPP linear model cannot be used. In Figure 9d, it can be seen that the PV-Full-bridge system maintains D_{min} at 0.1 when $n < 2D_{L2}\sqrt{R_{sM}}/\sqrt{R_L}$, while D_{max} remains at 0.45 under the condition of $n > 2D_{U2}\sqrt{R_{sM}}/\sqrt{R_L}$, which implies that the MPP does not exist outside the range of n, and the MPP linear model cannot be used.

Comparing Figure 9, the D_{max} of the PV system varies with n when n is within the MCCs. In this case, the MPP always exists, and the MPP linear model can be used for these four PV systems. The simulation results shown in Figure 9 are consistent with the corresponding data in Table 8, whereas the D_{max}-n curves of PV systems under different weather conditions differ significantly. Therefore, it can be concluded that the practical expressions of MCCs for various isolated PV systems in Table 4 are accurate for the PV-Forward, PV-Flyback, PV-Half-bridge and PV-Full-bridge systems.

3.2.2. Comparison of MCCs

The fifteen PV systems studied in this section can be applied under a wide range of practical requirements. However, the choice of the right PV system is complex. Therefore, it is essential to compare their MCCs. Here, it is assumed that the values of D_{L1}, D_{L2}, D_{U1}, D_{U2}, n, R_L and V_{Dbus} are the same as in Section 3.2.1. In this case, Table 9 shows the calculated values according to Table 5.

Table 9. Calculated values of MCCs.

PV System	Calculated MCC Values	
PV-Forward	$0.28\sqrt{R_{sM}} \le n \le 1.13\sqrt{R_{sM}}$	$4R_{sM} \le R_L \le 64R_{sM}$
PV-Flyback	$0.35\sqrt{R_{sM}} \le n \le 5.66\sqrt{R_{sM}}$	$6.25R_{sM} \le R_L \le 1600R_{sM}$
PV-Half-bridge	$0.14\sqrt{R_{sM}} \le n \le 0.636\sqrt{R_{sM}}$	$R_{sM} \le R_L < 20.25R_{sM}$
PV-Full-bridge	$0.28\sqrt{R_{sM}} \le n \le 1.272\sqrt{R_{sM}}$	$4R_{sM} \le R_L < 81R_{sM}$
PV-Forward-Dbus	$0.01V_{sM} \le n \le 0.04V_{sM}$	$V_{sM} \le V_{Dbus} \le 4V_{sM}$
PV-Flyback-Dbus	$0.013V_{sM} \le n \le 0.2V_{sM}$	$1.25V_{sM} \le V_{Dbus} \le 20V_{sM}$
PV-Half-bridge-Dbus	$0.005V_{sM} \le n \le 0.02V_{sM}$	$0.5V_{sM} \le V_{Dbus} < 2.25V_{sM}$
PV-Full-bridge-Dbus	$0.01V_{sM} \le n \le 0.045V_{sM}$	$V_{sM} \le V_{Dbus} \le 4.5V_{sM}$
PV-Forward-INV	$0.16\sqrt{R_{sM}} \le n \le 0.64\sqrt{R_{sM}}$	$1.28R_{sM} \le R_L \le 20.48R_{sM}$
PV-Flyback-INV	$0.2\sqrt{R_{sM}} \le n \le 3.2\sqrt{R_{sM}}$	$2R_{sM} \le R_L \le 512R_{sM}$
PV-Half-bridge-INV	$0.08\sqrt{R_{sM}} \le n \le 0.36\sqrt{R_{sM}}$	$0.32R_{sM} \le R_L < 6.48R_{sM}$
PV-Full-bridge-INV	$0.16\sqrt{R_{sM}} \le n \le 0.72\sqrt{R_{sM}}$	$1.28R_{sM} \le R_L < 25.92R_{sM}$

Some simulations based on Table 9 were performed to further analyze the MCCs. The simulation results are shown in Figure 10. In Figure 10, R_{LminFD} and R_{LmaxHB} denote the maximum and minimum values of R_L for the PV-Forward system and PV-Half-bridge system, respectively. Other circuit parameter boundaries are also presented in Figure 10.

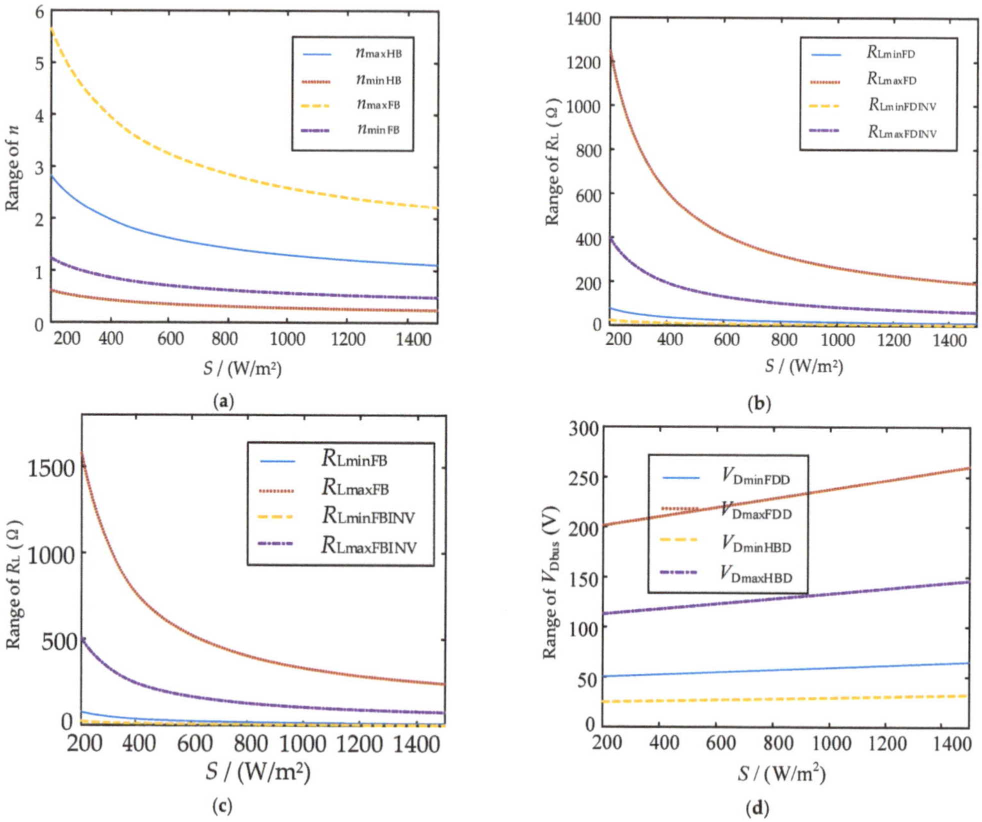

Figure 10. Comparison of curves of MCCs. (**a**) PV-Half-bridge compared with PV-Full-bridge system; (**b**) PV-Forward compared with PV-Forward-INV system; (**c**) PV-Full-bridge compared with PV-Full-bridge-INV system; (**d**) PV-Forward-Dbus compared with PV-Half-bridge-Dbus system.

Some conclusions can be drawn from Figure 10 and Table 9. Take the load resistance output as an example. When the weather parameters and the ratio are certain, the maximum value of the load resistance for the PV system using the forward converter as the MPPT circuit is about three times that for the PV system using the half-bridge converter. Meanwhile, the maximum value of the load resistance for the PV system using the full-bridge converter is about four times that for the PV system using the half-bridge converter. However, when $R_L < 4R_{sM}$, only the PV system using the half-bridge converter can successfully realize MPPT control. When the weather parameters and load resistance are certain, the maximum value of load resistance for the PV system using the full-bridge converter is about two times that for the PV system using the half-bridge converter. Meanwhile, only the PV system using the half-bridge converter can successfully realize MPPT control when $n < 0.28$. When an inverter is connected to the PV system, no matter what kind of converter is used as the MPPT control circuit, the range of circuit parameters is reduced to a certain extent. When the flyback converter is used, the load resistance, transformer ratio or bus voltage range is much larger than that of other isolated PV systems. Since both R_{sM} and V_{sM} are functions of S and T, the load, transformer ratio or bus voltage range changes with S and T. In addition, Figure 10 not only shows the range of variation in R_L, n and V_{Dbus} but also verifies the accuracy of the boundary values given in Table 9. The MPP linear model can be used only if the MPP is always present in the isolated PV system within this range.

In conclusion, both the different choices of isolated DC/DC converters and changing weather parameters may lead to changes in the MCCs.

3.3. Simulation Analysis of RMPPT

In order to verify the practicality of RMPPT and test its MPPT capability, the PV-Flyback system model was built by using Simulink. In this case, the MPP linear model in Section 2.1 is used. Meanwhile, n and R_L are equal to 2 and 1.7 Ω, respectively. In addition, the capacitors, inductors and transformers in the circuit are ideal components, the switching components are MOSFETs, and the PWM wave frequency is 15 kHz.

Simulation experiments on the practicality of RMPPT were conducted, and the results are shown in Table 10. D_{max} and D_{max1} denote D values at the MPP when the RMPPT and P&O methods are used, respectively. P_{omax} and P_{omax1} denote the maximum output power values of the PV cell when the RMPPT and P&O methods are used, respectively. P_{omax2} denotes the maximum output power of the PV system. The parameter settings are $n = 1/10$ and $R_L = 500$ Ω. The P&O method step size is set to 0.005.

Table 10. Experimental results for practicability of RMPPT.

(S,T)/(W/m², °C)	D_{max}	D_{max1}	P_{omax}	P_{omax1}	P_{omax2}
(750, 15)	0.4865	0.4821	152.19	152.13	149.63
(1000, 15)	0.5175	0.5204	214.7	214.89	212.5
(1250, 15)	0.54	0.5373	281.77	281.72	279.92
(750, 25)	0.4929	0.5007	151.29	151.22	148.79
(1000, 25)	0.524	0.5221	213.4	213.69	211.2
(1250, 25)	0.547	0.5455	280.87	280.83	277.93
(750, 35)	0.4994	0.5013	150.39	150.51	148.43
(1000, 35)	0.5308	0.5269	212.9	212.69	210.09
(1250, 35)	0.5539	0.5520	279.97	280.14	277.64

In Table 10, it can be seen that the values of D_{max} and P_{omax} calculated by RMPPT are basically equal to D_{max1} and P_{max1}, respectively. This proves the practicality of RMPPT. In addition, it can be seen from P_{omax1} and P_{omax2} that there is a difference between them due to the loss of the circuit components, the average value of which is the circuit loss, which is calculated to be about 2.41W.

Two sets of simulation experiments were performed for RMPPT using Simulink. And the MPPT methods were judged on the basis of stability and speed.

(1) Simulation experiment of irradiance change

In order to simulate a sudden weather change situation, it is assumed that at 0~0.3 s, S = 800 W/m² and T = 25 °C; at 0.3~0.7 s, S = 1200 W/m² and T = 25 °C; and at 0.7~1 s, S = 400 W/m² and T = 25 °C. Figure 11 shows the simulation results.

In Figure 11b, it can be seen that the tracking time and numerical stability of the MPP are much better than in the traditional P&O method when RMPPT is used in the isolated PV system with sudden changes in weather conditions (S). In Figure 11c, it can be seen that the P&O method itself has a step-length limitation, which causes D to oscillate around D_{max}, which is the reason why the output power of the P&O method oscillates at the MPP, while the RMPPT stabilizes at the MPP. It can also be seen in Figure 11 that, after the sudden change in S, D is actively adjusted to the new D_{max}, and the P_{omax} of the PV cell is also stabilized to the new P_{omax} after a rapid adjustment, which also proves the correctness of the conclusion in Section 2.1.

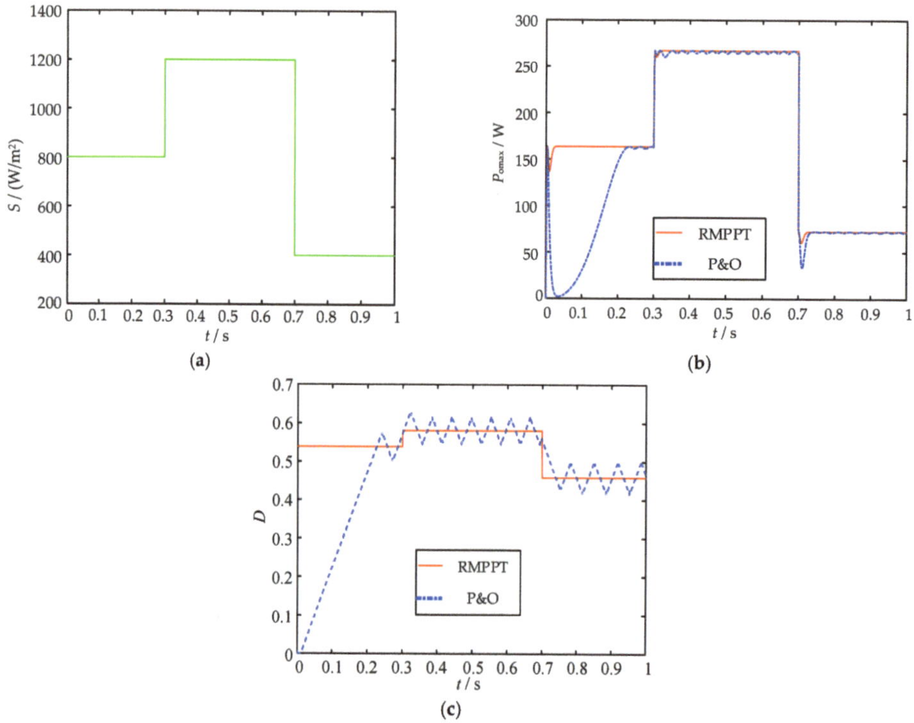

Figure 11. Simulation experiment of irradiance change. (**a**) S curve variation with t; (**b**) comparison of P_{omax}-t curves of RMPPT and P&O methods; (**c**) comparison of D-t curves of RMPPT and P&O methods.

(2) Simulation experiment of R_L change

Figure 12 shows the simulation results. In Figure 12b, it can be seen that the tracking time and numerical stability of the MPP are much better than those of the P&O method when RMPPT is used with sudden changes in R_L. It can also be seen in Figure 12 that D_{max} is actively adjusted to the new D_{max} after a sudden change in R_L, but P_{omax} remains at the same value after a short transient adjustment.

Therefore, it can be concluded that RMPPT outperforms the conventional P&O method in terms of MPPT rapidity and stability, regardless of changing weather conditions or circuit parameters.

Although only the MPPT method for the PV-Flyback system based on Equation (62) is proposed and verified in this section, the remaining MPPT methods for different isolated PV systems can be proposed analogously, which makes it easy for researchers and users of PV systems to select the corresponding MPPT methods.

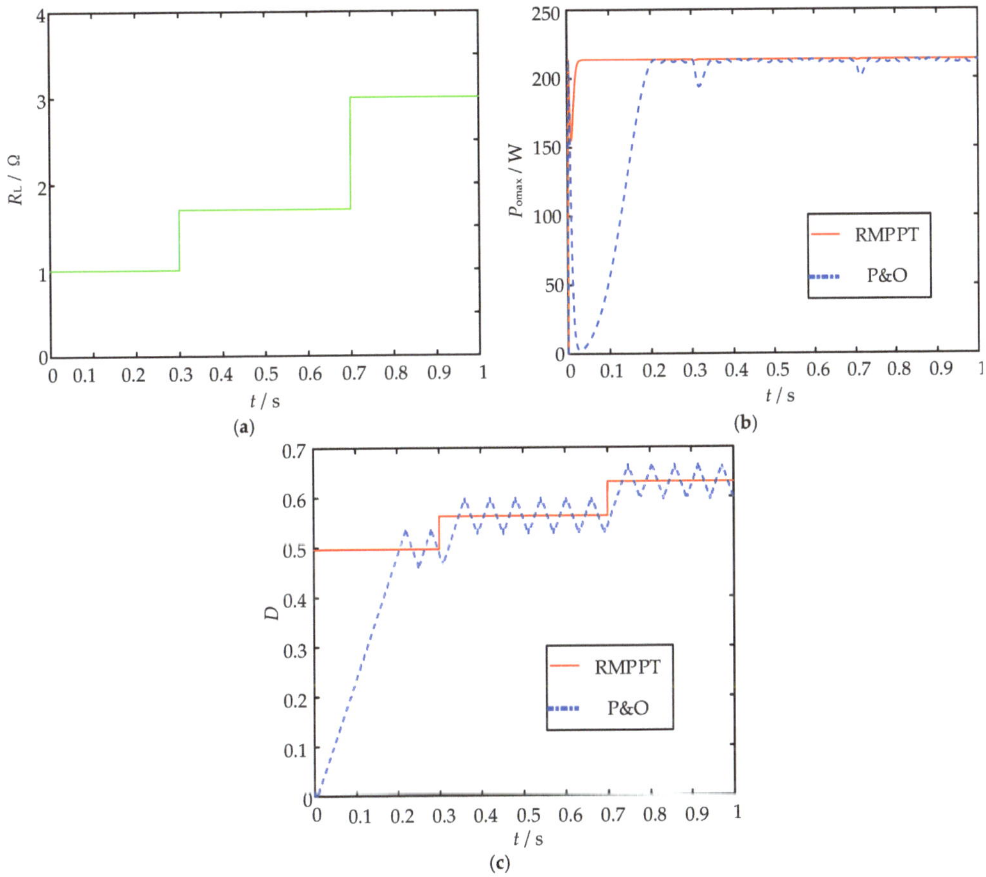

Figure 12. Simulation experiment of R_L change. (**a**) R_L curve variation with t; (**b**) comparison of P_{omax}-t curves of RMPPT and P&O methods; (**c**) comparison of D-t curves of RMPPT and P&O methods.

3.4. Simulation Analysis of BMPPT

In order to verify the practicality of BMPPT and test its MPPT capability, the PV-Flyback-Dbus system model shown in Figure 8 was built by using Simulink. The parameter settings are n = 2 and V_{Dbus} = 25 V, the capacitors, inductors and transformers in the circuit are ideal components, the switching components are MOSFETs, and the PWM wave frequency is 15 kHz. The simulation experiment results under varying temperature and DC bus voltage conditions are shown in Figure 13.

In Figure 13b,e, it can be seen that the tracking time and numerical stability of the MPP are much better than in the P&O method when BMPPT is used in isolated PV systems with sudden changes in T or V_{Dbus}. It can also be seen in Figure 13b,c that, after a sudden change in T, D_{max} is actively adjusted to the new D_{max}, and P_{omax} is also stabilized to the new P_{omax} after a rapid stepwise adjustment.

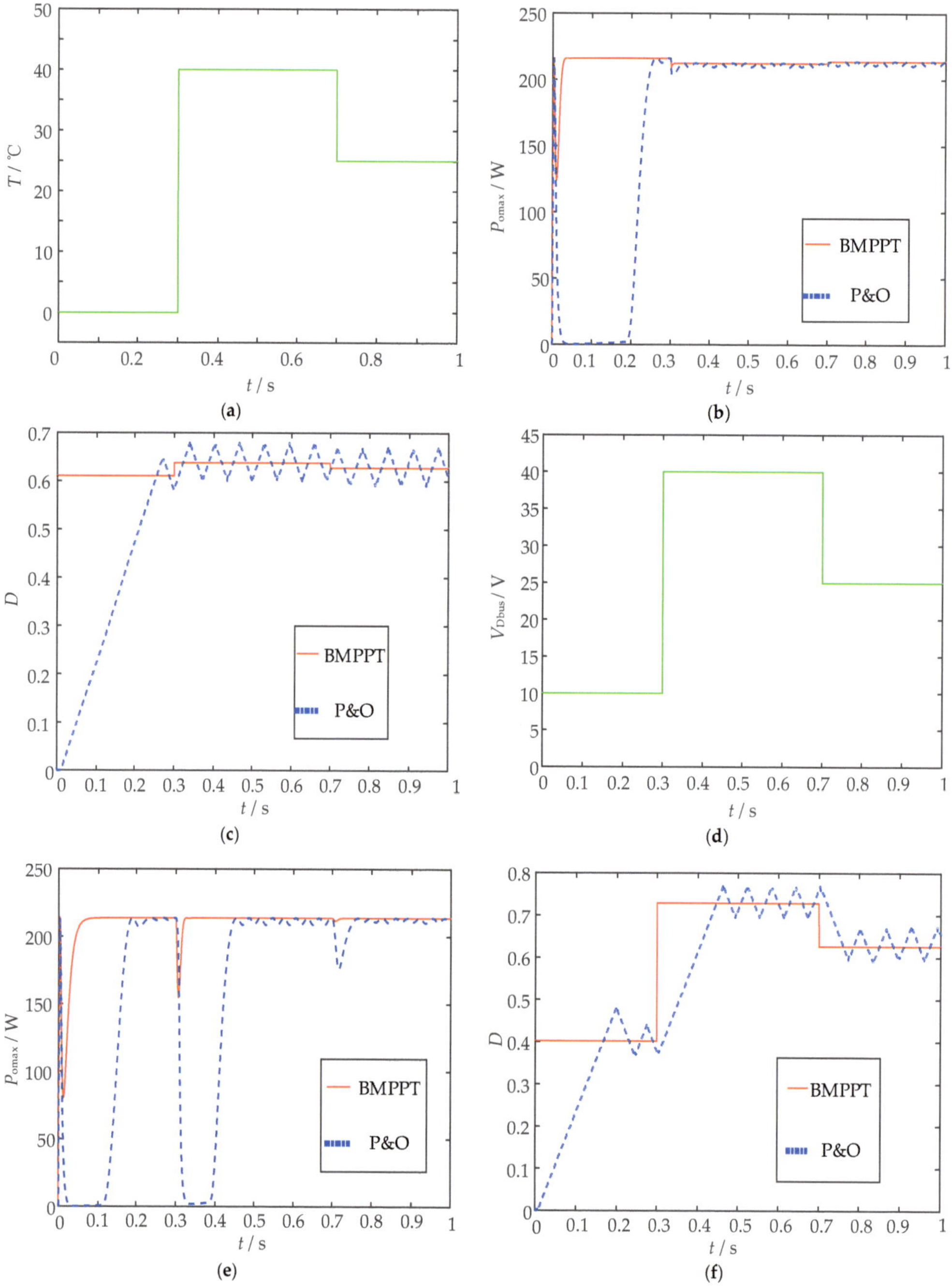

Figure 13. Simulation experiment of T and V_{Dbus} changes. (**a**) T curve variation with t; (**b**) comparison of P_{omax}-t curves of T change; (**c**) comparison of D-t curves of T change; (**d**) V_{Dbus} curve variation with t; (**e**) comparison of P_{omax}-t curves of V_{Dbus} change; (**f**) comparison of D-t curves of V_{Dbus} change.

Therefore, it can be concluded that BMPPT is far superior to the traditional P&O method in terms of rapidity and stability under changing weather conditions or circuit parameters.

Although only the MPPT method for the PV-Flyback-Dbus system based on the theory of Equation (63) is proposed and validated in this section, the remaining MPPT methods for different isolated PV systems can also be proposed analogously.

In this section, two MPPT methods (RMPPT and BMPPT) are verified when a load resistance and DC bus are selected as the output of the PV system, respectively. The conventional P&O method is compared with two MPPT methods implemented in Matlab/Simulink under varying weather conditions (irradiance and temperature) and circuit parameters (DC bus voltage and load resistance). The experimental results verify the high speed and accuracy of the two proposed MPPT methods and show the advantages of a simple program, small computational volume and low cost of hardware and software. They also verify the correctness and practicability of the MPP linear model established in Section 2.1.

4. Discussion

Tables 1–5 show the constraint conditions that enable the successful realization of MPPT control for isolated PV systems on the basis of the PV cell engineering model and MPP linear model. However, in practical applications, these constraint conditions usually play an important role in the hardware design, theoretical study and product installation of the PV system. On the one hand, since the boundaries of these constraints always change with the weather parameters, it is difficult to adjust the operating system in real time based on whether the load (or bus voltage) varies within the MCC range. For hardware designers, the MCCs can be utilized to select system configurations and circuit components. For the theoretical researcher, the MCCs can be used as a basis for ensuring the usability of the proposed control method. For the system installer, the MCCs can be used to estimate the MPPT effect based on solar irradiance and temperature recordings in the installation area. On the other hand, in practical applications, the maximum selected value of the load (or bus voltage) can be reflected by the MCCs. In other words, for a PV system, if the selected value of the load (or bus voltage) is not within the corresponding interval, the MPP cannot be successfully tracked, regardless of the used MPPT method, in which case, of course, the MPP linear model cannot be used. In addition, the MCCs can provide a theoretical basis when the MPP linear model is used to study the overall linearized model of the PV system.

However, in practice, the MPPT control of PV systems is usually affected by some other factors, such as the installed PV power, non-ideal DC/DC converter, non-ideal inverter and transmission efficiency. Therefore, the conclusions of this paper will be influenced by these factors to some extent. However, these factors are negligible. The reasons are as follows. On the one hand, the use of ideal isolated DC/DC converters and inverters can greatly simplify the theoretical study, just like in other studies. On the other hand, the aim of this work is to reveal the governing relationships between PV cell parameters and the load resistance or bus voltage when the MPP of the PV system is always present. Obviously, obtaining these relationships is very beneficial for the study of MPPT control methods using both PV cell models. Finally, the two constraint conditions in this paper represent the key results on the basis of which other factors can be easily considered and involved in practical applications.

5. Conclusions

For isolated PV systems, this paper solves the problem of when to apply the MPP linear model of the PV cell and proposes two faster and more accurate MPPT methods on the basis of MCCs, which are important for studying the overall linearization of isolated PV systems. In practical applications, the MCCs are a good guide for the circuit design, theoretical derivation and product selection of isolated PV systems. Theoretical researchers, hardware circuit designers and PV equipment installers can select the suitable isolated PV system according to different load and DC bus range requirements and make a preliminary estimation of MPPT effects. The main work in this paper is summarized as follows:

(1) The overall mathematical models of twenty isolated PV systems are established. And the relationships between the output power of isolated PV systems, the parameters of the PV cell and circuit parameters are found.
(2) The MCCs are found for isolated PV systems with different topologies and outputs on the basis of the PV cell engineering model and MPP linear model, respectively. They are a good guide for the circuit design, theoretical derivation and product selection of PV systems.
(3) Based on the MPP linear model and MCC, two MPPT methods (RMPPT and BMPPT) applicable to different output conditions are proposed. The experimental results verify the speed and accuracy of the two proposed MPPT methods. The MPPT time is improved from 0.23 s to 0.03 s. These two methods have the advantages of a simple program, small computational volume and low hardware and software costs.

Although this thesis finds some direct mathematical relationships between weather parameters (irradiance and temperature), circuit parameters (load resistance, transformer ratio and bus voltage) and control signals (PWM wave duty cycle) for isolated PV systems and proposes two MPPT methods applicable to different topologies and load types, there is still a lot of follow-up work to be carried out.

(1) The theoretical derivation in this paper makes some idealized assumptions. However, there may be more complicated situations in the practical circuit, and determining how to establish the MCCs and MPPT methods for more complicated situations is an important research direction.
(2) The two MPPT methods proposed put forward higher requirements on the speed, accuracy and economy of the irradiance and temperature sensors. If irradiance and temperature sensors with lower costs, higher accuracy and faster speed can be developed, the MPPT control method proposed in this paper can be more widely used.
(3) The MCCs proposed in this paper are based on the premise that the irradiance of all PV cells is uniform, but due to the environmental changes that may occur in the case of the partial shading of PV cells, it is also an important direction to consider the MCCs and the MPPT method in the case of non-uniform irradiance.

Author Contributions: Conceptualization, J.F. and S.L. (Shaowu Li); methodology, J.F.; software, J.F.; validation, J.F., S.L. (Shaowu Li), S.L. (Sanjun Liu) and X.Z.; formal analysis, J.F.; investigation, J.F., S.L. (Shaowu Li), S.L. (Sanjun Liu), X.D. and X.Z.; resources, S.L. (Shaowu Li) and S.L. (Sanjun Liu); data curation, J.F.; writing—original draft preparation, J.F. and S.L. (Shaowu Li); writing—review and editing, J.F., S.L. (Shaowu Li) and S.L. (Sanjun Liu); visualization, J.F. and S.L. (Shaowu Li); supervision, S.L. (Shaowu Li), S.L. (Sanjun Liu), X.D. and X.Z.; project administration, S.L. (Shaowu Li) and S.L. (Sanjun Liu); funding acquisition, S.L. (Shaowu Li), S.L. (Sanjun Liu) and X.D. All authors have read and agreed to the published version of the manuscript.

Funding: This research was funded by the National Natural Science Foundation of China (Nos. 61961016 and 61963014).

Data Availability Statement: All relevant data are within the paper.

Acknowledgments: The authors would like to sincerely thank the editor and anonymous reviewers for their valuable comments and suggestions to improve the quality of the article.

Conflicts of Interest: Author J.F. was employed by the company Jiangmen Pengjiang Power Supply Bureau of Guangdong Power Grid Co., Ltd. Author X.D. was employed by the company Hubei Chuangsinuo Electrical Technology Corp. Author X.Z. was employed by the company Guilin Power Supply Bureau of Guangxi Power Grid Co., Ltd. The remaining authors declare that the research was conducted in the absence of any commercial or financial relationships that could be construed as a potential conflict of interest.

Abbreviations

MPP	Maximum power point	STC	Standard test conditions
PV	Photovoltaic	PWM	Pulse-width modulation
MCC	MPPT constraint conditions	DC	Direct current
MPPT	Maximum power point tracking	AC	Alternating current
VWP	Variable-weather parameter		

Nomenclature

I	Output current of PV cell (A)	n	Transformer ratio of isolated DC/DC converter
V	Output voltage of PV cell (V)	M	SPWM wave modulation ratio
S	Solar irradiance (W/m^2)	V_r	Output voltage of inverter (V)
T	Cell temperature (°C)	I_r	Output current of inverter (A)
I_o	Output current of isolated DC/DC converter (A)	R_{sM}	Internal resistance of linear cell model (Ω)
V_o	Output voltage of isolated DC/DC converter (V)	V_{sM}	Open-circuit voltage of MPP linear model (V)
D	Duty cycle of the PWM signal of converter	D_{max}	D at the MPP
I_{sc}	Short-circuit current of PV cell under STC (A)	P_{omax}	Output power at MPP (W)
V_{oc}	Open-circuit voltage of PV cell under STC (V)	R_{iMPP}	Value of R_i at MPP (Ω)
I_m	MPP current of PV cell under STC (A)	V_{MPP}	Value of V at MPP (Ω)
V_m	MPP voltage of PV cell under STC (V)	I_{MPP}	Value of I at MPP (Ω)
R_i	Input resistance of isolated DC/DC converter (Ω)	V_{Dbus}	Voltage of DC bus (V)
R_L	Load or equivalent load resistance of PV system (Ω)	V_{Abus}	Voltage of AC bus (V)
D_{L1}	Minimum D for forward and flyback converters	D_{L2}	Minimum D for half-bridge, full-bridge, push–pull converter
D_{U1}	Maximum D for forward and flyback converters	D_{U2}	Maximum D for half-bridge, full-bridge, push–pull converter
P_o	Output power of PV system (W)		

References

1. Zhao, D.; Qian, M.; Ma, J.; Yamashita, K. Photovoltaic generator model for power system dynamic studies. *Sol. Energy* **2020**, *210*, 101–114. [CrossRef]
2. Koohi Kamali, S.; Rahim, N.A.; Mokhlis, H.; Tyagi, V.V. Photovoltaic electricity generator dynamic modeling methods for smart grid applications: A review. *Renew. Sustain. Energy Rev.* **2016**, *57*, 131–172. [CrossRef]
3. Moreira, H.S.; Silva, J.L.D.S.; Reis, M.V.G.D.; Mesquita, D.B.; Paula, B.H.K.; Villalva, M.G. Experimental comparative study of photovoltaic models for uniform and partially shading conditions. *Renew. Energy* **2021**, *164*, 58–73. [CrossRef]
4. Li, Q.; Zhao, S.; Wang, M.; Zou, Z.; Wang, B.; Chen, Q. An improved perturbation and observation maximum power point tracking algorithm based on a PV module four-parameter model for higher efficiency. *Appl. Energy* **2017**, *195*, 523–537. [CrossRef]
5. Yin, O.W.; Babu, B.C. Simple and easy approach for mathematical analysis of photovoltaic (PV) module under normal and partial shading conditions. *Optik* **2018**, *169*, 48–61. [CrossRef]
6. Zsiborács, H.; Baranyai, N.H.; Vincze, A.; Pintér, G. An economic analysis of the shading effects of transmission lines on photovoltaic power plant investment decisions: A case study. *Sensors* **2021**, *21*, 4973. [CrossRef]
7. Hsieh, Y.; Yu, L.; Chang, T.; Liu, W.; Wu, T.; Moo, C. Parameter identification of one-diode dynamic equivalent circuit model for photovoltaic panel. *IEEE J. Photovolt.* **2020**, *10*, 219–225. [CrossRef]
8. Si, X.; Chen, Q.; Gao, Z. A piecewise linearization based equivalent model of photovoltaic array. *Power Syst. Technol.* **2014**, *38*, 947–951.
9. Wang, Y.J.; Hsu, P.C. Modelling of solar cells and modules using piecewise linear parallel branches. *IET Renew. Power Gener.* **2011**, *5*, 215–222. [CrossRef]
10. Bharadwaj, P.; John, V. Linearised model for PV panel power output variation with changes in ambient conditions. *Sādhanā* **2017**, *42*, 2183–2187. [CrossRef]
11. Golshani, A.; Bathaee, S.M.T.; Moghaddas-Tafreshi, S.M. Small signal stability analysis of photovoltaic array based on averaged switch modeling technique. *J. Renew. Sustain. Energy* **2012**, *4*, 043117. [CrossRef]
12. Li, S. Linear equivalent models at the maximum power point based on variable weather parameters for photovoltaic cell. *Appl. Energy* **2016**, *182*, 94–104. [CrossRef]
13. Amir, A.; Amir, A.; Che, H.S.; Khateb, A.E.; Rahim, N.A. Comparative analysis of high voltage gain DC-DC converter topologies for photovoltaic systems. *Renew. Energy* **2019**, *136*, 1147–1163. [CrossRef]
14. Başoğlu, M.E.; Çakır, B. Comparisons of MPPT performances of isolated and non-isolated DC-DC converters by using a new approach. *Renew. Sustain. Energy Rev.* **2016**, *60*, 1100–1113. [CrossRef]
15. Brito, M.A.G.; Galotto, L.; Sampaio, L.P.; Melo, G.A.; Canesin, C.A. Evaluation of the main MPPT techniques for photovoltaic applications. *IEEE Trans. Ind. Electron.* **2013**, *60*, 1156–1167. [CrossRef]

16. Seyedmahmoudian, M.; Horan, B.; Soon, T.K.; Rahmani, R.; Oo, A.M.T.; Mekhilef, S.; Stojcevski, A. State of the art artificial intelligence-based MPPT techniques for mitigating partial shading effects on PV systems—A review. *Renew. Sustain. Energy Rev.* **2016**, *64*, 435–455. [CrossRef]
17. Hassan, T.U.; Abbassi, R.; Jerbi, H. A novel algorithm for MPPT of an isolated PV system using push pull converter with fuzzy logic controller. *Energies* **2020**, *13*, 4007. [CrossRef]
18. Islam, H.; Mekhilef, S.; Shah, N.B.M.; Soon, T.K.; Seyedmahmousian, M.; Horan, B.; Stojcevski, A. Performance evaluation of maximum power point tracking approaches and photovoltaic systems. *Energies* **2018**, *11*, 365. [CrossRef]
19. Bollipo, R.B.; Mikkili, S.; Bonthagorla, P.K. Hybrid, optimization, intelligent and classical PV MPPT techniques: Review. *CSEE J. Power Energy Syst.* **2021**, *7*, 25.
20. Mohapatra, A.; Nayak, B.; Das, P.; Mohanty, K.B. A review on MPPT techniques of PV system under partial shading condition. *Renew. Sustain. Energy Rev.* **2017**, *80*, 854–867. [CrossRef]
21. Li, S.; Ping, A.; Liu, Y.; Ma, X.; Li, C. A variable-weather-parameter MPPT method based on a defined characteristic resistance of photovoltaic cell. *Sol. Energy* **2020**, *199*, 673–684. [CrossRef]
22. Li, S. Circuit parameter range of photovoltaic system to correctly use the mpp linear model of photovoltaic cell. *Energies* **2021**, *14*, 3997. [CrossRef]
23. Chen, J.; Kang, Y. *Power Electronics: Power Electronic Conversion and Control Technology*, 3rd ed.; Higher Education Press: Beijing, China, 2011; pp. 81–93.
24. Qiu, G.; Luo, X. *Circuits*, 5th ed.; Higher Education Press: Beijing, China, 2013.

processes

Article

A Novel Hybrid Deep Learning Model for Forecasting Ultra-Short-Term Time Series Wind Speeds for Wind Turbines

Jianzan Yang [1], Feng Pang [1], Huawei Xiang [1], Dacheng Li [1] and Bo Gu [2,*]

1 Powerchina Guiyang Engineering Corporation Limited, Guiyang 550081, China; yangjz_gyy@powerchina.cn (J.Y.); pangf_gyy@powerchina.cn (F.P.); lakeonpeak@126.com (H.X.); ldchwhu@163.com (D.L.)
2 School of Electrical Engineering, North China University of Water Resources and Electric Power, Zhengzhou 450011, China
* Correspondence: gubo@ncwu.edu.cn

Abstract: Accurate forecasting of ultra-short-term time series wind speeds (UTSWS) is important for improving the efficiency and safe and stable operation of wind turbines. To address this issue, this study proposes a VMD-AOA-GRU based method for UTSWS forecasting. The proposed method utilizes variational mode decomposition (VMD) to decompose the wind speed data into temporal mode components with different frequencies and effectively extract high-frequency wind speed features. The arithmetic optimization algorithm (AOA) is then employed to optimize the hyperparameters of the model of the gated recurrent unit (GRU), including the number of hidden neurons, training epochs, learning rate, learning rate decay period, and training data temporal length, thereby constructing a high-precision AOA-GRU forecasting model. The AOA-GRU forecasting model is trained and tested using different frequency temporal mode components obtained from the VMD, which achieves multi-step accurate forecasting of the UTSWS. The forecasting results of the GRU, VMD-GRU, VMD-AOA-GRU, LSTM, VMD-LSTM, PSO-ELM, VMD-PSO-ELM, PSO-BP, VMD-PSO-BP, PSO-LSSVM, VMD-PSO-LSSVM, ARIMA, and VMD-ARIMA are compared and analyzed. The calculation results show that the VMD algorithm can accurately mine the high-frequency components of the time series wind speed, which can effectively improve the forecasting accuracy of the forecasting model. In addition, optimizing the hyperparameters of the GRU model using the AOA can further improve the forecasting accuracy of the GRU model.

Keywords: variational mode decomposition; arithmetic optimization algorithm; gated recurrent unit; ultra-short-term forecasting; time series wind speed

Citation: Yang, J.; Pang, F.; Xiang, H.; Li, D.; Gu, B. A Novel Hybrid Deep Learning Model for Forecasting Ultra-Short-Term Time Series Wind Speeds for Wind Turbines. *Processes* **2023**, *11*, 3247. https://doi.org/10.3390/pr11113247

Academic Editors: Krzysztof Rogowski, Chenyu Wu, Zhongkai Yi and Chenhui Lin

Received: 30 September 2023
Revised: 1 November 2023
Accepted: 9 November 2023
Published: 18 November 2023

1. Introduction

With the gradual depletion of fossil fuels and the continuous increase in energy demand arising from human societal development, wind power has been extensively developed and utilized as a renewable energy source in recent years [1]. According to the "Global Wind Report 2023" released by the Global Wind Energy Council (GWEC), global wind power added 78 GW of installed capacity in 2022, reaching a total installed capacity of 906 GW [2]. Wind speed is the primary factor affecting the operational characteristics of wind turbines, and its randomness, intermittency, and volatility pose a significant challenge for the optimal operation of wind turbines [3,4]. Accurate forecasting of the distribution characteristics of short-term wind speeds is an effective means of improving the safety and economic efficiency of wind turbine operations.

Based on the different timescales of wind speed forecasting, time series wind speed forecasting can be divided into three categories: ultra-short-term forecasting, short-term forecasting, and medium-to-long-term forecasting [5]. Time series wind speed forecasting methods can be classified into statistical forecasting methods, artificial intelligence forecasting methods, and hybrid forecasting methods based on the forecasting mechanism employed [6,7]. Statistical forecasting methods use statistical models to extract the historical features of time series wind speeds and forecast future wind speeds based on these features. Commonly used statistical forecasting models include the AR model [8,9], ARMA model [10–12], and ARIMA [13–15]. Statistical forecasting models exhibit good forecasting performance when the historical data are sufficient and relatively stable. However, their forecasting capability tends to degrade when dealing with incomplete historical data or non-stationary data. Especially with time series data with random distribution characteristics, the predictive performance of statistical prediction models is difficult to meet the prediction needs.

With the rapid development of artificial intelligence technology, its application to time series wind speed forecasting has received considerable attention in terms of research and applications [16,17]. Machine learning, with its strong capability for nonlinear mapping, has demonstrated excellent performance in wind speed forecasting [18–20]. In a previous study [21], a novel hybrid model was developed for short-term wind speed forecasting, and the calculation results illustrated that the proposed hybrid model outperformed single and recently developed forecasting models. A complete ensemble empirical mode decomposition with adaptive noise-least absolute shrinkage and selection operator-quantile regression neural network (CEEMDAN-LASSO-QRNN) model was developed for multistep wind speed probabilistic forecasting [22], and the experimental results indicated a higher accuracy and robustness of the proposed model in multistep wind speed probabilistic forecasting. Yang et al. [23] developed an innovative ensemble system based on mixed-frequency modeling to perform wind speed point and interval forecasting, and a multi-objective optimizer-based ensemble forecaster was proposed to provide deterministic and uncertain information regarding future wind speeds. The calculation results demonstrated that the system outperformed the benchmark techniques and could be employed for data monitoring and analysis on wind farms. Dong et al. [24] proposed an ensemble system of decomposition, adaptive selection, and forecasting to simulate the actual wind speed data of a wind farm. The calculation results indicated that the proposed forecasting model had outstanding forecasting accuracy for time series wind speeds. Machine learning methods can explore the nonlinear relationships of data, but they cannot explore the spatiotemporal distribution characteristics between data, resulting in low accuracy of machine learning methods in predicting time series data.

Deep learning exhibits exceptional capabilities in nonlinear mapping and time series data analysis, leading to superior performance in time series wind speed forecasting [25,26]. For instance, Lv et al. [27] applied a deep learning model to forecast time series wind speeds after removing data noise. The computational results demonstrated that the proposed method could accurately forecast time series wind speed sequences. A dynamic adaptive spatiotemporal graph neural network (DASTGN) was proposed for forecasting wind speed, and extensive experiments on real wind speed datasets in the Chinese seas showed that DASTGN improved the performance of the optimal baseline model by 3.05% and 3.69% in terms of MAE and RMSE, respectively [28]. A wind speed forecasting model based on hybrid variational mode decomposition (VMD), improved complete ensemble empirical mode decomposition with additive noise (ICEEMDAN), and a long short-term memory (LSTM) neural network has also been proposed [29]. By comparing and analyzing seven other forecasting models, the proposed model was found to have the best forecasting accuracy. Other studies [26,30–33] have presented a novel transformer-based deep neural network architecture integrated with a wavelet transform for forecasting wind speed and wind energy (power) generation for 6 h in the future, and the calculation results demonstrated that the integration of the transformer model with wavelet decomposition improved the forecasting accuracy. A single deep learning model can only explore the distribution characteristics (spatial or temporal distribution characteristics) of a certain aspect of data, and cannot simultaneously explore the spatiotemporal distribution and coupling characteristics of data. Therefore, a hybrid forecasting model with spatial and temporal feature mining capabilities is an effective method to improve the accuracy of ultra-short-term wind speed forecasting.

Integrating the advantages of different deep learning models to construct hybrid forecasting models can effectively overcome the limitations of individual deep learning forecasting models and further improve wind speed forecasting accuracy [34,35]. A hybrid forecasting model was developed by combining the strengths of a convolutional neural network (CNN) model and a bidirectional long short-term memory (BiLSTM) model, and was then applied to short-term wind speed forecasting [36,37]. The computational results demonstrated that the hybrid CNN-BiLSTM forecasting model could accurately capture the spatiotemporal distribution characteristics of wind speed information, leading to a higher forecasting accuracy than that of individual deep learning models. Assigning different weights to multiple deep learning models for ensemble construction in short-term wind speed forecasting allows each model's strengths to be fully utilized, thereby enhancing the overall forecasting accuracy of the ensemble model and expanding its practical applicability [38].

Based on an existing study on wind speed forecasting, this study presents a forecasting model based on VMD-AOA-GRU for ultra-short-term time series wind speed (UTSWS). In the proposed model, historical wind speed data from wind turbines were first decomposed into different frequency sub-sequences using variational mode decomposition (VMD). Then, the arithmetic optimization algorithm (AOA) was utilized to optimize the temporal length of the training data for each sub-sequence and the hyperparameters of the forecasting model of the gated recurrent unit (GRU), such as the number of hidden neurons, training epochs, learning rate, and learning rate decay period, to improve the convergence speed and forecasting accuracy of the GRU model. Finally, the forecasting accuracies of various hybrid forecasting models were compared to demonstrate the superiority of the proposed algorithm.

Compared with existing wind speed forecasting models, our proposed model has the following advantages:

(1) A UTSWS forecasting model based on VMD-AOA-GRU is proposed.
(2) VMD is employed to extract high-frequency wind speed features from time series wind speeds.
(3) The hyperparameters of the GRU model are optimized using the AOA to construct a hybrid AOA-GRU model.
(4) The proposed model outperforms other models for the four wind speed datasets.

The remainder of this paper is organized as follows: In Section 2, the principles of time series wind speed forecasting are introduced. The computational principles and decomposition process of VMD are presented in Section 3. The computational principles and process of the AOA are described in Section 4, including the calculation process for the accelerated function of the math optimizer, the global exploration phase, and the local exploitation phase. Section 5 describes the construction process for the AOA-GRU hybrid model, including the principles of GRU and the construction of the AOA-GRU model. The construction and validation processes for the VMD-AOA-GRU model are outlined in Section 6. In Section 7, the multi-step forecasting results of various models are compared, including GRU, VMD-GRU, VMD-AOA-GRU, LSTM, VMD-LSTM, PSO-ELM, VMD-PSO-ELM, PSO-BP, VMD-PSO-BP, PSO-LSSVM, VMD-PSO-LSSVM, ARIMA, and VMD-ARIMA. The results of this study are summarized in Section 8.

2. Principles of Time Series Wind Speed Forecasting

Wind speed sequences are essentially time series data, and for a specific wind farm, the wind speed time series can vary significantly under different climatic conditions. We assume that a set of real wind speed time series is denoted as $A_k = \{a_1, a_2, a_3, \ldots, a_k\}$, where

a_k represents the wind speed at the kth time step. Then, the wind speed, a_{k+1}, at the $(k+1)^{th}$ time step can be calculated using Equation (1).

$$a_{k+1} = f(A_k) = f(a_1, a_2, \cdots, a_{k-1}, a_k), \tag{1}$$

where $f(\cdot)$ represents the forecasting model, and k represents the length of the sliding time window.

Equation (1) expresses single-step wind speed forecasting. The principle of multi-step forecasting of time series wind speeds is similar to that of single-step forecasting. Taking two-step forecasting as an example, the wind speed, a_{k+2}, at time $k+2$ can be calculated using Equation (2).

$$a_{k+2} = f(a_2, a_3, \cdots, a_k, a_{k+1}) \tag{2}$$

In Equation (2), the wind speed sequence $\{a_1, a_2, a_3, \ldots, a_k\}$ is shifted one step to the left, and the forecasting value a_{k+1} at time $k+1$ is then added to the wind speed sequence as an input to the forecasting model. Similarly, the predicted value a_{k+n} at time $k+n$ can be calculated using Equation (3).

$$a_{k+n} = f\left(a_n, a_{n+1}, \cdots, a_{k+(n-2)}, a_{k+(n-1)}\right) \tag{3}$$

As indicated by the above equations, in time series wind speed forecasting, as the forecasting step increases, the forecasting error will accumulate with each step in the forecasting model, $f(\cdot)$, leading to a decrease in forecasting accuracy.

The principle of time series wind speed forecasting is illustrated in Figure 1.

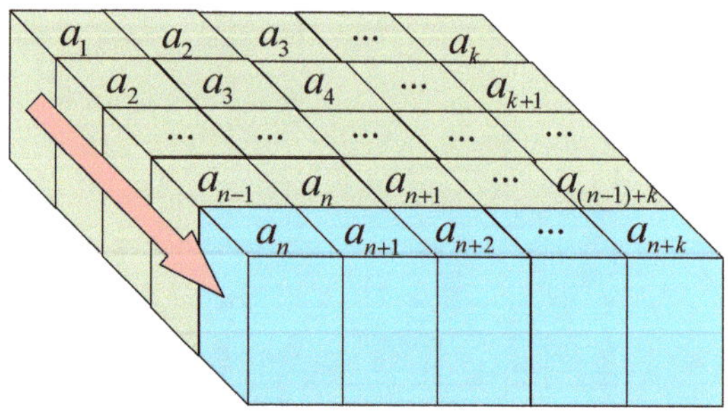

Figure 1. Principle of time series wind speed forecasting.

3. Variational Mode Decomposition

VMD is a non-stationary signal decomposition method introduced by Dragomiretskiy et al. in 2014 [39]. VMD achieves signal decomposition by introducing variational constraints, thus effectively overcoming the mode mixing and endpoint effects issues present in traditional empirical mode decomposition (EMD) methods.

VMD assumes that any signal $f(t)$ can be represented by a series of sub-signals u_k with specific center frequencies and finite bandwidths, and the sum of the bandwidths of all of the sub-signals is equal to the original signal. Then, the variational constraint equation shown in Equation (4) can be obtained.

$$\begin{cases} \min\limits_{\{u_k\},\{\omega_k\}} \left\{ \sum\limits_k \left|\left|\partial_t[(\delta(t) + j/\pi \cdot t) \otimes u_k(t)]e^{-i\omega_k t}\right|\right|_2^2 \right\} \\ \text{s.t. } \sum\limits_{k=1}^{K} u_k = f(t) \end{cases} \tag{4}$$

In Equation (4), K represents the number of variational modes, $\{u_k, \omega_k\}$ represents the kth decomposed mode component with its corresponding center frequency, ∂_t is the partial derivative with respect to time t, $\delta(t)$ represents the Dirac function, and $\otimes$ represents the convolution operator.

To transform a constrained variational problem into a non-constrained variational problem, VMD introduces quadratic penalty factors, α, and Lagrange multipliers, $\lambda(t)$, to construct the augmented Lagrangian function $L(\{u_k\}, \{\omega_k\}, \lambda)$, thereby converting the constrained problem into an unconstrained one. The constructed Lagrangian function is expressed in Equation (5).

$$\begin{aligned} L(\{u_k\},\{\omega_k\},\lambda) &= \alpha \sum_{k=1}^{K} \left\|\partial_t[(\delta(t)+j/\pi\cdot t)\otimes u_k(t)]e^{-i\omega_k t}\right\|_2^2 \\ &+\left\|f(t)-\sum_{k=1}^{K} u_k(t)\right\|_2^2+\left\langle \lambda(t), f(t)-\sum_{k=1}^{K} u_k(t)\right\rangle \end{aligned} \tag{5}$$

The augmented Lagrangian function $L(\{u_k\}, \{\omega_k\}, \lambda)$ is solved using the alternative direction method of multipliers (ADMM); the specific steps are as follows:

Step 1: Initialize $\{u_k^1\}$, $\{\omega_k^1\}$, λ^1 and the iteration number $n = 0$.

Step 2: Execute the iteration $n = n + 1$.

Step 3: Update $\{u_k^n\}$, $\{\omega_k^n\}$, and λ^n based on Equations (6)–(8), respectively.

$$u_k^n(\omega) = \frac{f(\omega) - \sum_{i=1}^{k-1} u_i^n(\omega) - \sum_{i=k+1}^{K} u_i^{n-1}(\omega) + \frac{\lambda^{n-1}(\omega)}{2}}{1+2\alpha(\omega-\omega_k^{n-1})^2} \tag{6}$$

$$\omega_k^n = \frac{\int_0^\infty \omega |u_k^n(\omega)|^2 d\omega}{\int_0^\infty |u_k^n(\omega)|^2 d\omega} \tag{7}$$

$$\lambda^n(\omega) = \lambda^{n-1}(\omega) + \tau[f(\omega) - \sum_{k=1}^{K} u_k^n(\omega)] \tag{8}$$

In the above equations, $f(\omega)$ represents the Fourier transform of signal $f(t)$, and τ is the noise tolerance parameter.

Step 4: If the convergence condition is satisfied, the iteration stops; otherwise, return to Step 2 for further refinement.

$$\sum_{k=1}^{K} \left\|u_k^n - u_k^{n-1}\right\|_2^2 \Big/ \left\|u_k^{n-1}\right\|_2^2 < e \tag{9}$$

In Equation (9), e is the convergence condition for stopping the iteration.

The above calculation process demonstrates that VMD transforms the signal from the time domain to the frequency domain for decomposition. For non-stationary time series signals, this approach effectively preserves the non-stationary information while ensuring the robustness of the decomposition process.

4. Arithmetic Optimization Algorithm (AOA)

The arithmetic optimization algorithm (AOA) is a metaheuristic optimization algorithm based on the concept of mixed arithmetic operations; it was proposed by Abualigah et al. in 2021 [40]. The AOA is characterized by its fast convergence speed and high precision. The AOA consists of three parts: the mathematical optimizer acceleration function, global exploration stage, and local exploitation stage. A mathematical optimizer acceleration function is employed to select the optimization strategy. In the global exploration stage, multiplication and division strategies are utilized for global search, enhancing the dispersion of solutions and improving the global optimization ability of the AOA. In the local exploitation stage, addition and subtraction strategies are used to reduce the dispersion and strengthen the local optimization ability of the AOA.

(1) Mathematical Optimizer Acceleration Function

At the beginning of each iteration loop, the global exploration and local exploitation stages are selected using a mathematical optimizer acceleration function. Suppose that there are N candidate solutions for the problem to be solved in the solution space, Z, and the position of the ith candidate solution in the Z-dimensional solution space is $X_i(x_{i1}, x_{i2}, \ldots, x_{iZ})$, where $i = 1, 2, \ldots, N$. The solution set can then be represented by Equation (10).

$$X = \begin{bmatrix} x_{1,1} & \cdots & \cdots & x_{1,j} & x_{1,Z-1} & x_{1,Z} \\ x_{2,1} & \cdots & \cdots & x_{2,j} & \cdots & x_{2,Z} \\ \cdots & \cdots & \cdots & \cdots & \cdots & \cdots \\ \vdots & \vdots & \vdots & \vdots & \vdots & \vdots \\ x_{N-1,1} & \cdots & \cdots & x_{N-1,j} & \cdots & x_{N-1,Z} \\ x_{N,1} & \cdots & \cdots & x_{N,j} & x_{N,Z-1} & x_{N,Z} \end{bmatrix} \tag{10}$$

The AOA selects the search stage using the mathematical optimizer acceleration function. When $r_1 \geq$ MOA, the AOA performs the global exploration stage, and when $r_1 <$ MOA, the AOA performs the local exploitation stage. Here, r_1 is a random number in the range [0, 1]. The MOA is calculated using Equation (11).

$$\text{MOA}(t) = \text{Min} + t \cdot \left(\frac{\text{Max} - \text{Min}}{T}\right) \tag{11}$$

Min and Max represent the minimum and maximum values of the mathematical optimizer acceleration function, which are typically set to 0.2 and 1, respectively; t is the current number of iterations, and T is the overall number of iterations.

(2) Global Exploration Stage

In the global exploration stage, the AOA employs two search strategies: multiplication and division. When $r_2 \geq 0.5$, the multiplication search strategy is executed, and when $r_2 < 0.5$, the division search strategy is executed. The formulas for the multiplication and division search strategies are given in Equation (12).

$$x(t+1) = \begin{cases} x(t) \cdot \text{MOP} \cdot ((\text{UB} - \text{LB}) \cdot \mu + \text{LB}), r_2 \geq 0.5 \\ x(t)/(\text{MOP} + \varepsilon) \cdot ((\text{UB} - \text{LB}) \cdot \mu + \text{LB}), r_2 < 0.5 \end{cases} \tag{12}$$

In Equation (12), UB and LB represent the upper and lower bounds of the solution space, respectively, r_2 is a random number between [0, 1], and μ is the control parameter for adjusting the search process, with a typical value of 0.499. MOP is the mathematical optimizer probability, which is calculated as shown in Equation (13).

$$\text{MOP}(t) = 1 - \frac{t^{\frac{1}{\alpha}}}{T^{\frac{1}{\alpha}}}, \tag{13}$$

where α represents the sensitive parameter, which defines the local exploitation accuracy during the iteration process; it typically has a value of 5.

(3) Local Exploitation Stage

In the local exploitation stage, addition and subtraction operations are employed by the AOA to fine-tune the solutions obtained during the global exploration stage. The formulae for the addition and subtraction operations are expressed in Equation (14).

$$X(t+1) = \begin{cases} x_b(t) - \text{MOP} \cdot ((\text{UB} - \text{LB}) \cdot \mu + \text{LB}),\ r_3 < 0.5 \\ x_b(t) + \text{MOP} \cdot ((\text{UB} - \text{LB}) \cdot \mu + \text{LB}),\ r_3 \geq 0.5 \end{cases} \tag{14}$$

In Equation (14), r_3 is a random variable with a value in the range of [0, 1]. When $r_3 < 0.5$, the subtraction operation is performed; when $r_3 \geq 0.5$, the addition operation is executed.

5. AOA-GRU Hybrid Model

5.1. GRU Algorithm Principles

The gated recurrent unit (GRU) is a variant of the long short-term memory (LSTM) neural network proposed by Cho et al. in 2014 [41,42]. The GRU not only effectively solves the problems of gradient vanishing and gradient explosion in recurrent neural networks (RNNs), but also avoids the problems of a large number of parameters, low training efficiency, and slow convergence speed in LSTM models. Since its inception, the GRU model has been widely used in the study of time series problems.

The internal structure of the GRU model is illustrated in Figure 2. Figure 2 shows that the GRU model has two gate control units, namely, the update and reset gates. The update gate determines how much information from the previous and current time steps is continuously transmitted to the future, whereas the reset gate determines how much information from the past should be forgotten.

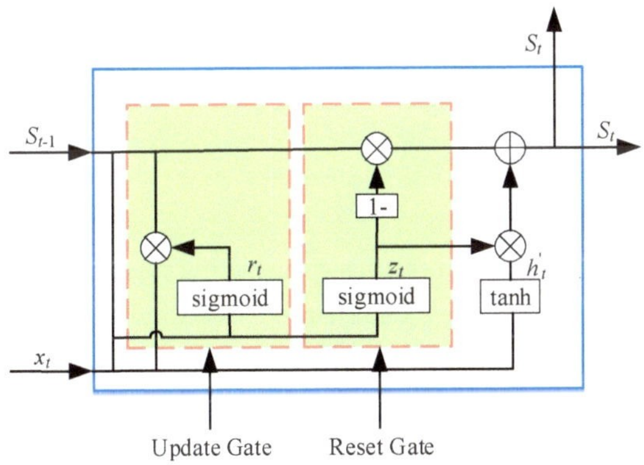

Figure 2. Internal structure of the GRU.

(1) Reset Gate

The calculation formula for the reset gate is given by Equation (15).

$$r_t = \text{sigmoid}\left(W_r^T x_t + U_r^T S_{t-1} + B_r\right) \tag{15}$$

In Equation (15), x_t represents the input vector at time step t, S_{t-1} represents the hidden state at time step $t-1$, W_t and U_t are the weight matrices of the reset gate, B_r is the bias matrix of the reset gate, and r_t is the output of the reset gate at time step t.

The output range of the sigmoid activation function is [0, 1], and its mathematical expression is as follows:

$$\text{sigmoid}(x) = \frac{1}{1+e^{-x}}. \tag{16}$$

(2) Update Gate

The calculation formula for the update gate is given by Equation (17).

$$z_t = \text{sigmoid}\left(W_z^T X_t + U_z^T S_{t-1} + B_z\right) \tag{17}$$

In Equation (17), W_z and U_z are the weight matrices of the update gate, B_z is the bias matrix of the update gate, and z_t is the output of the update gate at time step t.

(3) Candidate Hidden State of the GRU Model

The candidate hidden state of the GRU model is calculated using Equation (18).

$$S_t' = \tanh\left(W_{h_t'}^T X_t + U_{h_t'}^T (r_t * S_{t-1}) + B_{h_t'}\right), \tag{18}$$

where $W_{h_t'}^T$ and $U_{h_t'}^T$ are the weight matrices of the candidate hidden state in the GRU model, $B_{h_t'}$ is the bias matrix of the candidate hidden state, S_t' represents the candidate hidden state of the GRU model, and the tanh function is the hyperbolic tangent activation function, which has an output range of $[-1, 1]$.

(4) Hidden State of the GRU Model

The hidden state of the GRU model is calculated using Equation (19).

$$S_t = (1 - z_t) \odot S_{t-1} + z_t \odot S_t', \tag{19}$$

where S_t represents the hidden state of the GRU model.

(5) GRU Model Output

The output of the GRU model can be calculated using Equation (20).

$$y_t = \text{sigmoid}\left(W_y^T S_t + B_y\right), \tag{20}$$

where y_t represents the output vector of the GRU model, W_y^T is the weight matrix, and B_y is the bias matrix.

5.2. Hyperparameters Affecting the Forecasting Performance of GRU Models

When the hyperparameters of the GRU model are reasonably set, the model exhibits good forecasting accuracy. The hyperparameters affecting the forecasting performance of the GRU model include the number of hidden layers, number of hidden layer neurons, training epochs, initial learning rate, and learning rate decay period. To elucidate the relationship between model hyperparameters and forecasting accuracy, the hyperparameters of the GRU model are adjusted using the grid search method, and the impact of changes in model hyperparameters on the forecasting accuracy of the wind speed series Dataset1 are observed (relevant information on the wind speed series Dataset1 can be found in Section 6.1). The variation ranges of each parameter adjusted using the grid search method are listed in Table 1.

Table 1. Grid search method for tuning GRU model hyperparameters.

Hyperparameter	Search Scope	Optimal Parameter Value
Number of Hidden Layers	[1, 2, 3, 4]	2
Number of Hidden Layer Neurons	[10, 20, 30, 40, 50, 60]	20
Number of Training Epochs	[30, 40, 50, 60, 70, 80]	70
Initial Learning Rate	[0.02, 0.04, 0.06, 0.08, 0.1]	0.06
Learning Rate Decay Period	[10, 20, 30, 40, 50, 60]	30

(1) Impact of the number of hidden layers on the model forecasting performance

In theory, the greater the number of hidden layers the GRU model has, the higher the GRU model forecasting accuracy will be. However, the training time and computational memory requirements of the GRU model increase rapidly with the number of hidden layers. Additionally, a higher number of hidden layers may lead to issues such as overfitting and gradient vanishing. Therefore, selecting the appropriate number of hidden layers is critical for setting the hyperparameters of the GRU model.

Figure 3 shows the relationship between the model forecasting error (RMSE), model training time, and number of hidden layers (1, 2, 3, or 4) for the time series wind speed sequence in Dataset1. The solid blue line with squares represents the model forecasting error for different numbers of hidden layers, and the solid red line with dots represents the model training time. It can be observed that when the number of hidden layers is set to one, the GRU model has a relatively large forecasting error but requires less training time. As the number of hidden layers increases, the model forecasting accuracy improves; however, the training time also increases. When the number of hidden layers is two, the GRU model achieves the best balance between forecasting accuracy and training time.

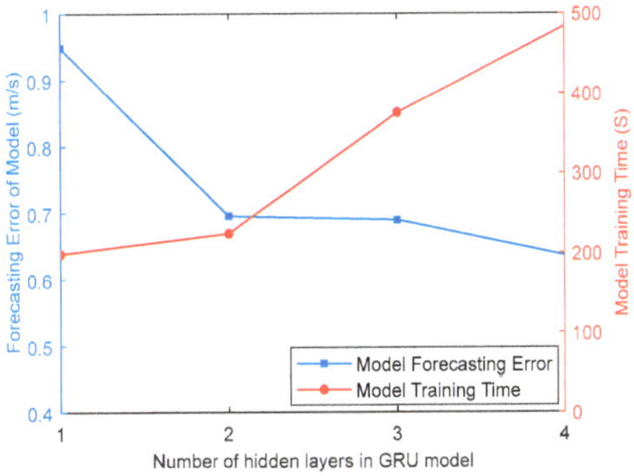

Figure 3. Relationships among the number of hidden layers, forecasting error, and training time.

(2) Influence of the number of hidden layer neurons on the model forecasting performance

The number of hidden layer neurons also has a significant impact on the forecasting performance of the GRU model. When the number of hidden layer neurons is small, the GRU model is prone to underfitting; when there are too many neurons, the GRU model is prone to overfitting. Therefore, selecting a reasonable number of hidden layer neurons is also an important aspect of the hyperparameter settings in the GRU model. Figure 4 shows the correlations between the number of hidden layer neurons, model forecasting error, and training time. As shown in Figure 4, when the number of hidden layer neurons is 20, the forecasting accuracy and training time of the GRU model reach an optimal balance.

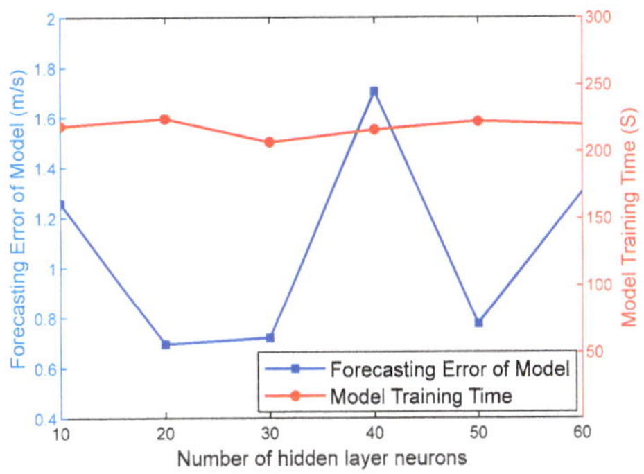

Figure 4. Relationship among the number of hidden layer neurons, forecasting error, and training time.

(3) Impact of training epochs on the model forecasting performance

One training epoch represents a complete operation on the data in a neural network. The weight matrix in the GRU network must be updated every time the data are fully trained. Too few training epochs cannot accurately extract temporal features from the training data, resulting in underfitting of the GRU model. Excessive training epochs increase the training time of the GRU model. Therefore, determining a reasonable number of training epochs and accurately extracting temporal features from the data are important aspects to be considered in the hyperparameter setting of the GRU model.

As shown in Figure 5, when the number of training epochs for the GRU model is set to 70, the model achieves the lowest forecasting error, and the increase in training time is relatively minimal.

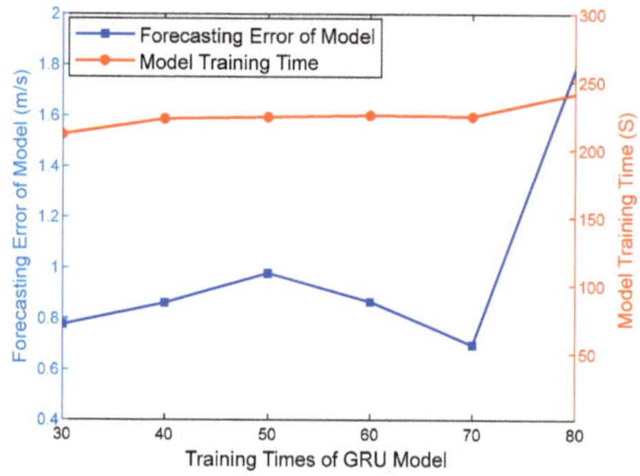

Figure 5. Relationship among the number of training epochs, forecasting error, and training time.

(4) Impact of the learning rate and learning rate decay period on the model forecasting performance

The learning rate is a crucial hyperparameter that controls the learning progress of the GRU model. Setting a learning rate that is too high or too low can affect both the model training time and forecasting accuracy. Typically, a higher learning rate is used in the initial stages of model training to enhance the initial learning speed. After a certain number of training iterations, it is necessary to decrease the learning rate to improve the model forecasting accuracy.

Figure 6a illustrates the impact of the learning rate on the model forecasting accuracy and training time. As shown in Figure 6a, the GRU model achieves the best forecasting performance when the initial learning rate is set to 0.06.

Figure 6b shows the influence of the learning rate decay period on the model forecasting accuracy and training time. As shown in Figure 6b, the GRU model performs optimally when the learning rate decay period is set to 30.

Based on the information in Figures 3–6, it is evident that setting the number of hidden layers, number of neurons in the hidden layer, number of training epochs, learning rate, and learning rate decay period appropriately can effectively improve the forecasting accuracy of the GRU model.

5.3. AOA Optimized Hyperparameters of the GRU Model

Based on the discussion in Section 5.2, it is clear that setting appropriate hyperparameters for the GRU model can improve the forecasting accuracy while reducing the model training time. However, in practical engineering applications, the training dataset is continually changing, and the GRU model hyperparameters must be adjusted accordingly. When the grid search method is used to adjust the GRU model, the hyperparameters are inefficient and may not meet the demands of real-world applications. Therefore, in this study, the AOA is utilized to optimize the GRU model hyperparameters, including the training data sequence length, number of hidden layer neurons, training epochs, initial learning rate, and learning rate decay period. The number of hidden layers in the GRU model is determined using a grid search method and remains constant throughout the optimization process. The construction process for the AOA-GRU model is illustrated in Figure 7.

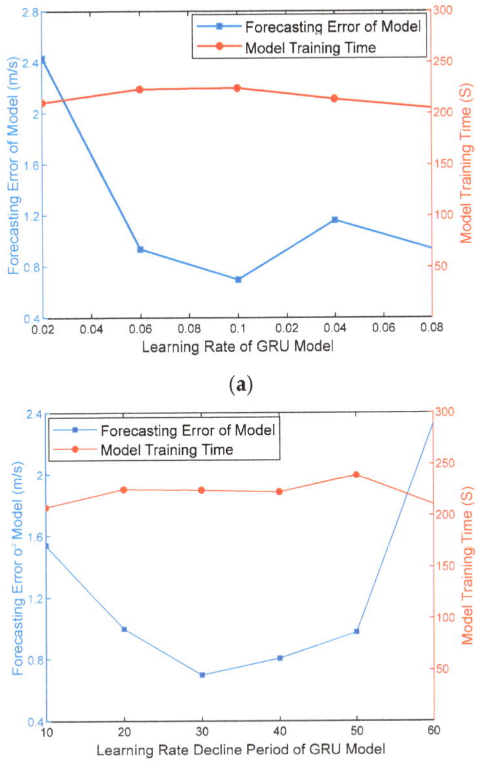

(**a**)

(**b**)

Figure 6. Relationship among the learning rate, learning rate decay period, forecasting error, and training time. (**a**) Impact of the learning rate on the model forecasting performance. (**b**) Impact of the learning rate decay period on the model forecasting performance.

The specific AOA optimization process for the hyperparameters of the GRU model is as follows.

Step 1: Construct an AOA candidate solution structure, $X_i(x_{i1}, x_{i2}, x_{i3}, x_{i4}, x_{i5})$, including the training data time series length, x_{i1}, number of hidden layer neurons, x_{i2}, training iterations, x_{i3}, initial learning rate, x_{i4}, and learning rate decay period, x_{i5}. Determine the range and number of candidate solutions.

Step 2: Based on the position coordinates $X_i(x_{i1}, x_{i2}, x_{i3}, x_{i4}, x_{i5})$ of each candidate solution, construct their respective GRU forecasting models and forecast the wind speed sequence.

Step 3: Calculate the forecasting error of each GRU model based on its forecasting results and use the forecasting error as the fitness value for each candidate solution of the AOA.

Step 4: Save the coordinates of the candidate solution with the best fitness value as $X^*(x_1, x_2, x_3, x_4, x_5)$, representing the optimal model hyperparameters.

Step 5: Utilize the AOA to update the coordinates of each candidate solution $X_i(x_{i1}, x_{i2}, x_{i3}, x_{i4}, x_{i5})$ and rebuild each GRU model based on the updated coordinates to forecast the wind speed.

Step 6: Based on the forecasting results of the GRU model, recalculate the forecasting errors of each GRU model and determine the position coordinates $X^*(x_1, x_2, x_3, x_4, x_5)$ of the candidate solution with the best fitness value, i.e., the optimal training data length and model hyperparameters.

Step 7: Check whether the termination condition is satisfied. If not, return to Step 5 and continue the process. If the termination condition is satisfied, proceed to Step 8.

Step 8: Build the final GRU forecasting model based on the coordinates of the global optimal candidate solution, $X^*(x_1, x_2, x_3, x_4, x_5)$, to forecast the wind speed.

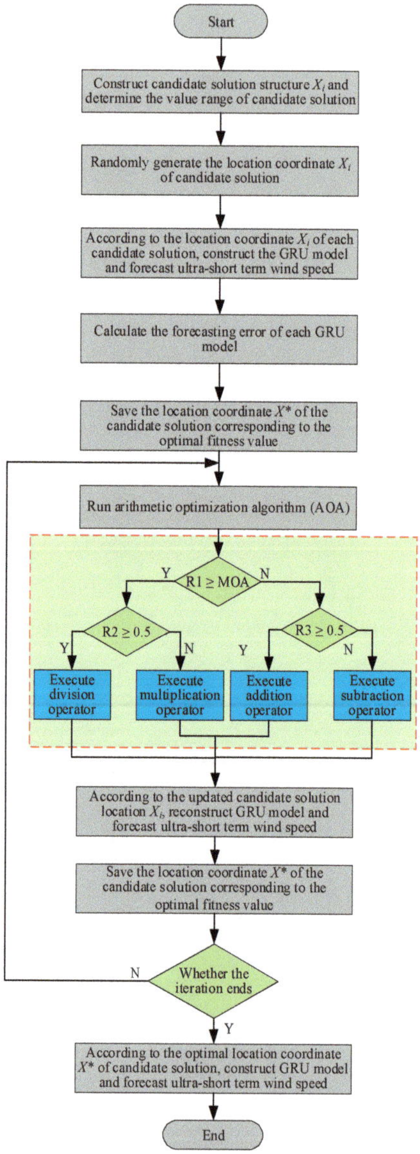

Figure 7. Construction process for the AOA-GRU model.

6. Construction and Verification of the VMD-AOA-GRU Model

6.1. Data Sources and Sample Set Partition

(1) Data Sources

The wind speed data used in this study were derived from the Supervisory Control and Data Acquisition (SCADA) system of a wind farm located in the northwestern region of China that includes 134 wind turbines. The distribution of the wind turbines is shown in Figure 8. The relative coordinates of each turbine are shown in the figure, with the x- and y-axis units expressed in meters. The collection period for the wind speed data is 180 d, and the data time resolution is 10 min. Among them, the 58th and 123rd wind turbines (with red borders) are the research objects of this paper.

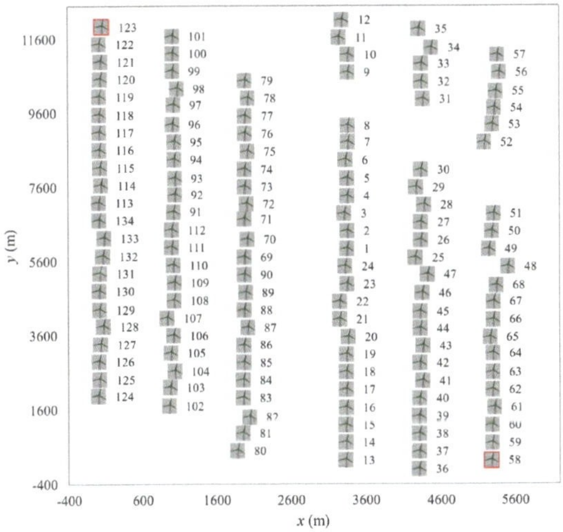

Figure 8. Wind turbine location distribution of wind farm.

The ultra-short-term forecasting of wind speed mainly refers to the forecasting of one or several time steps in the future. In order to verify the high forecasting accuracy of the model proposed in this paper at different forecasting time scales, the ultra-short-term wind speeds of one time step, two time steps, and three time steps were forecasted, with each time step being 10 min.

(2) Sample Set Partitioning

To validate the effectiveness of the proposed algorithm, the wind speed data from the 58th and 123rd wind turbines, located far apart, were selected for calculation and analysis. Additionally, to verify the forecasting performance of the proposed algorithm under different climatic conditions, four datasets from the 10th to the 19th day and from the 153rd to the 162nd day were chosen for model training and testing (two datasets per turbine). The datasets belonging to the 58th wind turbine were denoted as Dataset1 (from the 10th to the 19th day) and Dataset2 (from the 153rd to the 162nd day), while the datasets from the 123rd wind turbine were denoted as Dataset3 (from the 10th to the 19th day) and Dataset4 (from the 153rdto the 162nd day). Each dataset contained 1440 data points, with the first 1320 data points designated as the training dataset and the last 120 data points designated as the testing dataset.

6.2. VMD-AOA-GRU Model Construction

When forecasting the UTSWS, VMD was first used to decompose the time series wind speed data into different frequency time series data, thus effectively extracting the high-frequency data features from the wind speed data while removing noise and thereby enhancing the forecasting accuracy of the model. Subsequently, the constructed AOA-GRU model was used to train and test the temporal wind speed data at different frequencies. The implementation process for the entire VMD-AOA-GRU model is illustrated in Figure 9.

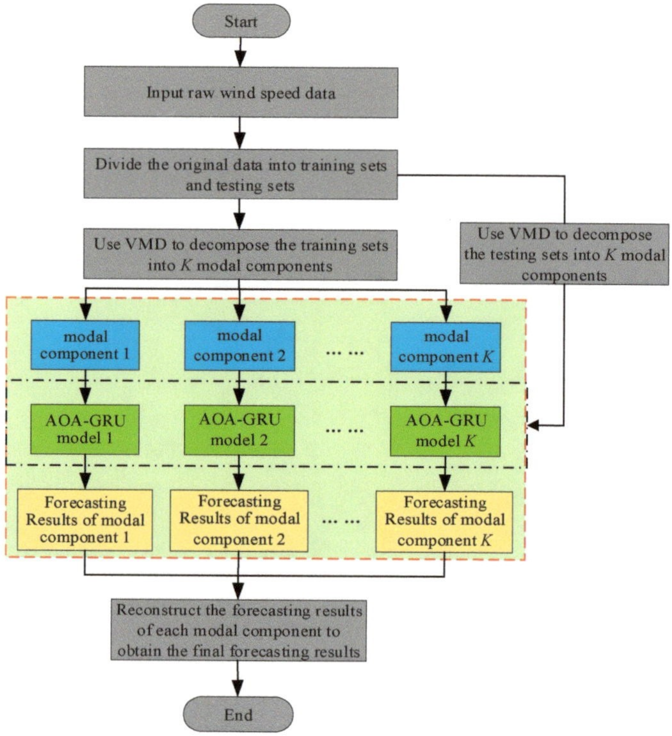

Figure 9. Implementation process for the VMD-AOA-GRU model.

The specific implementation steps are as follows:

(1) Utilize VMD to decompose the training and testing datasets and obtain K modal components of different frequencies.
(2) Input each modal component derived from the decomposed training dataset into the AOA-GRU model separately and train the AOA-GRU model.
(3) Input each modal component derived from the decomposed testing dataset into the trained AOA-GRU model to achieve ultra-short-term forecasting for each modal component.
(4) Reconstruct the UTSWS based on the ultra-short-term forecasting results of each modal component of the testing dataset.

6.3. Verification of the VMD-AOA-GRU Model

6.3.1. Determination of the Number of VMD Modal Components

When the number of modal components in VMD is small, some important information in the original time series wind speed will be filtered out, reducing the forecasting accuracy of the VMD-AOA-GRU model. When the number of modal components is large, modal overlap can easily occur or additional noise may be generated. Therefore, the selection of an appropriate number of VMD modal components has become a key concern in VMD research and application. To determine the optimal number of temporal wind speed modal components, the number of temporal wind speed decomposed modal components was gradually increased in this study, and the Pearson correlation coefficient method was used to calculate the correlation coefficients between the decomposed modal components. The optimal number of VMD modal components can be determined by calculating their correlation coefficients.

Table 2 presents the center frequency distribution of the time series wind speed for Dataset1 after decomposition under different numbers of modal components. Table 2 demonstrates that as the number of modal components increases, the spacing between the center frequencies of each mode gradually decreases.

Table 2. Central frequency distribution of the decomposed time series wind speed.

Number of Modes (K)	Central Frequency							
2	0.1491 Hz	0.0004 Hz						
3	0.3401 Hz	0.0776 Hz	0.0003 Hz					
4	0.4367 Hz	0.1792 Hz	0.0551 Hz	0.0003 Hz				
5	0.4494 Hz	0.2573 Hz	0.1002 Hz	0.0156 Hz	0.0002 Hz			
6	0.3768 Hz	0.2106 Hz	0.1094 Hz	0.0529 Hz	0.0098 Hz	0.0001 Hz		
7	0.3875 Hz	0.2525 Hz	0.1540 Hz	0.1007 Hz	0.0511 Hz	0.0092 Hz	0.0001 Hz	
8	0.4255 Hz	0.3154 Hz	0.2231 Hz	0.1543 Hz	0.1033 Hz	0.0514 Hz	0.0088 Hz	0.0001 Hz

Table 3 presents the Pearson correlation coefficients between the various modal components of the original time series wind speed after decomposition under different numbers of modal components. The results reveal that when the number of modal components is two or three, the Pearson correlation coefficient between each modal component is low. However, because of the small number of modal components, important temporal wind speed information is easily lost. When the number of modal components is five, six, or greater, the Pearson correlation coefficient between some modal components is higher (bold font in Table 3), indicating the existence of overlap between these modal components. When the number of modal components is four, the Pearson correlation coefficient between each modal component is low, and the number of modal components is relatively small. Thus, it can be concluded that the most reasonable choice for the VMD calculation of the time series wind speed is four modal components. In the subsequent VMD calculation of the time series wind speed in this study, four modal components were used.

Table 3. Pearson correlation coefficients between adjacent modal components.

Number of Modal Components (K)	C_{12}	C_{23}	C_{34}	C_{45}	C_{56}	C_{67}	C_{78}
2	0.015163						
3	0.025336	0.04361					
4	0.022502	0.061709	0.053189				
5	0.035944	0.048752	**0.069011**	**0.157038**			
6	0.040904	0.064714	**0.109889**	**0.079833**	**0.144653**		
7	0.061023	0.072363	**0.117903**	**0.132261**	**0.074168**	**0.14186**	
8	0.068861	0.085539	**0.118399**	**0.116491**	**0.113842**	**0.066931**	**0.144964**

6.3.2. Training of VMD-AOA-GRU Model

It is necessary to train the VMD-AOA-GRU model to achieve ultra-short-term forecasting. When training the VMD-AOA-GRU model in this study, the training datasets used were the four training datasets introduced in Section 6.1 (Training Dataset 1, Training Dataset 2, Training Dataset 3, and Training Dataset 4). As described in Sections 6.2 and 6.3.1, during the training of the VMD-AOA-GRU model using the training datasets, each group of training data needed to be decomposed into four modal components (resulting in 16 modal components from four datasets). Subsequently, the four VMD-AOA-GRU models were trained using the four decomposed modal components of each training dataset. In order to solve the problem of overfitting and bias during the model training process, the model was continuously trained 10 times, and the average hyperparameter value obtained from 10 times training was taken as the hyperparameter of the proposed model. The training process was executed on a laptop equipped with an Intel i7-6500U CPU running at a main frequency of 2.29 GHz and memory of 16 GB. The calculation software was MATLAB 2021a.

Table 4 presents a detailed list of the hyperparameter values of the VMD-AOA-GRU model optimized by AOA during the one-step forecasting process. Table 4 reveals that (1) the hyperparameter values of the GRU model trained by different modal components obtained using the same training dataset are different, and (2) the hyperparameter values of the GRU model trained by different modal components obtained from different training datasets are different.

Table 4. Hyperparameter values of the GRU model optimized by AOA.

Sequence	Training Dataset 1				Training Dataset 2			
Modal Components	1	2	3	4	1	2	3	4
Temporal Length of Training Data	21	17	9	47	14	11	10	20
Number of Neurons in Hidden Layers	28	50	21	50	50	30	50	44
Number of Training Epochs	91	81	75	100	100	34	71	89
Learning Rate	0.0478	0.0355	0.079	0.0572	0.0766	0.0581	0.0269	0.0574
Learning Rate Decay Period	15	30	9	16	10	25	29	30
MAE	3.46%	4.7%	4.32%	3.96%	7.54%	4.4%	3.23%	2.74%
RMSE	4.98%	6.59%	5.36%	4.82%	9.71%	5.81%	3.93%	3.35%
MAPE	0.53%	0.71%	0.85%	0.01%	0.62%	0.30%	0.21%	0.01%
Training Time	187.4 s	190.1 s	153.9 s	232.4 s	233.9 s	88.6 s	172.1 s	201.9 s
Sequence	**Training Dataset 3**				**Training Dataset 4**			
Modal Components	1	2	3	4	1	2	3	4
Temporal Length of Training Data	9	10	9	56	9	16	11	20
Number of Neurons in Hidden Layers	39	44	16	24	30	33	34	48
Number of Training Epochs	67	83	62	96	58	96	35	80
Learning Rate	0.1	0.0391	0.1	0.0589	0.087	0.0544	0.0409	0.0564
Learning Rate Decay Period	30	30	30	30	26	25	30	17
MAE	4.51%	4.08%	2.12%	2.58%	3.66%	3.4%	3.14%	2.74%
RMSE	6.52%	5.53%	2.92%	3.13%	4.59%	4.41%	4.13%	3.47%
MAPE	1.60%	0.32%	0.12%	0.01%	2.51%	0.32%	0.15%	0.01%
Training Time	157.2 s	189.7 s	127.3 s	187.2 s	133 s	201.1 s	92.1 s	186.6 s

There are three reasons for this outcome.

(a) The wind speed data features contained in the four modal components decomposed from the same training dataset differ. Therefore, after using the four modal components to train the AOA-GRU model, the hyperparameter values of the GRU model optimized by AOA are different.

(b) Owing to the large distance between the 58th and 123rd wind turbines, there are certain differences in the time series wind speed data for these two turbines during the same time period, and the decomposed modal components also differ. After using these modal components to train the AOA-GRU model, the hyperparameter values of the GRU model optimized by AOA are different.

(c) The time series wind speed data from the same wind turbine differ significantly during different time periods, and the decomposed modal components also exhibit significant differences. After training the AOA-GRU model using modal components, the hyperparameter values of the GRU model optimized by AOA are different.

From the above analysis, it can be observed that the hyperparameter values of the GRU model change with changes in the training dataset. Therefore, in the process of real-time forecasting of ultra-short-term wind speeds, it is necessary to use the AOA to optimize the hyperparameters of the GRU model in real time to obtain the optimal forecasting performance.

6.3.3. Forecasting Analysis of the VMD-AOA-GRU Model

To validate the performance of the proposed VMD-AOA-GRU model for ultra-short-term wind speed forecasting, the forecasting results of the GRU, VMD-GRU, AOA-GRU, and VMD-AOA-GRU models were compared at different forecasting time steps. The hyperparameter values of GRU and VMD-GRU were adjusted using the grid search method, which ensured that these models were under fair comparison conditions during the comparison process. The training dataset of these models was derived from the training data described in Section 6.1, and the forecasting results of the four testing datasets (Testing Dataset1, Testing Dataset2, Testing Dataset3, and Testing Dataset4) are shown in Figure 10.

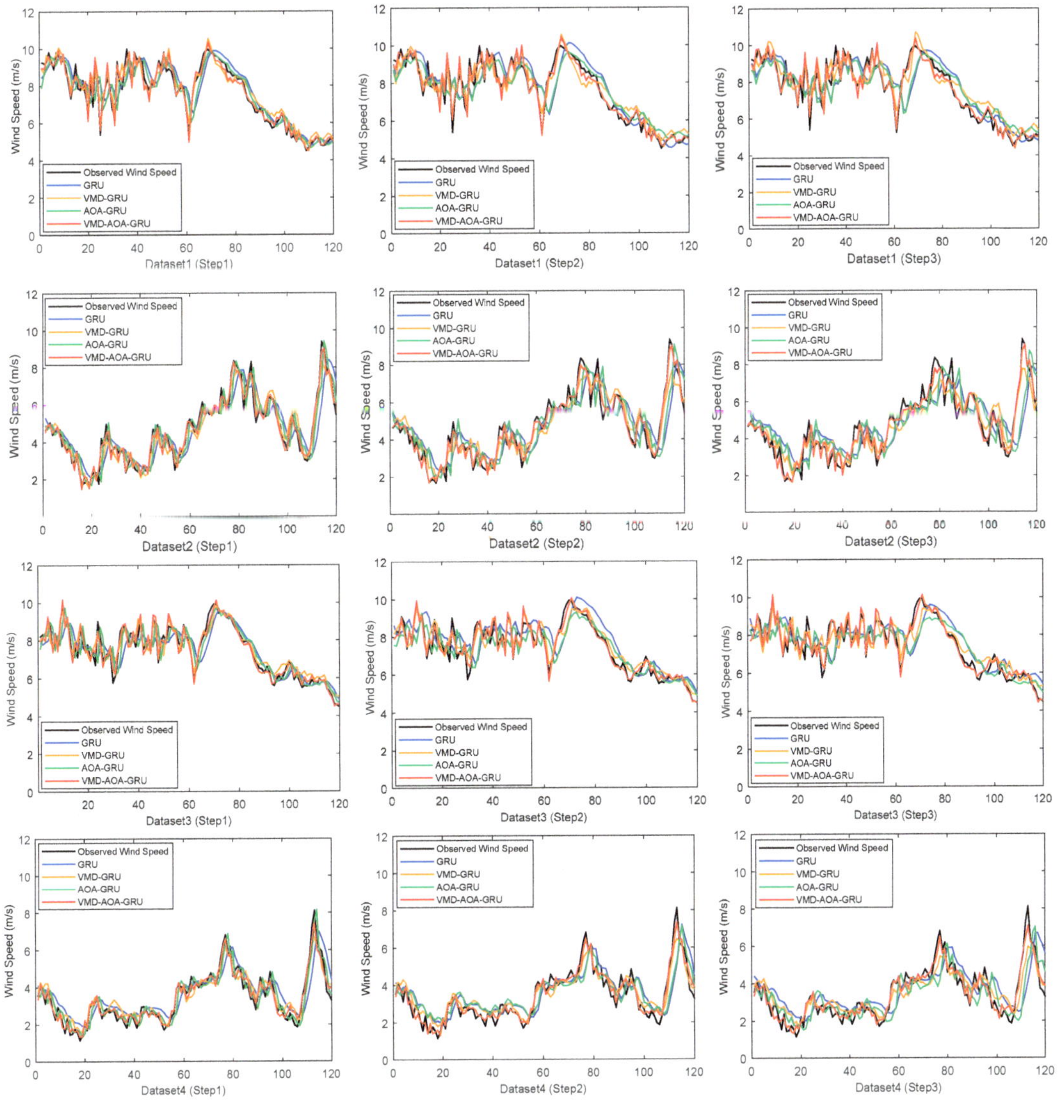

Figure 10. Multi-step forecasting results for the GRU model and its improved model.

In Figure 10, the black solid line represents the actual observed wind speed, the blue solid line represents the forecasting of the GRU model, the yellow solid line represents the forecasting of the VMD-GRU model, the green solid line represents the forecasting of the AOA-GRU model, and the red solid line represents the forecasting of the VMD-AOA-GRU model.

Based on the forecasting results in Figure 10, the following conclusions can be drawn.

(1) Under different forecasting time steps, all four forecasting models can accurately reflect the trend of actual wind speed changes, confirming that the GRU model and its hybrid model perform well in time series wind speed forecasting.
(2) The wind speed inflection point in Figure 10 reveals that the forecasting results of the GRU and AOA-GRU models lag behind those of the VMD-GRU and VMD-AOA-GRU models. This is because the VMD algorithm can effectively extract high-frequency component features (corresponding to the rapidly changing part of the wind speed) from the wind speed sequence, and these high-frequency component features are used as a component of the input data of the forecasting model, enabling the VMD-GRU and VMD-AOA-GRU models to accurately forecast the sudden changes in the actual wind speed.

Table 5 presents the forecasting error values for the GRU, VMD-GRU, AOA-GRU, and VMD-AOA-GRU models at different forecasting time steps. The following conclusions are drawn based on the results in Table 5:

(1) The VMD-GRU model outperforms the GRU model in terms of forecasting accuracy, and the VMD-AOA-GRU model exhibits a higher forecasting accuracy than the AOA-GRU model. This demonstrates that the VMD algorithm effectively captures the high-frequency components of the wind speed data, thereby enhancing the forecasting accuracy of the forecasting models.
(2) The AOA-GRU model outperforms the GRU model in terms of forecasting accuracy, and the VMD-AOA-GRU model outperforms the VMD-GRU model in terms of forecasting accuracy. This indicates that utilizing the AOA to optimize the hyperparameters of the GRU model can effectively improve its forecasting accuracy.
(3) For the different forecasting models, as the forecasting time step increases, the forecasting error also increases.

Table 5. Forecasting error values at different forecasting time steps.

	Time Steps	1			2			3		
	Error Type	MAE	RMSE	MAPE	MAE	RMSE	MAPE	MAE	RMSE	MAPE
Testing Dataset 1	GRU	0.5499	0.7527	0.0734	0.7272	0.9096	0.0964	0.7524	0.9413	0.0994
	VMD-GRU	0.4159	0.5091	0.0576	0.5143	0.6375	0.0710	0.6018	0.7176	0.0846
	AOA-GRU	0.5178	0.7378	0.0696	0.6743	0.8652	0.0912	0.7420	0.9236	0.1007
	VMD-AOA-GRU	0.2280	0.2990	0.0292	0.2536	0.3382	0.0323	0.2704	0.3585	0.0349
Testing Dataset 2	GRU	0.7005	0.9290	0.1615	0.8503	1.0866	0.2025	0.9865	1.2427	0.2413
	VMD-GRU	0.3967	0.5232	0.0927	0.5400	0.6781	0.1274	0.5542	0.6989	0.1354
	AOA-GRU	0.5259	0.7174	0.1227	0.7793	1.0050	0.1807	0.9585	1.2020	0.2235
	VMD-AOA-GRU	0.2463	0.3001	0.0615	0.2727	0.3286	0.0641	0.3411	0.4422	0.0843
Testing Dataset 3	GRU	0.5426	0.7111	0.0728	0.7227	0.8794	0.0982	0.7178	0.8949	0.0977
	VMD-GRU	0.3390	0.4373	0.0483	0.3697	0.4769	0.0508	0.4685	0.6097	0.0660
	AOA-GRU	0.4937	0.6592	0.0659	0.6291	0.8042	0.0838	0.6862	0.8508	0.0918
	VMD-AOA-GRU	0.1988	0.2576	0.0263	0.2027	0.2729	0.0269	0.2363	0.2923	0.0301
Testing Dataset 4	GRU	0.6143	0.7849	0.2029	0.6785	0.8854	0.2218	0.8397	1.0956	0.2713
	VMD-GRU	0.3970	0.4965	0.1346	0.4283	0.5355	0.1489	0.5287	0.6590	0.1756
	AOA-GRU	0.4425	0.5597	0.1387	0.6373	0.8183	0.2145	0.7011	0.9456	0.2094
	VMD-AOA-GRU	0.2170	0.2779	0.0701	0.2608	0.3351	0.0855	0.3102	0.3908	0.1054

7. Comparison of Different Forecasting Models

To further demonstrate the superiority of the proposed VMD-AOA-GRU forecasting model, a comparative analysis was conducted for the multiple-step forecasting of various models. The compared models included the LSTM, GRU, PSO-BP, PSO-ELM, PSO-LSSVM, VMD-GRU, VMD-LSTM, VMD-PSO-BP, VMD-PSO-ELM, VMD-PSO-LSSVM, and VMD-AOA-GRU models. In order to ensure that these models were under fair comparison conditions during the comparison process, except for the VMD-AOA-GRU model, the hyperparameters of all other models were optimized using the grid search method. The forecasting results for these models are shown in Figure 11. The training and testing datasets used for the comparative analysis were consistent with those presented in Section 6.1.

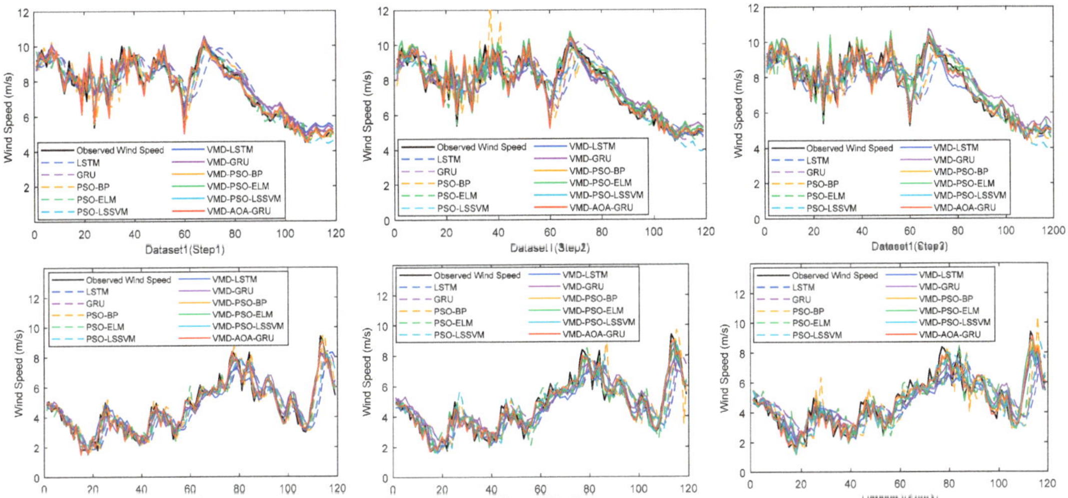

Figure 11. Wind speed forecasting accuracy of different forecasting models.

In Figure 11, the black solid line represents the observed wind speed, the blue dashed line represents the forecasting of the LSTM model, the purple dashed line represents the forecasting of the GRU model, the orange dashed line represents the forecasting of the

PSO-BP model, the green dashed line represents the forecasting of the PSO-ELM model, the cyan dashed line represents the forecasting of the PSO-LSSVM model, and the red solid line represents the forecasting of the VMD-AOA-GRU model. Based on the observations and analysis shown in Figure 11, the following conclusions can be drawn.

(1) All of the machine learning models accurately capture the trends of the actual wind speed, demonstrating that the use of machine learning models for ultra-short-term wind speed forecasting is feasible.
(2) At the wind speed inflection points, the VMD-AOA-GRU, VMD-LSTM, VMD-GRU, VMD-PSO-BP, VMD-PSO-ELM, and VMD-PSO-LSSVM models accurately forecast the positions of the inflection points (the highest accuracy in single-step forecasting). In contrast, the other models exhibit a lag in forecasting the positions of the inflection points compared with the actual wind speed. This further validates that the VMD algorithm can accurately extract the high-frequency components of the time series wind speed, thereby enhancing the accuracy of wind-speed forecasting.
(3) Among the forecasting models, the VMD-AOA-GRU model shows the closest similarity to the distribution characteristics of the actual time series wind speed. This demonstrates that the forecasting performance of the VMD-AOA-GRU model is superior to that of the other models.

Table 6 lists the forecasting errors of each forecasting model under the four testing wind speed sequences at one, two, and three steps. The following conclusions are drawn based on the results in Table 6:

(1) The forecasting accuracy of the hybrid VMD models is higher than that of the non-hybrid VMD models, indicating that deep mining of high-frequency features in the time series wind speed through VMD can effectively improve the forecasting accuracy of the forecasting model.
(2) The forecasting accuracy of the LSTM and GRU models is lower than that of some machine learning models, indicating that although the LSTM and GRU models have the theoretical potential to achieve high forecasting accuracy by mining temporal correlations in the data, their forecasting accuracy is affected by improper hyperparameter settings.
(3) The forecasting accuracy of the VMD-AOA-GRU model is higher than that of all other models, demonstrating that optimizing the hyperparameters of the GRU model through AOA effectively enhances the forecasting accuracy of the GRU model.
(4) As the forecasting time step increases, the forecasting accuracy of all models gradually decreases, which aligns with the inherent characteristics of forecasting models.

Table 6. Wind speed forecasting errors of different models.

	Time Step		**1**			**2**			**3**	
	Error Type	**MAE**	**RMSE**	**MAPE**	**MAE**	**RMSE**	**MAPE**	**MAE**	**RMSE**	**MAPE**
Testing Dataset 1	LSTM	0.7122	0.8927	0.0948	0.7605	0.9483	0.1004	0.8092	1.0211	0.1049
	GRU	0.5499	0.7527	0.0734	0.7272	0.9096	0.0964	0.7524	0.9413	0.0994
	PSO-BP	0.5722	0.8016	0.0760	0.7276	0.9305	0.0971	0.7565	1.0682	0.1002
	PSO-ELM	0.2937	0.3873	0.0399	0.4859	0.6034	0.0654	0.5159	0.6478	0.0698
	PSO-LSSVM	0.5241	0.7260	0.0716	0.6662	0.8312	0.0919	0.7286	0.8855	0.0995
	VMD-LSTM	0.4140	0.5131	0.0593	0.4148	0.5472	0.0545	0.4637	0.6218	0.0596
	VMD-GRU	0.4159	0.5091	0.0576	0.5143	0.6375	0.0710	0.6018	0.7176	0.0846
	VMD-PSO-BP	0.2306	0.2987	0.0292	0.2896	0.3994	0.0367	0.3467	0.4775	0.0444
	VMD-PSO-ELM	0.2371	0.3172	0.0302	0.4253	0.5602	0.0560	0.4832	0.6103	0.0638
	VMD-PSO-LSSVM	0.2295	0.3045	0.0293	0.2888	0.3955	0.0367	0.3162	0.4365	0.0406
	VMD-AOA-GRU	**0.2280**	**0.2990**	**0.0292**	**0.2536**	**0.3382**	**0.0323**	**0.2704**	**0.3585**	**0.0349**

Table 6. *Cont.*

	Time Step Error Type	MAE	1 RMSE	MAPE	MAE	2 RMSE	MAPE	MAE	3 RMSE	MAPE
Testing Dataset 2	LSTM	0.7528	1.0541	0.1643	0.8739	1.1703	0.1995	1.0092	1.3030	0.2428
	GRU	0.7005	0.9290	0.1615	0.8503	1.0866	0.2025	0.9865	1.2427	0.2413
	PSO-BP	0.5447	0.7329	0.1265	0.7984	1.0600	0.1791	0.9295	1.2103	0.2099
	PSO-ELM	0.4124	0.5301	0.1029	0.5670	0.7277	0.1419	0.6451	0.8573	0.1566
	PSO-LSSVM	0.6130	0.8520	0.1380	0.7448	0.9983	0.1696	0.9032	1.2067	0.2007
	VMD-LSTM	0.4073	0.5381	0.0953	0.4659	0.6060	0.1026	0.5551	0.7177	0.1266
	VMD-GRU	0.3967	0.5232	0.0927	0.5400	0.6781	0.1274	0.5542	0.6989	0.1354
	VMD-PSO-BP	0.2496	0.3046	0.0632	0.3523	0.3535	0.0649	0.4273	0.5705	0.1002
	VMD-PSO-ELM	0.2564	0.3070	0.0643	0.2618	0.4604	0.0820	0.4124	0.5346	0.0971
	VMD-PSO-LSSVM	0.2474	0.3066	0.0622	0.2848	0.3730	0.0677	0.4589	0.6011	0.1071
	VMD-AOA-GRU	**0.2463**	**0.3001**	**0.0615**	**0.2727**	**0.3286**	**0.0641**	**0.3411**	**0.4422**	**0.0843**
Testing Dataset 3	LSTM	0.6011	0.7619	0.0798	0.7163	0.8916	0.0954	0.8098	0.9865	0.1089
	GRU	0.5426	0.7111	0.0728	0.7227	0.8794	0.0982	0.7178	0.8949	0.0977
	PSO-BP	0.4963	0.6558	0.0648	0.6969	0.8854	0.0921	0.7414	0.9364	0.0981
	PSO-ELM	0.3469	0.4394	0.0462	0.4447	0.5677	0.0597	0.5260	0.6446	0.0713
	PSO-LSSVM	0.6398	0.7872	0.0854	0.7180	0.8766	0.0957	0.7375	0.8947	0.0981
	VMD-LSTM	0.2927	0.3846	0.0399	0.4904	0.5983	0.0691	0.4918	0.6181	0.0696
	VMD-GRU	0.3390	0.4373	0.0483	0.3697	0.4769	0.0508	0.4685	0.6097	0.0660
	VMD-PSO-BP	0.2078	0.2755	0.0277	0.2502	0.2790	0.0337	0.2645	0.3690	0.0350
	VMD-PSO-ELM	0.2006	0.2716	0.0265	0.2085	0.3570	0.0277	0.2253	0.3981	0.0405
	VMD-PSO-LSSVM	0.2031	0.2690	0.0269	0.2040	0.2835	0.0270	0.2978	0.3371	0.0311
	VMD-AOA-GRU	**0.1988**	**0.2576**	**0.0263**	**0.2027**	**0.2729**	**0.0269**	**0.2363**	**0.2923**	**0.0301**
Testing Dataset 4	LSTM	0.7196	0.9250	0.2394	0.8377	1.0775	0.2821	0.8782	1.1983	0.2640
	GRU	0.6143	0.7849	0.2029	0.6785	0.8854	0.2218	0.8397	1.0956	0.2713
	PSO-BP	0.4324	0.5544	0.1331	0.6957	0.9454	0.2054	0.7131	0.9583	0.2135
	PSO-ELM	0.3143	0.4076	0.1008	0.4435	0.5777	0.1375	0.5032	0.6291	0.1586
	PSO-LSSVM	0.4349	0.5586	0.1345	0.6184	0.8034	0.1848	0.7263	0.9627	0.2155
	VMD-LSTM	0.4238	0.5176	0.1448	0.4586	0.5561	0.1585	0.5844	0.7049	0.2096
	VMD-GRU	0.3970	0.4965	0.1346	0.4283	0.5355	0.1489	0.5287	0.6590	0.1756
	VMD-PSO-BP	0.2177	0.2902	0.0730	0.3526	0.3358	0.1138	0.3906	0.5082	0.1267
	VMD-PSO-ELM	0.2239	0.2860	0.0728	0.2620	0.4415	0.0872	0.4021	0.5043	0.1239
	VMD-PSO-LSSVM	0.2250	0.2785	0.0704	0.2778	0.3551	0.0904	0.3554	0.4476	0.1145
	VMD-AOA-GRU	**0.2170**	**0.2779**	**0.0701**	**0.2608**	**0.3351**	**0.0855**	**0.3102**	**0.3908**	**0.1054**

8. Conclusions

This study proposes an ultrashort-term forecasting model for time series wind speeds based on VMD-AOA-GRU. The model first uses VMD to decompose the time series wind speed data into different frequency modal components, effectively extracting high-frequency wind speed features. Then, the AOA is employed to optimize the hyperparameters of the GRU model to construct a high-accuracy AOA-GRU forecasting model. The time series modal components decomposed by VMD are then employed to train and test the AOA-GRU model, achieving multi-step forecasting of ultra-short-term time series wind speeds. The forecasting results for the GRU, VMD-GRU, VMD-AOA-GRU, LSTM, PSO-BP, PSO-ELM, PSO-LSSVM, VMD-LSTM, VMD-PSO-BP, VMD-PSO-ELM, and VMD-PSO-LSSVM models were compared, and the results are as follows:

(1) The forecasting accuracies of the hybrid VMD models (VMD-AOA-GRU, VMD-LSTM, VMD-PSO-BP, VMD-PSO-ELM, and VMD-PSO-LSSVM) are higher than those of the non-hybrid VMD models (GRU, LSTM, PSO-BP, PSO-ELM, and PSO-LSSVM), indicating that the VMD can deeply explore high-frequency components in time series

wind speed, particularly the high-frequency features at inflection points, effectively improving the accuracy of the forecasted time series wind speed.
(2) Although the LSTM and GRU deep learning models can capture the temporal correlations in time series wind speeds, their forecasting accuracy may be lower than that of some commonly used machine learning models (PSO-BP, PSO-ELM, and PSO-LSSVM) when their hyperparameter settings are improper. This indicates that a reasonable setting of hyperparameters for deep learning models significantly affects the forecasting accuracy.
(3) The forecasting accuracy of the GRU model can be effectively improved by using the AOA to optimize the hyperparameters of the GRU model. The calculation results show that the forecasting accuracy of the VMD-AOA-GRU model constructed in this study is higher than that of the other models.
(4) As the forecasting time step increases, the forecasting accuracy of the model gradually decreases.

The study results in this paper can be widely used for the optimization control of wind turbines, thereby improving the operational efficiency of wind turbines and reducing their fatigue losses. However, during the research process, this paper did not consider the accuracy of wind direction forecasting, a topic that will need to be a focus of subsequent research.

Author Contributions: Conceptualization, J.Y.; Methodology, J.Y.; Software, F.P.; Validation, F.P.; Formal analysis, F.P.; Investigation, H.X., D.L. and B.G.; Resources, H.X., D.L. and B.G.; Data curation, H.X., D.L. and B.G.; Writing – original draft, J.Y.; Writing – review & editing, J.Y.; Visualization, D.L.; Project administration, J.Y. All authors have read and agreed to the published version of the manuscript.

Funding: This work was supported by the National Key Research and Development Program of China (2019YFE0104800), and the Henan Natural Science Foundation (232300420152).

Data Availability Statement: Data are contained within the article.

Acknowledgments: The authors thank the editors and reviewers for their helpful comments regarding the manuscript and other individuals who contributed but are not listed as authors of this study.

Conflicts of Interest: Jianzan Yang, Feng Pang, Huawei Xiang, Dacheng Li were employed by the company Powerchina Guiyang Engineering Corporation Limited, Bo Gu declares no conflict of interest.

Nomenclature

AOA	arithmetic optimization algorithm
AR	autoregressive model
ARIMA	autoregressive integrated moving average model
ARMA	auto-regressive moving average model
BiLSTM	bidirectional long short-term memory
BP	back propagation neural network
CNN	convolutional neural network
CNN-BiLSTM	CNN and BiLSTM hybrid model
DASTGN	dynamic adaptive spatiotemporal graph neural network
ELM	extreme learning machine
GRU	gated recurrent unit
ICEEMDAN	improved complete ensemble empirical mode decomposition with additive noise
LSSVM	least squares support vector machines
LSTM	long short-term memory networks
PSO	particle swarm optimization
PSO-BP	PSO and BP hybrid model
PSO-ELM	PSO and ELM hybrid model

PSO-LSSVM	PSO and LSSVM hybrid model
UTSWS	ultra-short-term time series wind speeds
VMD	variational mode decomposition
VMD-AOA-GRU	VMD, AOA and GRU hybrid model
VMD-GRU	VMD and GRU hybrid model
VMD-LSTM	VMD and LSTM hybrid model
VMD-PSO-BP	VMD, PSO and BP hybrid model
VMD-PSO-ELM	VMD, PSO and ELM hybrid model
VMD-PSO-LSSVM	VMD, PSO and LSSVM hybrid model

References

1. IRENA. Renewable Capacity Statistics 2023. Available online: https://www.irena.org/Publications/2023/Mar/Renewable-capacity-statistics-2023 (accessed on 29 September 2023).
2. GWEC. Global Wind Report 2023. Available online: https://gwec.net/globalwindreport2023/ (accessed on 29 September 2023).
3. Saini, V.K.; Kumar, R.; Al-Sumaiti, A.S.; Sujil, A.; Heydarian-Forushani, E. Learning Based Short Term Wind Speed Forecasting Models for Smart Grid Applications: An Extensive Review and Case Study. *Electr. Power Syst. Study* **2023**, *222*, 109502. [CrossRef]
4. Wang, Y.; Zou, R.; Liu, F.; Zhang, L.; Liu, Q. A Review of Wind Speed and Wind Power Forecasting with Deep Neural Networks. *Appl. Energy* **2021**, *304*, 117766. [CrossRef]
5. Shahram, H.; Liu, X.L.; Lin, Z.; Saeid, L. A Critical Review of Wind Power Forecasting Methods-Past, Present and Future. *Energies* **2020**, *13*, 3764.
6. Wang, J.; Song, Y.; Liu, F.; Hou, R. Analysis and Application of Forecasting Models in Wind Power Integration: A Review of Multi-Step-Ahead Wind Speed Forecasting Models. *Renew. Sustain. Energy Rev.* **2016**, *60*, 960–981. [CrossRef]
7. Han, Y.; Mi, L.; Shen, L.; Cai, C.; Liu, Y.; Li, K.; Xu, G. A Short-Term Wind Speed Prediction Method Utilizing Novel Hybrid Deep Learning Algorithms to Correct Numerical Weather Forecasting. *Appl. Energy* **2022**, *312*, 118777. [CrossRef]
8. Lydia, M.; Kumar, S.S.; Selvakumar, A.I.; Kumar, G.E.P. Linear and Non-Linear Autoregressive Models for Short-Term Wind Speed Forecasting. *Energy Convers. Manag.* **2016**, *112*, 115–124. [CrossRef]
9. Srihari, P.; Kiran, T.; Vishalteja, K. A Hybrid VMD Based Contextual Feature Representation Approach for Wind Speed Forecasting. *Renew. Energy* **2023**, *219*, 119391.
10. Zhang, Y.; Zhao, Y.; Kong, C.; Chen, B. A New Prediction Method Based on VMD-PRBF-ARMA-E Model Considering Wind Speed Characteristic. *Energy Convers. Manag.* **2020**, *203*, 112254. [CrossRef]
11. Zhu, K.; Mu, L.; Yu, R.; Xia, X.; Tu, H. Probabilistic Modelling of Surface Drift Prediction in Marine Disasters Based on the NN-GA and ARMA Model. *Ocean Eng.* **2023**, *281*, 114804. [CrossRef]
12. Zhu, Y.; Xie, S.; Xie, Y.; Chen, X. Temperature Prediction of Aluminum Reduction Cell Based on Integration of Dual Attention LSTM for Non-Stationary Sub-Sequence and ARMA for Stationary Sub-Sequences. *Control Eng. Pract.* **2023**, *138*, 105567. [CrossRef]
13. Aasim Singh, S.N.; Mohapatra, A. Repeated Wavelet Transform Based ARIMA Model for Very Short-Term Wind Speed Forecasting. *Renew. Energy* **2019**, *136*, 758–768. [CrossRef]
14. Liu, X.L.; Lin, Z.; Feng, Z.M. Short-term Offshore Wind Speed Forecast by Seasonal ARIMA-A Comparison against GRU and LSTM. *Energy* **2021**, *227*, 120492. [CrossRef]
15. Akshita, G.; Arun, K. Two-Step Daily Reservoir Inflow Prediction Using ARIMA-Machine Learning and Ensemble Models. *J. Hydro-Environ. Res.* **2022**, *45*, 39–52.
16. Hu, Y.H.; Guo, Y.S.; Fu, R. A Novel Wind Speed Forecasting Combined Model Using Variational Mode Decomposition, Sparse Auto-Encoder and Optimized Fuzzy Cognitive Mapping Network. *Energy* **2023**, *278 Pt A*, 127926. [CrossRef]
17. Li, M.; Yang, Y.; He, Z.; Guo, X.; Zhang, R.; Huang, B. A Wind Speed Forecasting Model Based on Multi-objective Algorithm and Interpretability Learning. *Energy* **2023**, *269*, 126778. [CrossRef]
18. Parmaksiz, H.; Yuzgec, U.; Dokur, E.; Erdogan, N. Mutation Based Improved Dragonfly Optimization Algorithm for a Neuro-fuzzy System in Short Term Wind Speed Forecasting. *Knowl.-Based Syst.* **2023**, *268*, 110472. [CrossRef]
19. Dokur, E.; Erdogan, N.; Salari, M.E.; Karakuzu, C.; Murphy, J. Offshore Wind Speed Short-Term Forecasting Based on a Hybrid Method: Swarm Decomposition and Meta-Extreme Learning Machine. *Energy* **2022**, *248*, 123595. [CrossRef]
20. Yang, Y.; Zhou, H.; Wu, J.; Ding, Z.; Wang, Y.-G. Robustified Extreme Learning Machine Regression with Applications in Outlier-Blended Wind-Speed Forecasting. *Appl. Soft Comput.* **2022**, *122*, 108814. [CrossRef]
21. Sun, S.; Wang, Y.; Meng, Y.; Wang, C.; Zhu, X. Multi-Step Wind Speed Forecasting Model Using a Compound Forecasting Architecture and an Improved QPSO-Based Synchronous Optimization. *Energy Rep.* **2022**, *8*, 9899–9918. [CrossRef]
22. He, Y.; Wang, Y.; Wang, S.; Yao, X. A Cooperative Ensemble Method for Multistep Wind Speed Probabilistic Forecasting. *Chaos Solitons Fractals* **2022**, *162*, 112416. [CrossRef]
23. Yang, W.; Hao, M.; Hao, Y. Innovative Ensemble System Based on Mixed Frequency Modeling for Wind Speed Point and Interval Forecasting. *Inf. Sci.* **2023**, *622*, 560–586. [CrossRef]
24. Dong, Y.; Li, J.; Liu, Z.; Niu, X.; Wang, J. Ensemble Wind Speed Forecasting System Based on Optimal Model Adaptive Selection Strategy: Case Study in China. *Sustain. Energy Technol. Assess.* **2022**, *53 Pt B*, 102535. [CrossRef]

25. Hao, Y.; Yang, W.D.; Yin, K.D. Novel Wind Speed Forecasting Model Based on a Deep Learning Combined Strategy in Urban Energy Systems. *Expert Syst. Appl.* **2023**, *219*, 119636. [CrossRef]
26. Wang, Y.; Xu, H.; Song, M.; Zhang, F.; Li, Y.; Zhou, S.; Zhang, L. A Convolutional Transformer-based Truncated Gaussian Density Network with Data Denoising for Wind Speed Forecasting. *Appl. Energy* **2023**, *333*, 120601. [CrossRef]
27. Lv, M.; Li, J.; Niu, X.; Wang, J. Novel Deterministic and Probabilistic Combined System Based on Deep Learning and Self-improved Optimization Algorithm for Wind Speed Forecasting. *Sustain. Energy Technol. Assess.* **2022**, *52 Pt B*, 102186. [CrossRef]
28. Gao, Z.; Li, Z.; Xu, L.; Yu, J. Dynamic Adaptive Spatio-temporal Graph Neural Network for Multi-node Offshore Wind Speed Forecasting. *Appl. Soft Comput.* **2023**, *141*, 110294. [CrossRef]
29. Sibtain, M.; Bashir, H.; Nawaz, M.; Hameed, S.; Azam, M.I.; Li, X.; Abbas, T.; Saleem, S. A Multivariate Ultra-Short-Term Wind Speed Forecasting Model by Employing Multistage Signal Decomposition Approaches and a Deep Learning Network. *Energy Convers. Manag.* **2022**, *263*, 115703. [CrossRef]
30. Nascimento, E.G.S.; Talison, A.C.; Davidson, M.M. A Transformer-Based Deep Neural Network with Wavelet Transform for Forecasting Wind Speed and Sind Energy. *Energy* **2023**, *278*, 127678. [CrossRef]
31. Bentsen, L.Ø.; Warakagoda, N.D.; Stenbro, R.; Engelstad, P. Spatio-Temporal Wind Speed Forecasting Using Graph Networks and Novel Transformer Architectures. *Appl. Energy* **2023**, *333*, 120565. [CrossRef]
32. Bala, S.B.; Kiran, T.; Vishalterja, K. Hybrid Wind Speed Forecasting Using ICEEMDAN and Transformer Model with Novel Loss Function. *Energy* **2023**, *265*, 126383.
33. Wu, H.; Meng, K.; Fan, D.; Zhang, Z.; Liu, Q. Multistep Short-Term Wind Speed Forecasting Using Transformer. *Energy* **2022**, *261 Pt A*, 125231. [CrossRef]
34. Liu, G.; Wang, Y.; Qin, H.; Shen, K.; Liu, S.; Shen, Q.; Qu, Y.; Zhou, J. Probabilistic Spatiotemporal Forecasting of Wind Speed Based on Multi-Network Deep Ensembles Method. *Renew. Energy* **2023**, *209*, 231–247. [CrossRef]
35. Lv, S.X.; Wang, L. Multivariate Wind Speed Forecasting Based on Multi-Objective Feature Selection Approach and Hybrid Deep Learning Model. *Energy* **2023**, *263 Pt E*, 126100. [CrossRef]
36. Zheng, L.; Lu, W.S.; Zhou, Q.Y. Weather Image-Based Short-Term Dense Wind Speed Forecast with a ConvLSTM-LSTM Deep Learning Model. *Build. Environ.* **2023**, *239*, 110446. [CrossRef]
37. Zhang, Y.M.; Wang, H. Multi-Head Attention-Based Probabilistic CNN-BiLSTM for Day-Ahead Wind Speed Forecasting. *Energy* **2023**, *278 Pt A*, 127865. [CrossRef]
38. Wang, J.Z.; An, Y.N.; Lu, H.Y. A Novel Combined Forecasting Model Based on Neural Networks, Deep Learning Approaches, and Multi-Objective Optimization for Short-Term Wind Speed Forecasting. *Energy* **2022**, *251*, 123960. [CrossRef]
39. Dragomiretskiy, K.; Zosso, D. Variational Mode Decomposition. *IEEE Trans. Signal Process.* **2014**, *62*, 531–544. [CrossRef]
40. Abualigah, L.; Diabat, A.; Mirjalili, S.; Abd Elaziz, M.; Gandomi, A.H. The Arithmetic Optimization Algorithm. *Comput. Methods Appl. Mech. Eng.* **2021**, *376*, 113609. [CrossRef]
41. Cho, K.; van Merriënboer, B.; Gulcehre, C.; Bahdanau, D.; Bougares, F.; Schwenk, H.; Bengio, Y. Learning Phrase Representations Using RNN Encoder-Decoder for Statistical Machine Translation. *arXiv* **2014**, arXiv:1406.1078. [CrossRef]
42. Bahdanau, D.; Cho, K.; Bengio, Y. Neural Machine Translation by Jointly Learning to Align and Translate. *arXiv* **2014**, arXiv:1409.0473. [CrossRef]

processes MDPI

Article

Bi-Level Inverse Robust Optimization Dispatch of Wind Power and Pumped Storage Hydropower Complementary Systems

Xiuyan Jing [1], Liantao Ji [2,*] and Huan Xie [3]

1 State Grid Corporation of China, Beijing 100031, China; xiuyan-jing@sgcc.com
2 China Electric Power Research Institute, Beijing 100192, China
3 Electric Power Research Institute State Grid Jibei Electric Power Co., Ltd., Beijing 100054, China; xiaosan_616@aliyun.com
* Correspondence: jiliantaoget@outlook.com

Abstract: This paper presents a bi-level inverse robust economic dispatch optimization model consisting of wind turbines and pumped storage hydropower (PSH). The inner level model aims to minimize the total generation cost, while the outer level introduces the optimal inverse robust index (OIRI) for wind power output based on the ideal perturbation constraints of the objective function. The OIRI represents the maximum distance by which decision variables in the non-dominated frontier can be perturbed. Compared to traditional methods for quantifying the worst-case sensitivity region using polygons and ellipses, the OIRI can more accurately quantify parameter uncertainty. We integrate the grid multi-objective bacterial colony chemotaxis algorithm and the bisection method to solve the proposed model. The former is adopted to solve the inner level problem, while the latter is used to calculate the OIRI. The proposed approach establishes the relationship between the maximum forecast deviation and the minimum generation cost associated with each non-dominated solution in the optimal load allocation. To demonstrate its economic viability and effectiveness, we simulate the proposed approach using real power system operation data and conduct a comparative analysis.

Keywords: wind power; pumped storage hydropower; economic dispatch

Citation: Jing, X.; Ji, L.; Xie, H. Bi-Level Inverse Robust Optimization Dispatch of Wind Power and Pumped Storage Hydropower Complementary Systems. *Processes* **2024**, *12*, 729. https://doi.org/10.3390/pr12040729

Academic Editors: Davide Papurello, Chenyu Wu, Zhongkai Yi and Chenhui Lin

Received: 21 February 2024
Revised: 21 March 2024
Accepted: 29 March 2024
Published: 4 April 2024

1. Introduction

The global power industry has placed significant emphasis on net zero emission, proposed during the 21st UN Climate Change Conference. The future is poised for a substantial surge in the utilization of renewable energy for power generation. Nevertheless, the inherent unpredictability of renewable resources poses a challenge, leading to potential imbalances between supply and demand. To mitigate the impact of renewable energy generators on the power system, the paramount focus is on advancing energy storage systems. Pumped storage hydropower (PSH) has gained widespread popularity due to its substantial capacity and cost-effectiveness [1]. Javed et al. [2] discussed the economic, environmental, and technical aspects of solar–wind–PHS systems, affirming the positive role of PHS in integrating renewable energy into the power systems. The integration of PSH has increased the flexibility and efficiency of renewable power generation.

With the large-scale integration of wind power into the grids, the impact of wind power's uncertainty on the scheduling and control operations of the power system has become increasingly pronounced [3]. Therefore, it is crucial to address the uncertainties associated with wind power when studying the coordinated scheduling of wind power–PSH systems. Many researchers have explored the relationship between uncertainty and wind power distribution using probabilistic models, including interval prediction [4], quantile regression prediction [5], the scenario generation method [6], two-stage stochastic programming [7], chance-constrained programming [8], and interval forecasting [9]. Interval prediction, quantile regression prediction, and the scenario generation method primarily focus on characterizing uncertain variables. However, these methods face challenges, such as high computational complexity and the inability to guarantee computational accuracy and security, limiting their further development. The last three methods rely on deterministic probability distribution models and require significant sampling computations. However, due to the impact of wind power uncertainty [10], these optimal dispatch models are complex, high-dimensional, non-convex, and non-differentiable multi-objective optimization problems. As a result, traditional methods often yield inaccurate density functions for wind power output, limiting their practical application.

Robust optimization methods are widely used because they do not require specific probability distributions to deal with wind power uncertainty. Since the 1970s, Soyster et al. [11] pioneered the use of linear robust optimization methods for solving uncertain linear programming problems. To overcome the limitations of this approach, a constraint protecting the nominal parameter level was introduced in the literature [12]. A maximum perturbation is added to the left side of the constraint equation, and an optimization dualization technique is used to integrate the equations, resulting in a linear robust model. Jin et al. [13] provide a more comprehensive definition of robustness in optimization problems. Robustness refers to the ability of the non-dominated solution set to maintain certain performance characteristics when decision variables are perturbed. It is often used to characterize the sensitivity of parameters to disturbances.

References [14,15] tackle supply and demand uncertainties by constructing a stochastic robust optimization model. The Benders' decomposition method is then employed to solve this model. However, the solutions obtained through this method may fall outside of the specified range during parameter perturbations. References [16,17] introduce a bi-level interval robust scheduling optimization model that accounts for the unpredictable nature of wind power. This model is utilized to formulate wind power scheduling plans for the next few hours. By applying strong duality theory, the model is transformed into a quadratic programming problem for the solution. However, interval robust optimization methods tend to search for overly conservative solutions that meet the fluctuation constraints in the objective functions. Ji et al. [18] improve upon conventional fuzzy scheduling optimization problems by incorporating fuzzy numbers to account for the inherent uncertainty in wind power ramping time. The methodology generates an uncertain set that encompasses worst-case scenarios, presenting a scheduling model based on fuzzy robust optimization methods. This multi-objective robust design based on fuzzy theory commonly employs the weighted analysis method for formulating objective functions, introducing a notably subjective element into the methodology. Gunawan et al. [19] propose a multi-objective robust optimization method by introducing the concept of the worst-case sensitivity region (WCSR). The WCSR's radius acts as a robust indicator and is included in the constraints, transforming the problem into a constrained multi-objective optimization. However, this method has some limitations, such as the subjective construction of uncertainty and the lack of capability to search for a globally optimal solution. To overcome these issues, we propose a bi-level inverse robust optimization dispatch model for wind power–PSH complementary systems. The main contributions of this paper are summarized as follows.

Firstly, this paper presents a novel approach for improving the integration of wind power into the electricity grid while achieving economic efficiency. The proposed approach is based on the characteristics of pumped storage hydropower stations and develops a bi-level inverse robust economic dispatch optimization model that considers the uncertainties of wind power and the participation of pumped storage hydropower stations in scheduling. Secondly, the proposed model incorporates ideal disturbance constraints on the objective function, which are based on decision makers' preferences. We use hypersphere contraction and expansion to explore decision variable points with the best robustness in the non-dominated solution set. The OIRI is introduced to represent the maximum distance at which decision variables in the non-dominated frontiers can be perturbed. It provides the worst-case perturbation for each point in each dimension of the high-dimensional space, which enhances the generating units' robustness against disturbances and establishes the optimal relationship between wind power integration and the objective function. Thirdly, we propose a novel approach combining the grid multi-objective bacterial colony chemotaxis algorithm with the bisection method. The inner level problem is addressed by the grid multi-objective bacterial colony chemotaxis algorithm, and the OIRI is calculated using the bisection method. The effectiveness of the proposed model and algorithm is validated through simulations using actual grid-connected data. This approach can be useful for dispatchers in optimizing the schedule of wind power–PSH complementary systems.

2. Inverse Robust Optimization Preliminaries

This subsection briefly introduces the general bi-level inverse robust optimization model. Inner level optimization aims to obtain superior decision vectors and feasible solutions within the feasible domain. The outer level seeks to derive the maximum range of robustness among the results obtained from the inner layer by imposing ideal disturbance constraints. According to the conventional robust optimization model and the worst-case sensitivity region, we have the following bi-level inverse robust model.

Outer level problem:

$$find\{OIRI(\boldsymbol{x}) = \max\|\boldsymbol{x} - \boldsymbol{x}_0\|_2, \boldsymbol{f}\} \tag{1}$$

$$s.t.\, \boldsymbol{f}(\boldsymbol{x}_0) \in \overline{U(\boldsymbol{f}(\boldsymbol{x}), \boldsymbol{d})} \cap \Phi \tag{2}$$

$$\boldsymbol{d} = \boldsymbol{f}(\boldsymbol{x}) - \boldsymbol{f}(\boldsymbol{x}_0),\ u(\boldsymbol{f}) = \{\boldsymbol{d} | \boldsymbol{d} \in U(0, \boldsymbol{\varepsilon})\},\ \boldsymbol{d} \in u(\boldsymbol{f}) \tag{3}$$

Inner level problem:

$$\text{Minimize}\, \boldsymbol{f} = [f_1, f_2, \cdots, f_m]^T \tag{4}$$

$$g_j(\boldsymbol{x}),\ \max_{d \in u(f)} g_j(\boldsymbol{x}_0) \le 0, j = 1, 2, \ldots, J \tag{5}$$

$$h_k(\boldsymbol{x}),\ \max_{d \in u(f)} h_k(\boldsymbol{x}_0) = 0, k = 1, 2, \ldots, K \tag{6}$$

where $\boldsymbol{x} = [x_1, x_2, \ldots, x_n]^T$ is the n-dimensional decision vector and $\boldsymbol{x}_0$ is the decision vector satisfying all constraints in the inner and outer levels. The optimal inverse robust indicator OIRI will be defined in Definition 1. $\boldsymbol{f}$ is the m-dimensional objective function vector; $\boldsymbol{d}$ denotes the ideal disturbance vector; $\boldsymbol{\varepsilon} = [\varepsilon_1, \varepsilon_2, \ldots, \varepsilon_m]^T$ is the m-dimensional vector consisting of ideal disturbance coefficients; and all elements are positive. $U(0, \boldsymbol{\varepsilon})$ represents the neighborhood of the zero point composed of all points with a distance less than ε from the circle; $\overline{U(\cdot)}$ is the closure of the neighborhood of $U(\cdot)$; Φ denotes the inner decision vector space; $g_j(\boldsymbol{x})/h_k(\boldsymbol{x})$ represent inequality/equality constraints, respectively; J and K are the number of inequality constraints and equality constraints, respectively; and m and n are the number of objective functions and decision variables, respectively.

The Pareto front of this model (1)–(6) is referred to as the inverse robust Pareto frontier. The following definitions are given:

Definition 1. *Assuming Ω is the feasible region of the inner level problem, for any $\boldsymbol{x} \in \Omega$ and any given ideal disturbance coefficients $\boldsymbol{\varepsilon} > 0$, there exists a convex neighborhood of $\Delta_{max} \subset \Omega$ so that for any $\boldsymbol{x}_0 \in \Delta_{max}$, $\boldsymbol{y} = \boldsymbol{f}(\boldsymbol{x}_0)$ satisfies the inner and outer constraints. Δ_{max} is referred to as the optimal inverse robust indicator (OIRI) if and only if the following two conditions are met:*

(1) *The mapping function $\boldsymbol{f}$ satisfies $\boldsymbol{f}: \Delta_{max} \longmapsto U(\boldsymbol{f}(\boldsymbol{x}), \varepsilon) \bigcap \Phi$.*

(2) *There is no region Λ that satisfies both $\Lambda \subset \Omega$ and $\Delta_{max} \subseteq \Lambda$, so that $\boldsymbol{f}: \Lambda \longmapsto \overline{U(\boldsymbol{f}(\boldsymbol{x}), \varepsilon)} \bigcap \Phi$.*

Equation (2) indicates that when solving the outer level problem, it is necessary to ensure that the objective function values of boundary points fall within the feasible domain and the range of ideal disturbance constraints. The optimal inverse robust indicator represents the maximum distance by which a decision vector x, satisfying Equations (2)–(4) in the non-dominated frontier, can be perturbed outward.

3. Bi-Level Inverse Robust Model for Wind Power–PSH Complementary Systems

The scheduling model developed in this section is a bi-level inverse robust optimization model that takes into account thermal power plants, pumped storage hydropower stations, and wind power. The model is structured as Equations (1)–(6) and considers the decision makers' expected deviation range from the total generation cost and forecast results as ideal disturbance constraints. Additionally, the model defines the OIRI as the maximum wind power fluctuation that the optimization results can tolerate. The inner level of the model minimizes the total generation cost, while the outer level solves for the OIRI and the corresponding total generation cost f, thus determining the Pareto frontier.

The objective function of the outer level problem is defined as

$$\text{OIRI} = \max \sum_{t=1}^{T} \left\| \boldsymbol{P}_t^{w} - \boldsymbol{P}_t^{w0} \right\|_2 \tag{7}$$

where $\boldsymbol{P}_t^{w} = \left[P_{j,t}^{w} \right]_{N_w \times 1}$ is the non-dominated decision vector of wind power obtained from the inner level problem and $\boldsymbol{P}_t^{w0} = \left[P_{j,t}^{w0} \right]_{N_w \times 1}$ is the decision vector within $U(\boldsymbol{P}_t^{w}, \varepsilon)$. A larger OIRI indicates greater robustness in the inner level optimization results for $\boldsymbol{P}_t^{w}$, implying a stronger ability of the wind power generation plan to withstand fluctuations.

The total generation cost f is defined as

$$f = \sum_{t=1}^{T} \left(\sum_{i=1}^{N_c} C_{i,t}(P_{i,t}) + \sum_{j=1}^{N_w} \varphi_j \left(\left(P_{j,t}^{w} \ or \ P_{j,t}^{w0} \right) - P_{j,t}^{prw} \right)^2 + \sum_{k=1}^{N_p} \left[SC_{k,t}(h_{k,t}) + \phi_k \left(P_{k,t}^{ps} - P_{k,t}^{prps} \right)^2 \right] \right) \tag{8}$$

where T is the number of dispatch time intervals. $N_c/N_w/N_p$ represent the number of thermal units, wind farms, and PSHs, respectively. $P_{j,t}^{w}/P_{j,t}^{prw}$ denote the output power/forecasted output power of wind farm j in the inner level at period t, respectively; $P_{j,t}^{w0}$ represents the output power of wind farm j in the outer level at period t; $P_{j,t}^{ps}/P_{j,t}^{prps}$ denote the output power/forecasted output power of PSH j in the inner level at period t, respectively; and the fuel cost $C_{i,t}(P_{i,t})$ of the thermal unit i can be expressed as a quadratic function in relation to its output power $P_{i,t}$. $\varphi_j\left(\left(P_{j,t}^{w} \ or \ P_{j,t}^{w0}\right), P_{j,t}^{prw}\right)$ denotes the penalty cost of wind farm j deviating from the planned output at period t. φ_j/ϕ_k are the penalty coefficients. $SC_{k,t}\left(P_{j,t}^{ps}\right)$ is the sum of start-up and shut-down costs. It can be expressed as

$$SC_{k,t}(h_{k,t}) = h_{k,t}(1 - h_{k,t-1})S_k^{ps} + h_{k,t-1}(1 - h_{k,t})D_k^{ps} \tag{9}$$

where $h_{k,t}$ denotes the PSH's working state, taking 1 for start and taking 0 for stop and $S_{k,t}^{ps}/D_{k,t}^{ps}$ are the start-up cost and shut-down cost of the PSH k.

As shown in (5) and (6), the constraints of the inner level are listed as follows.

(1) Power balance constraints:

$$\sum_{i=1}^{N_c} P_{i,t} + \sum_{j=1}^{N_w} P_{j,t}^{w} + \sum_{k=1}^{N_p} P_{k,t}^{ps} = D_t \tag{10}$$

where D_t is the total power load.

(2) Generation limit constraints:

$$\underline{P}_i \le P_i \le \overline{P}_i \tag{11}$$

$$\underline{P}_j^w \leq P_j^w \leq \overline{P}_j^w \tag{12}$$

where $\underline{P}_i/\overline{P}_i$ are the lower/upper limits of thermal unit i, respectively. $\underline{P}_j^w/\overline{P}_j^w$ are the lower/upper output limits of wind farm j, respectively.

(3) Climbing constraints for thermal units:

$$-rp_{i,t}^d \cdot \Delta T \leq P_{i,t} - P_{i,t-1} \leq rp_{i,t}^u \cdot \Delta T \tag{13}$$

where $rp_{i,t}^d/rp_{i,t}^u$ are the maximum upward/downward ramping rates, respectively. ΔT is the dispatch time interval.

(4) Spinning reserve constraints:

$$\begin{cases} \sum\limits_{i=1}^{N_c} h_{i,t}\left(\overline{P}_{i,t}^{spin} - P_{i,t}\right) + \sum\limits_{k=1}^{N_k} P_{k,t}^{ps} \geq SP_t^{up} \\ \overline{P}_{i,t}^{spin} = h_{i,t-1} \cdot \min\left(h_{i,t-1}\overline{P}_{i,t}, P_{i,t-1} + rp_{i,t}^u \cdot \Delta T\right) \\ SP_t^{up} = \beta^c \sum\limits_{i=1}^{N_c} P_{i,t} + \beta^w \sum\limits_{j=1}^{N_w} P_{j,t}^w \end{cases} \tag{14}$$

$$\begin{cases} \sum\limits_{i=1}^{N_c} h_{i,t}\left(P_{i,t} - \underline{P}_{i,t}^{spin}\right) + \sum\limits_{k=1}^{N_k} \underline{P}_{k,t}^{ps} \geq SP_t^{dn} \\ \underline{P}_{i,t}^{spin} = h_{i,t-1} \cdot \max\left(h_{i,t-1}\underline{P}_{i,t}, P_{i,t-1} - rp_{i,t}^d \cdot \Delta T\right) \\ SP_t^{dn} = \beta^w \sum\limits_{j=1}^{N_w} P_{j,t}^w \end{cases} \tag{15}$$

where $h_{i,t}$ is the binary variable and denotes the working state of thermal unit i at period t; $\underline{P}_{i,t}^{spin}/\overline{P}_{i,t}^{spin}$ are the minimum/maximum feasible outputs of thermal unit i at period t; and SP_t^{up}/SP_t^{dn} are the upward/downward requirements of power systems at period t. β^c/β^w are the spinning reserve rates of the thermal unit and the wind farm.

(5) Transmission constraints:

$$\underline{TP}_l \leq \sum_{i=1}^{N_c} TF_{l,i}P_{i,t} + \sum_{j=1}^{N_w} TF_{l,j}P_{j,t}^w + \sum_{k=1}^{N_p} TF_{l,k}P_{k,t}^{ps} \leq \overline{TP}_l \tag{16}$$

where $\underline{TP}_l/\overline{TP}_l$ are the lower/upper power flow limits of line l and $TF_{l,i}$ denotes the power transfer distribution factors from unit i to line l.

(6) Minimum on and off time constraints:

$$\begin{cases} \left(T_{i,t-1}^{on} - Z_i^{on}\right)(h_{i,t-1} - h_{i,t}) \geq 0 \\ \left(T_{i,t-1}^{off} - Z_i^{off}\right)(h_{i,t} - h_{i,t-1}) \geq 0 \end{cases} \tag{17}$$

where $T_{i,t-1}^{on}/T_{i,t-1}^{off}$ are the start-up/shut-down times of thermal unit i at period $t-1$, respectively and M_i^{on}/M_i^{off} represent the minimum start-up/shut-down times of the thermal power unit i.

(7) PSH's capacity constraints:

In actual operation, PSHs are required to satisfy constraints regarding reservoir water balance. We adopt a daily energy balance mode wherein the electricity consumed during pumping and the electricity generated during generation are equal throughout the day, thus ensuring that the reservoir water level remains relatively unchanged at the beginning and end of the day.

$$-\eta_k \sum_{t=1}^{T} \sum_{k=1}^{N_p} P_{k,t}^{ps} = \sum_{t=1}^{T} \sum_{k=1}^{N_p} P_{k,t}^{ps} \quad (18)$$

where η_k is the energy conversion efficiency of PSH k.

(8) Start–stop frequency constraints:

$$\sum_{t=1}^{T} |h_{k,t} - h_{k,t-1}| \leq \overline{TH}_k \quad (19)$$

where $\overline{TH}_k$ is the maximum daily start–stop times of PSH k. Frequent starts and stops can accelerate the deterioration of the generator, leading to increased maintenance requirements, higher repair costs, and a potentially shorter operational lifespan. Equation (19) refers to limitations imposed on how frequently the PSH can be start up or shut down within a given period. When the PSH k starts up or shuts down, $|h_{k,t} - h_{k,t-1}|$ is equal to 1. Summing up $|h_{k,t} - h_{k,t-1}|$ over an operating cycle T yields the total number of start-ups and shut-downs of PSH k.

(9) Ideal perturbation constraint:

$$\left\| f(\boldsymbol{P}_t^{w}) - f(\boldsymbol{P}_t^{w_0}) \right\|_2 \leq \varepsilon \quad (20)$$

Equation (20) ensures that the outer level optimization satisfies the ideal disturbance constraints. In this section, we present a dispatch model that utilizes bi-level inverse robust optimization theory. This model provides the optimal load allocations for conventional thermal power units, wind farms, and PSHs. Additionally, it will identify the maximum insensitivity regions relative to each optimal allocation point when there are uncertain parameters in the system undergoing perturbations.

4. Solution Procedure

The model developed in this paper has an inherent coupling between the inner and outer layers. Therefore, a singular intelligent optimization algorithm is not enough for obtaining effective solutions. To address the inner level optimization problem, the grid multi-objective bacterial colony chemotaxis algorithm is employed.

4.1. Brief Introduction of the Multi-Objective Bacterial Colony Chemotaxis Algorithm

The role of the multi-objective bacterial colony chemotaxis algorithm [20,21] is to solve the inner level optimization problem. The trajectory of a bacterium comprises consecutive linear paths interspersed with instantaneous changes in direction, with each path defined by its speed, direction, and duration. All paths are characterized by a uniform, constant speed. Upon changing direction, a bacterium selects a new trajectory based on a probability distribution that exhibits azimuthal symmetry relative to the preceding direction. The angle between consecutive trajectories follows a probability distribution, while the duration of each trajectory is determined by an exponentially declining probability distribution. Notably, both the angle and duration probability distributions are unaffected by the parameters of the preceding trajectory.

Here, x is used to represent decision variables $P_{i,t}$ and $P_{j,t}^{ps}$. Before applying the multi-objective bacterial colony chemotaxis algorithm, all constraints are added as penalty terms to the objective function. Figure 1 shows the flow chart of the multi-objective bacterial colony chemotaxis algorithm, which can be briefly explained as follows.

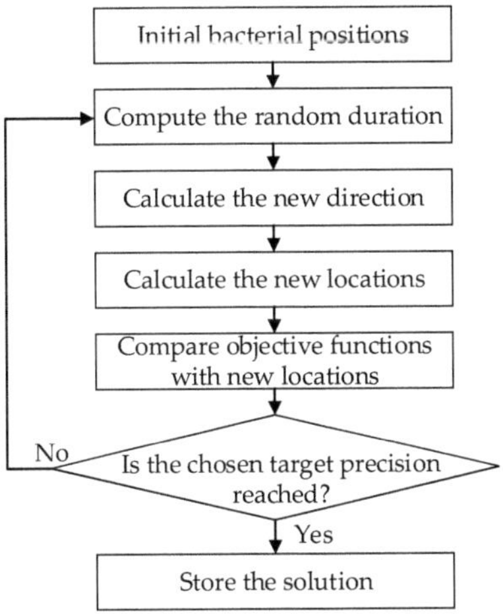

Figure 1. Flow chart of the multi-objective bacterial colony chemotaxis algorithm.

Step 1: Initialize bacterial positions. Generate N_{ba} bacteria randomly in this model. The velocity v is assumed to be a scalar constant value.

Step 2: Compute the random duration τ that follows an exponential probability density function:

$$P(X=\tau)=\frac{1}{\gamma}e^{-\tau/\gamma} \tag{21}$$

$$\gamma=\begin{cases}\gamma_0 \text{ if } \boldsymbol{x}_{pre}\succ \boldsymbol{x}_{cur}\\ \gamma_0\left(1+b*\min\left|\frac{f_{mpr}}{l_{pr}}\right|\right) \text{ if } \boldsymbol{x}_{pre}\prec \boldsymbol{x}_{cur}\end{cases} \tag{22}$$

$$\gamma_0=\xi^{0.03}\cdot 10^{-1.73}\ b=\gamma_0\cdot\left(\gamma_0^{-1.54}\cdot 10^{0.6}\right) \tag{23}$$

where γ_0 and b are the minimal mean time and dimensionless parameter, respectively. They can be calculated through (23) according to reference [21]. ξ is the calculation precision. $f_{ipr}, i=1,2,\ldots,m$ is the difference between the actual and the previous function values. l_{pr} is the vector connecting the previous and the actual positions in the parameter space. $\boldsymbol{x}_{pre}$ and $\boldsymbol{x}_{\mathrm{cur}}$ are the previous and current locations of bacteria.

The position and motion of bacteria is defined by a radius r_0 and $n-1$ angles $\boldsymbol{\theta}=\{\theta_1,\theta_2,\ldots,\theta_{n-1}\}$.

$$x_1=r_0\prod_{s=1}^{n-1}\cos(\theta_s)\ ,\ x_i=r_0\sin(\theta_{i-1})\prod_{s=1}^{n-1}\cos(\theta_s), i=2,3,\ldots,n \tag{24}$$

Step 3: Calculate the new direction. The angle θ_i between the previous and new directions follows a Gaussian distribution, governing both left and right turns, respectively.

$$P(X_i=\theta_i,v_i=\mu_i)=\frac{1}{\sigma_i\sqrt{2\pi}}\exp\left[-\frac{(\theta_i-v_i)}{2\sigma_i^2}\right]^2 \tag{25}$$

$$P(X_i=\theta_i,v_i=-\mu_i)=\frac{1}{\sigma_i\sqrt{2\pi}}\exp\left[-\frac{(\theta_i-v_i)}{2\sigma_i^2}\right]^2 \tag{26}$$

where $\theta_i\in[0,180^\circ]$. The expectation value μ_i, variance σ_i, and correlation time τ_C are determined by the formulation:

$$\mu_i=62^\circ(1-\exp(-\tau_C\tau_{pr}))\ \sigma_i=26^\circ(1-\exp(-\tau_C\tau_{pr})) \tag{27}$$

where τ_{pr} is the duration of the previous step.

Step 4: Calculate the new locations x_{new1} and x_{new2}. The new locations are calculated through the following equations:

$$x_{\text{new1}} = x_{\text{pre}} + v \cdot \tau \quad (28)$$

$$x_{\text{new2}} = x_{\text{pre}} - 2 \times \text{U}(0,1) \times \left(x_{\text{pre}} - x_{\text{cen}}\right) \quad (29)$$

$$x_{\text{cen}} = \frac{1}{N_{ba}} \sum_{i=1}^{N_{ba}} x_i \quad (30)$$

where $\text{U}(0,1)$ is a random number governed by the uniform distribution.

Step 5: Determine the new location x_{new}. By comparing x_{new1} with x_{new2}, all bacteria choose the better one as their new locations. Terminate iteration when the change in the objective function value is less than the target precision ξ; otherwise, proceed to Step 2.

4.2. Solution Method for the Bi-Level Inverse Robust Model

To calculate the OIRI for the outer level optimization problem, the hypersphere can be used to quantify maximum uncertainty. This involves partitioning the hyperspherical surface to facilitate the outward expansion of each small arc face. This results in a more accurate irregular super-enclosed surface characterized by the radial distance r and polar angle $\theta = [\theta_1, \theta_2, \ldots, \theta_{n-1}]$. Its maximum radius corresponds to the OIRI. Applying the boundary principle, we replace the discussion of changes across the entire domain with an examination of variations in domain boundary points.

Let x be equivalent to P_t^w. Assume x is a non-dominated solution obtained from the inner level problem. Consider a spherical neighborhood, denoted as $O(x)$, around $x = [x_1, x_2, \ldots, x_n]$, with a radius denoted as r_0. Every point on the boundary of $O(x)$, denoted as $\partial O(x)$, can be expressed using parametric equations. Let $x_0 = [x_{01}, x_{02}, \ldots, x_{0n}]$ and $x_0 \in \partial O(x)$, so we have

$$\begin{cases} x_{01} = x_1 + r_0 \cos\theta_1 \\ x_{02} = x_2 + r_0 \sin\theta_1 \cos\theta_2 \\ \cdots \\ x_{0n} = x_n + r_0 \sin\theta_1 \sin\theta_2 \cdots \cos\theta_{n-1} \end{cases} \quad (31)$$

where $\theta_1, \theta_2, \ldots, \theta_{n-2} \in [0, \pi]$, $\theta_{n-1} \in [0, 2\pi]$. We adopt a uniform sampling method to sample points on the boundary $\partial O(x)$. The boundary $\partial O(x)$ is divided into M parts, where $M = \lambda_1 \cdot \lambda_2, \ldots, \lambda_{n-1}$ ($\lambda_i \in N^+, i = 1, 2, \ldots, n-1$). Define a set $B = \{x_0 = x_{q_1 q_2 \cdots q_{n-1}} | 1 \le q_i \le \lambda_i, q_i \in N^+, i = 1,2,..,n-1\}$, so we have

$$x_{q_1 q_2 \cdots q_{n-1}} = \begin{bmatrix} x_1 + r_0 \cos\left(\frac{\pi}{\lambda_{n-1}} \eta_1\right), \\ x_2 + r_0 \sin\left(\frac{\pi}{\lambda_{n-1}} \eta_1\right) \cos\left(\frac{\pi}{\lambda_{n-2}} \eta_2\right), \ldots, \\ x_n + r_0 \sin\left(\frac{\pi}{\lambda_{n-1}} \eta_1\right) \ldots \cos\left(\frac{\pi}{\lambda_2} \eta_{n-2}\right) \cos\left(\frac{2\pi}{\lambda_1} \eta_{n-1}\right) \end{bmatrix}^T \quad (32)$$

After defining

$$D(x) = [\|f_1(x) - f_1(x_0)\|_2, \|f_2(x) - f_2(x_0)\|_2, \ldots, \|f_m(x) - f_m(x_0)\|_2]^T \quad (33)$$

the bisection method can be applied to approximate a larger hyperregion, ensuring that when decision variables are perturbed within this region, the function values are constrained within a certain range. In other words, adjusting each point $x_0 \in B$ ensures that x_0 satisfies $\|D(x_0)\|_\infty \in [\varepsilon - \delta_\varepsilon, \varepsilon + \delta_\varepsilon]$, and δ_ε is a predefined convergence accuracy. The specific steps of the bisection method are as follows.

Step 1: Initializing set $\boldsymbol{B}$.

(1) If there is a point $x_0 \in \boldsymbol{B}$ that satisfies $\|D(x_0)\|_\infty \in [\varepsilon - \delta_\varepsilon, \varepsilon + \delta_\varepsilon]$, then the values of $f(x_0)$ and $\langle r_0, \theta \rangle$ are stored, and the process proceeds to Step 2.

(2) If there are two points $x_1, x_2 \in \boldsymbol{B}$ and $x_1 \neq x_2$ that satisfy $\|D(x_1)\|_\infty < \varepsilon - \delta_\varepsilon$ and $\|D(x_2)\|_\infty > \varepsilon + \delta_\varepsilon$, the process proceeds to Step 2. Otherwise, adjust the radial radius through the bisection method.

(3) If for any $x_0 \in \boldsymbol{B}$ that satisfies $\|D(x_1)\|_\infty > \varepsilon - \delta_\varepsilon$, then $r_1 = 2r_0$ and $r_2 = r_0 + 0.5(r_1 - r_0)$. If for any $x_0 \in \boldsymbol{B}$ that satisfies $\|D(x_2)\|_\infty < \varepsilon + \delta_\varepsilon$, then $r_1 = 0.5r_0$ and $r_2 = r_0 + 0.5(r_1 - r_0)$. Calculate the size of $\|D(x_0)\|_\infty$ in the set $\boldsymbol{B}$ based on radial radius r_1 and r_2.

Step 2: Update the points in set $\boldsymbol{B}$.

Define $\boldsymbol{B}_1 = \{x_0 | \|D(x_0)\|_\infty < \varepsilon - \delta_\varepsilon \text{ or } \|D(x_0)\|_\infty > \varepsilon + \delta_\varepsilon\}$. As in the procedure in Step 1, scale the radial radius of all $x_0 \in \boldsymbol{B}_1$ until $\|D(x_0)\|_\infty \in [\varepsilon - \delta_\varepsilon, \varepsilon + \delta_\varepsilon]$ is satisfied. Record the angular and radial information of x_0.

Step 3: Calculate OIRI $= max_{x_0 \in \boldsymbol{B}} r_0$ and all angles and radial directions $\langle r_0, \theta \rangle$.

In traditional multi-objective optimization models, points within the decision space often exhibit better robustness, whereas feasible solutions near the boundaries of the decision space tend to have poorer robustness. When non-dominated solutions are located on or near the boundary, the perturbation of objective function values cannot extend beyond the super-neighborhood, and the corresponding perturbation of decision space vectors cannot expand outward. Here, an adaptive reduction of perturbation space is employed, limiting the range of perturbations of decision vectors within the feasible domain. This approach avoids redundant evaluations and reduces computational burden.

We integrate the grid multi-objective bacterial colony chemotaxis algorithm and the bisection method to solve the proposed model. Figure 2 illustrates the calculation process.

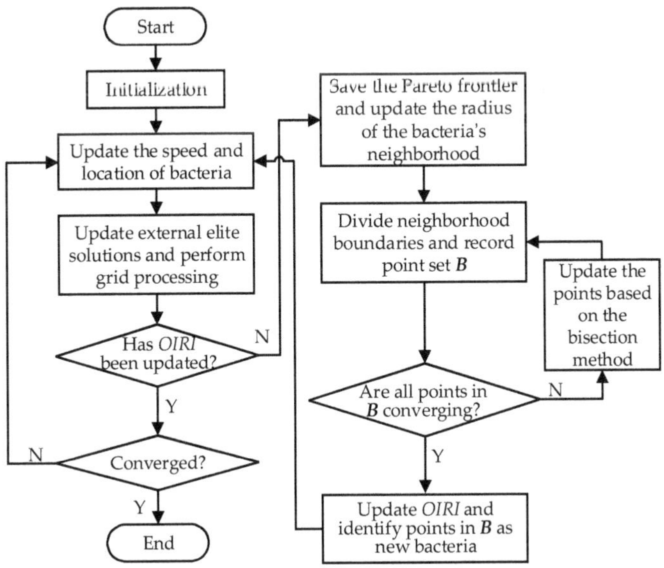

Figure 2. Flow chart of solution procedure.

Step 1: System parameter initialization. It involves setting parameters, such as calculation precision, bacteria migration rate, initial positions of the bacterial colony, ideal error constraints, error precision, and the number of boundary samples, among others.

Step 2: Update the speed and location of bacteria. According to the conventional bacterial colony chemotaxis algorithm [20,21], a bacterium will make a decision by comparing x_{new1} with x_{new2} and choose the better one as its new location x_{new}.

Step 3: Update the external Pareto set and perform grid processing. Guide the mutation of bacteria using non-dominance-based variation. Select non-dominated solutions from the new generation population and incorporate them into the Pareto set while simultaneously removing inferior solutions from the Pareto set.

Step 4: If the OIRI has been updated, proceed to Step 5. Otherwise, go to Step 6.

Step 5: Terminate the program if the convergence accuracy or the maximum number of iterations is reached. Otherwise, go back to Step 2.

Step 6: Save the Pareto frontier and update the radius of the bacteria's neighborhood.

Step 7: Divide neighborhood boundaries and record point set $\boldsymbol{B}$.

Step 8: If all points in set $\boldsymbol{B}$ have converged, go to Step 9. Otherwise, go to Step 6.

Step 9: Update the OIRI and identify points in $\boldsymbol{B}$ as new bacteria. Define the crowd distance of x_{new1}/x_{new2} as CD_1/CD_2.

The OIRI and f have been defined by (7) and (8). The new location can be calculated as follows:

(1) If two out of the three conditions $f(x_{new1}) \prec f(x_{new2})$, $\mathrm{OIRI}(x_{new1}) > \mathrm{OIRI}(x_{new2})$, and $CD_1 > CD_2$ are satisfied, then let $x_{new} = x_{new1}$;

(2) If two out of the three conditions $f(x_{new1}) \succ f(x_{new2})$, $\mathrm{OIRI}(x_{new1}) < \mathrm{OIRI}(x_{new2})$, and $CD_1 < CD_2$ are satisfied, then let $x_{new} = x_{new2}$;

(3) Otherwise, maintain the original position unchanged, $x_{new} = x$.

Go to Step 2.

5. Case Study

5.1. Modified 42-Bus Power System

In the modified 42-bus power system, the participating generating units include eight thermal power plants consisting of twenty-two units with capacities of 540 MW (four units), 315 MW (five units), 300 MW (five units), 270 MW (three units), and 180 MW (five units), with a total installed capacity of 6945 MW. There are two wind farms with a total capacity of 1080 MW, and the system wind power penetration rate ranges from approximately 12% to 20%. There are also two pumped storage power stations, each with a capacity of 270 MW. The maximum and minimum reservoir capacities are 2.0015×10^6 m^3 and 4.8388×10^5 m^3, respectively. The network topology is illustrated in Figure 3. The numbers in the Figure 3 represent the bus numbers.

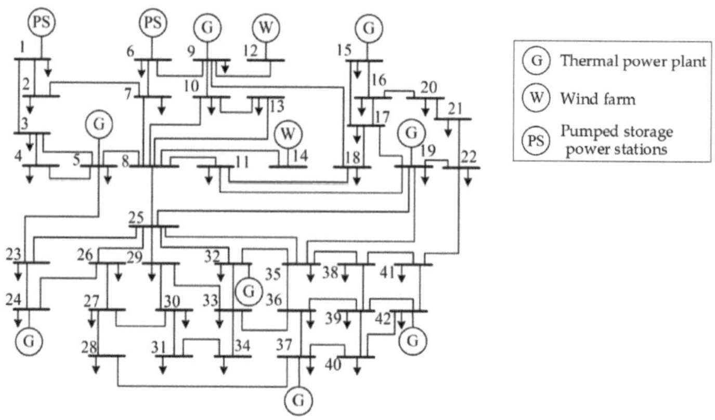

Figure 3. Schematic diagram of network topology.

The ramping rates for thermal power units are 5% of their rated capacity. The start-up and shut-down cost for each PSH is USD 585. The penalty factor for deviation from the planned output for PSHs is 0.5. The fuel cost $C_{i,t}(P_{i,t})$ of the thermal unit i can be expressed as a quadratic function, and its cost coefficients for thermal power units are provided in Table 1.

Table 1. Cost coefficient of thermal power units.

Rated Power/MW	P_{max}/MW	P_{min}/MW	a_i	b_i	c_i
540	540	270	0.00037	18.315	807.11
315	315	157.5	0.00051	17.435	843.80
297	297	148.5	0.00053	19.241	710.68
270	270	135	0.00063	22.789	521.00
180	180	90	0.00066	22.181	634.71

In Table 1, the units for a_i, b_i, and c_i are USD/MW^{2}h, USD/MWh, and USD/h, respectively. Based on the pre-scheduled plan, one PSH will pump water at its rated power during the low-load period, which is from 00:00 to 06:30. Additionally, during the high-demand periods from 09:00 to 11:45 and 16:15 to 18:30, one PSH produces electricity at rated power. The operational time horizon is 24 h, and it is divided into 96 periods, with each period lasting 15 min.

5.2. Inverse Robust Optimization Results Analysis

The daily power load and wind power output are depicted in Figures 4 and 5, respectively. It is observed that the predicted wind power exceeds the actual wind power, leading to wind curtailment during practical operation. The primary reason for wind curtailment is that during certain periods, such as from 00:00 to 06:30, the load demand is relatively low, and the power systems' capacity to accommodate wind power is insufficient.

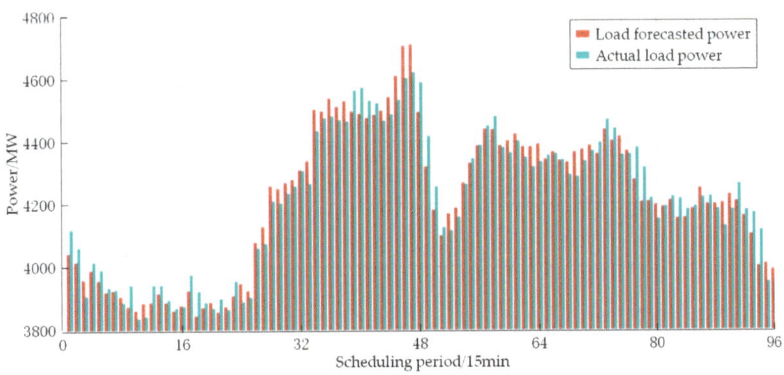

Figure 4. The predicted power load and actual power load.

Due to the requirement for PSHs to operate at rated power during pumping mode, only one or two PSHs can be in operation from 00:00 to 06:30. In light of these constraints, two scenarios are considered, and the inverse robust optimization approach is employed to find optimal solutions for them.

Scenario 1: Schedule one PSH to operate at full capacity during the specified time period (00:00 to 06:30). During other time intervals, the PSH can transition to power generation mode based on the specific optimization conditions.

Scenario 2: Flexibly schedule one or two PSHs to operate at full capacity based on the actual optimization conditions during the specified time period (00:00 to 06:30). During other time intervals, the PSHs can transition to power generation mode based on the specific optimization conditions.

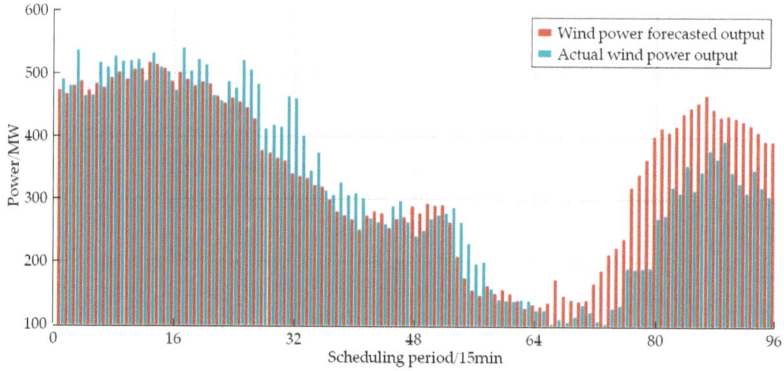

Figure 5. The predicted wind power and actual wind power.

As shown in Figures 6 and 7, a comparison is provided between Scenario 1 and Scenario 2 in terms of the total generation cost and the OIRI with the ideal disturbance coefficients $\varepsilon = 0.05$ and $\varepsilon = 0.10$, respectively. Each set includes 50 Pareto optimal solutions. When the ideal disturbance coefficient is set to 0.05, the non-dominated front obtained by Scenario 2 outperforms that of Scenario 1. It indicates that when the ideal disturbance coefficient is set to 0.10, its relatively loose constraints on the objective function enhance the solution set's robustness, thereby validating the rationality of the inverse robust optimization model proposed in this paper. The following analysis focuses solely on cases where the ideal disturbance coefficient is restricted to 0.10.

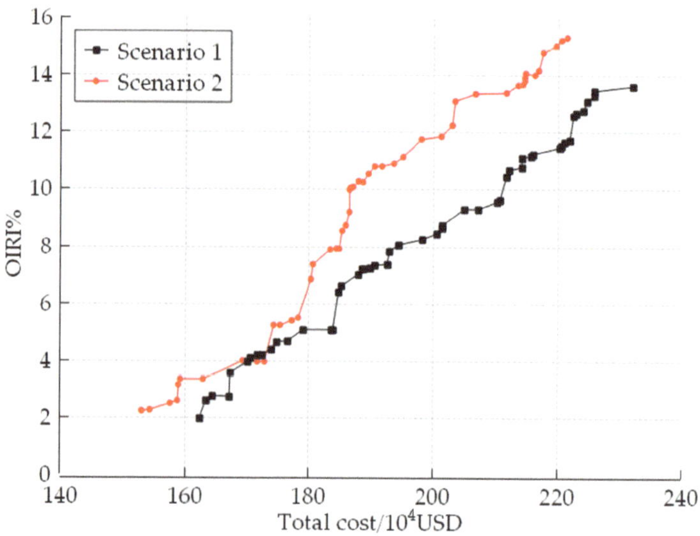

Figure 6. The results of OIRI and total generation cost (ideal disturbance constraint for 0.05).

In Table 2, the maximum wind power fluctuation that the system can withstand is OIRI (max), and the minimum total power generation cost is f (min). The results show the constraint relationships between the wind power accommodation and the total generation cost. In Scenario 1, when the ideal disturbance coefficient is 0.1, the wind power can be completely consumed as long as the error of wind power prediction is lower than 17.48%. The total generation cost under the maximum wind power forecasting deviation is USD 247.82 × 10^4. In Scenario 2, wind power can be fully consumed when its prediction deviation is within 20.01%. Additionally, at the maximum wind power prediction deviation, the total generation cost is USD 232.59 × 10^4.

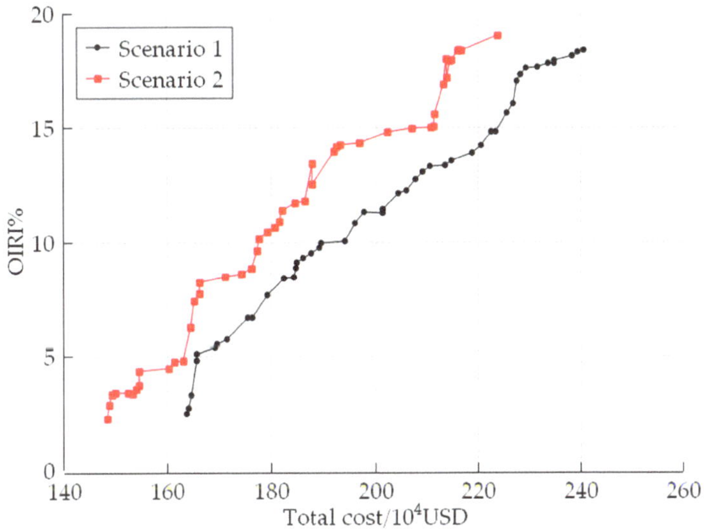

Figure 7. The results of OIRI and total generation cost (ideal disturbance constraint for 0.1).

Table 2. Comparison of optimum of OIRI and cost.

	OIRI (max)	f/10^4 USD	OIRI	f(min)/10^4 USD
Scenario 1	17.48%	247.82	0.018%	164.09
Scenario 2	20.01%	232.59	0.027%	147.20

The results of the optimization indicate that the maximum wind power prediction error that the system can tolerate can be determined based on the OIRI value, when operators have specific requirements for the total operating cost. If the wind power prediction error is large and the requirement for the total cost is low, the system may not be able to fully integrate wind power. For instance, when the wind prediction error is 0.15, both Scenario 1 and Scenario 2 require operating costs of around USD 225 × 10^4 to achieve complete wind power absorption. On the other hand, lower operating costs can be achieved when the wind power prediction deviation is smaller. If the wind power prediction deviation is 0.05, the minimum operating cost is USD 180 × 10^4. It is clear that if the grid follows Scenario 2 in scheduling, it will achieve complete integration of wind power with lower overall generation costs.

Figure 8 illustrates the daily working state curves of the PSHs under both scenarios. It is evident that the proposed inverse robust scheduling optimization strategy ensures the complete consumption of wind power. Additionally, Scenario 2 is more cost-effective than Scenario 1. The primary reason is that in Scenario 2, the PSHs absorb more wind power during the pumping stage, providing more flexibility in selecting output plans during the power generation periods.

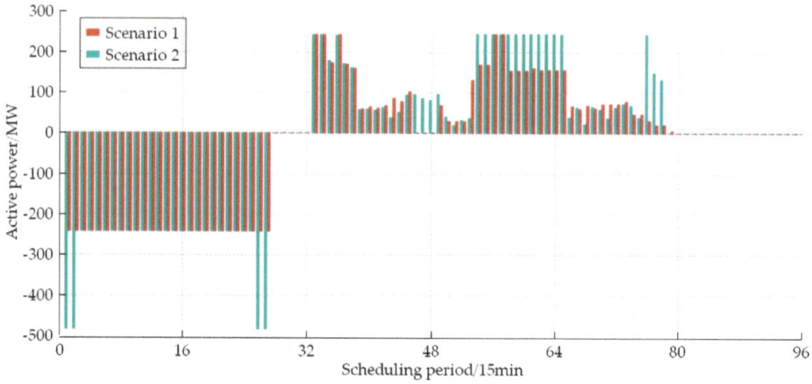

Figure 8. Working states of PSHs under different schemes.

5.3. Indictor Comparison Analysis

The WCSR is often used to represent the maximum hypersphere radius that ensures the robustness of the non-dominated solution set. The comparison between WCSR and OIRI is employed below to illustrate that compared to conventional interval quantification methods for robustness, OIRI can achieve better economy when addressing optimization scheduling problems affected by wind power uncertainties. WCSR can be calculated as follows:

$$WCSR = \min\sum_{t=1}^{T}\left\|\boldsymbol{P}_t^{w} - \boldsymbol{P}_t^{w_0}\right\|_2 \tag{34}$$

Based on the data presented in Table 3, the OIRI method achieves full integration of wind power regardless of whether Scheme 1 or Scheme 2 is adopted. When the WCSR method is employed under Scheme 1, full integration of wind power cannot be achieved, and the total generation costs are higher. Under Scheme 2, although the WCSR method can fully integrate wind power, its total generation costs are higher compared to the results obtained using the OIRI method. Traditional scheduling methods do not take into account uncertainty and usually need less computation time compared to the algorithm considered in OIRI. However, the proposed algorithm has the ability to identify schedules that are lower in cost and can integrate more wind power. Moreover, its computation time is less than 5 min, which is within an acceptable range.

Table 3. Comparison between OIRI and WCSR methods.

Classification	Conventional Method	Scenario 1		Scenario 2	
		OIRI	WCSR	OIRI	WCSR
the total cost/10^4 USD	276.41	225.60	229.04	192.11	203.91
coal consumption cost/10^4 USD	233.15	193.74	220.62	185.94	197.45
wind power penalty cost/10^4 USD	31.94	0	5.04	0	0
pumped storage start–stop cost/10^4 USD	11.32	2.9	2.9	5.76	5.76
pumped storage penalty cost/10^4 USD	0	0.29	0.48	0.41	0.69
wind curtailment/MWh	4487.16	0	708.45	0	0
CPU time (s)	153.1	219.9	253.9	228.4	261.7

5.4. Large-Scale System Testing

The proposed method was compared with some commonly used heuristic methods, such as the non-dominated sorting genetic algorithm (NSGA) [22], multi-objective particle swarm optimization (MOPSO) [23], niched Pareto genetic algorithm (NPGA) [24], and strength Pareto evolutionary algorithm (SPEA) [25]. These methods have been successfully applied to optimize economic dispatch. Their performances were evaluated by testing a modified IEEE 118-bus system whose detailed parameters are available online at http://motor.ece.iit.edu/data (accessed on 1 January 2023). The system consists of 118 buses, 33 generators, and 186 transmission lines. Based on the original IEEE 118-bus system, two wind farms and two pumped storage power stations were added at bus 12 and 61, respectively. Each pumped storage power station has a capacity of 270 MW. The forecasted wind power can be found in reference [14].

The results of the simulation are summarized in Table 4. To ensure a fair comparison between the different approaches, ten optimization runs were conducted, and the average values are shown. It was found that the proposed method has the lowest total cost and the shortest computation time. All of the methods tested were able to fully integrate wind power using PSHs and the OIRI index. Compared to the previous 42-bus test system, the number of thermal units in the modified IEEE 118-bus system increased from 22 to 33, and finding the optimal economic dispatch plan is a critical factor in reducing costs. The proposed algorithms were able to find better scheduling schemes for both thermal units and PSHs when wind power was fully integrated. This resulted in reduced coal generation costs of thermal units and start–stop costs of PSHs.

Table 4. Comparison between different heuristic methods and OIRI index.

	Proposed Method	NSGA	MOPSO	NPGA	SPEA
the total cost/10^4 USD	309.21	352.62	316.19	316.42	315.81
coal consumption cost/10^4 USD	299.28	319.24	306.03	306.21	306.61
wind power penalty cost/10^4 USD	0	0	0	0	0
pumped storage start–stop cost/10^4 USD	9.27	31.92	8.97	8.95	8.97
pumped storage penalty cost/10^4 USD	0.66	1.46	1.19	1.26	0.23
wind curtailment/MWh	0	0	0	0	0
CPU time (s)	542.4	549.8	590.7	560.1	571.3

6. Conclusions

To integrate wind power efficiently and minimize electricity generation costs, we propose a bi-level inverse robust wind power–PSHs optimization scheduling model. The model is solved using a combination of the grid multi-objective bacterial colony chemotaxis algorithm and the bisection method. This paper introduces an inverse robust indicator, which not only addresses the limitations of pumped storage in mitigating wind power fluctuations but also establishes the relationship between the maximum forecast deviation and the minimum generation cost associated with each non-dominated solution in the optimal load allocation. Case studies on operational data from a specific regional power grid validate the proposed method. Two optimization scenarios are developed and compared with the actual operation data. The case studies indicate that optimizing the pumping and generation operation of PSHs addresses the limitations of accommodating wind power. The proposed method enhances both economic efficiency and wind power consumption. It provides a clear relationship between the objective function and decision vectors, offering valuable insights for developing robust scheduling plans.

Author Contributions: Writing—original draft preparation, X.J.; writing—review and editing, L.J.; Supervision, H.X. All authors have read and agreed to the published version of the manuscript.

Funding: This research was funded by the Science and Technology Project of State Grid Corporation of China (No. 4000-202255061A-1-1-ZN).

Data Availability Statement: The data presented in this study are available on request from the corresponding author.

Conflicts of Interest: Author Xiuyan Jing was employed by the company State Grid Corporation of China. Author Liantao Ji was employed by the company China Electric Power Research Institute. Author Huan Xie was employed by the company Electric Power Research Institute State Grid Jibei Electric Power Co., Ltd. The authors declare that this study received funding from State

Grid Corporation of China. The funder had the following involvement with the study: analysis, interpretation of data, the writing of this article and the decision to submit it for publication.

Nomenclature

J	Number of inequality constraints
K	Number of equality constraints
T	Number of dispatch time intervals
ΔT	Dispatch time interval
ε	m-dimensional vector consisting of ideal disturbance coefficients
N_c	Number of thermal units
N_w	Number of wind farms
N_p	Number of pumped storage hydropower
$P_{j,t}^{w}$	Output power of wind farm j in the inner level at period t
$P_{j,t}^{prw}$	Forecasted power of wind farm j in the inner level at period t
$P_{j,t}^{w_0}$	Output power of wind farm j in the outer level at period t
$P_{j,t}^{ps}$	Output power of PSH j in the inner level at period t
$P_{j,t}^{prps}$	Forecasted power of PSH j in the inner level at period t
φ_j/ϕ_k	Penalty coefficients
$h_{i,t}$	Binary variable; it equals 1/0 if equipment i is ON/OFF at period t
$S_{k,t}^{ps}/D_{k,t}^{ps}$	Start-up/shut-down cost of the PSH k
$\underline{P}_i/\overline{P}_i$	Lower/upper limits of thermal unit i
$\underline{P}_j^{w}/\overline{P}_j^{w}$	Lower/upper output limits of wind farm j
$rp_{i,t}^{d}/rp_{i,t}^{u}$	Maximum upward/downward ramping rates of thermal unit i
$\underline{P}_{i,t}^{spin}/\overline{P}_{i,t}^{spin}$	Minimum/maximum feasible outputs of thermal unit i at period t
SP_t^{up}/SP_t^{dn}	Upward/downward requirement of power systems at period t
β^c/β^w	Spinning reserve rates of thermal unit and wind farm
$\underline{TP}_l/\overline{TP}_l$	Lower/upper power flow limits of line l
$TF_{l,i}$	Power transfer distribution factors from unit i to line l
$T_{i,t-1}^{on}/T_{i,t-1}^{off}$	Start-up/shut-down time of thermal unit i at period $t-1$
M_i^{on}/M_i^{off}	Minimum start-up/shut-down time of the thermal power unit i
η_k	Energy conversion efficiency of PSH k
$\overline{TH}_k$	Maximum daily start–stop times of PSH k
δ_ε	Predefined convergence accuracy
$C_{i,t}(P_{i,t})$	Cost function of thermal unit i at period t
$SC_{k,t}(h_{i,t})$	Sum of start-up and shut-down costs

References

1. Zhao, J.F.; Oh, U.J.; Park, J.C.; Park, E.S.; Im, H.B.; Lee, K.Y.; Choi, J.S. A Review of World-wide Advanced Pumped Storage Hydropower Technologies. *IFAC-Pap.* **2022**, *55*, 170–174. [CrossRef]
2. Javed, M.S.; Ma, T.; Jurasz, J.; Amin, M.Y. Solar and Wind Power Generation Systems with Pumped Hydro Storage: Review and Future Perspectives. *Renew. Energy* **2020**, *148*, 176–192. [CrossRef]
3. Yuan, W.Y.; Xin, W.P.; Su, C.G.; Cheng, C.; Yan, D.; Wu, Z. Cross-regional Integrated Transmission of Wind Power and Pumped-Storage Hydropower Considering the Peak Shaving Demands of Multiple Power Grids. *Renew. Energy* **2022**, *190*, 1112–1126. [CrossRef]
4. Hu, Z.J.; Zhang, M.L.; Wang, X.F.; Li, C.; Hu, M.Y. Bi-level Robust Dynamic Economic Emission Dispatch Considering Wind Power Uncertainty. *Electr. Power Syst. Res.* **2016**, *135*, 35–47. [CrossRef]
5. Bentsen, L.; Warakagoda, N.D.; Stenbro, R.; Engelstad, P. Relative Evaluation of Probabilistic Methods for Spatio-Temporal Wind Forecasting. *J. Clean. Prod.* **2024**, *434*, 139944. [CrossRef]
6. Tang, C.H.; Xu, J.; Tan, Y.S.; Sun, Y.Z.; Zhang, B.S. Lagrangian Relaxation with Incremental Proximal Method for Economic Dispatch with Large Numbers of Wind Power Scenarios. *IEEE Trans. Power Syst.* **2019**, *11*, 436–447. [CrossRef]
7. Xu, X.; Yan, Z.; Shahidehpour, M.; Li, Z.; Yan, M.; Kong, X. Data-Driven Risk-Averse Two-Stage Optimal Stochastic Scheduling of Energy and Reserve with Correlated Wind Power. *IEEE Trans. Sustain. Energy* **2020**, *27*, 206–215. [CrossRef]
8. Wu, C.Y.; Wei, G.; Zhou, S.Y.; Chen, X.G. Coordinated Optimal Power Flow for Integrated Active Distribution Network and Virtual Power Plants Using Decentralized Algorithm. *IEEE Trans. Power Syst.* **2021**, *36*, 3541–3551. [CrossRef]

9. Ran, X.H.; Zhang, J.H.; Zhu, W.J.; Liu, K.P.; Liu, Y.S. A Model of Correlated Interval–Probabilistic Conditional Value-at-Risk and Optimal Dispatch with Spatial Correlation of Multiple Wind Power Generations. *Int. J. Electr. Power Energy Syst.* **2024**, *155*, 109500. [CrossRef]
10. Wang, Z.N.; Fang, G.H.; Wen, X.; Tan, Q.F.; Zhang, P.; Liu, Z.H. Coordinated Operation of Conventional Hydropower Plants as Hybrid Pumped Storage Hydropower with Wind and Photovoltaic Plants. *Energy Convers. Manag.* **2023**, *277*, 116654. [CrossRef]
11. Soyster, A.L. Convex Programming and Set-Inclusion Constraints and Applications to Inexact Linear Programming. *Oper. Res.* **1973**, *21*, 1154–1157. [CrossRef]
12. Ahmed, S.; Sahinididis, N.V. Robust Process Planning under Uncertainty. *Ind. Eng. Chem. Res.* **1998**, *37*, 1883–1892. [CrossRef]
13. Jin, Y.C.; Branke, J. Evolutionary Optimization in Uncertain Environments-A Survey. *IEEE Trans. Evol. Comput.* **2005**, *9*, 303–317. [CrossRef]
14. Zhao, C.Y.; Wang, J.H.; Watson, J.P.; Guan, Y.P. Multi-Stage Robust Unit Commitment Considering Wind and Demand Response Uncertainties. *IEEE Trans. Power Syst.* **2013**, *28*, 2708–2717. [CrossRef]
15. Xiong, H.B.; Shi, Y.H.; Chen, Z.; Guo, C.X.; Ding, Y. Multi-Stage Robust Dynamic Unit Commitment Based on Pre-Extended—Fast Robust Dual Dynamic Programming. *IEEE Trans. Power Syst.* **2023**, *38*, 2411–2422. [CrossRef]
16. Wu, W.C.; Chen, J.H.; Zhang, B.M.; Sun, H.B. A Robust Wind Power Optimization Method for Look-Ahead Power Dispatch. *IEEE Trans. Sustain. Energy* **2014**, *5*, 507–515. [CrossRef]
17. Zhang, R.F.; Chen, Y.; Li, B.X.; Jiang, T.; Li, X.; Chen, H.H.; Ning, R.X. Adjustable Robust Interval Economic Dispatch of Integrated Electricity and District Heating Systems under Wind Power Uncertainty. *Energy Rep.* **2022**, *8*, 13138–13149. [CrossRef]
18. Xu, M.; Li, W.W.; Feng, Z.H.; Bai, W.W.; Jia, L.L.; Wei, Z.H. Economic Dispatch Model of High Proportional New Energy Grid-Connected Consumption Considering Source Load Uncertainty. *Energies* **2023**, *16*, 1696. [CrossRef]
19. Gunawan, S.; Azarm, S. Multi-objective Robust Optimization Using A Sensitivity Region Concept. *Struct. Multidiscip. Optim.* **2005**, *29*, 50–60. [CrossRef]
20. Lu, Z.; Zhao, H.; Xiao, H.F.; Wang, H.R.; Wang, H.J. An Improved Multi-Objective Bacteria Colony Chemotaxis Algorithm and Convergence Analysis. *Appl. Soft Comput.* **2015**, *31*, 274–292. [CrossRef]
21. Lu, Z.; Feng, T.; Li, X.P. Low-carbon Emission/Economic Power Dispatch Using the Multi-Objective Bacterial Colony Chemotaxis Optimization Algorithm Considering Carbon Capture Power Plant. *Int. J. Electr. Power Energy Syst.* **2013**, *53*, 106–112. [CrossRef]
22. Basu, M. A Nondominated Sorting Genetic Algorithm III with Three Crossover Strategies for the Combined Heat and Power Dynamic Economic Emission Dispatch with or without Prohibited Operating Zones. *Eng. Appl. Artif. Intell.* **2023**, *123*, 106443.
23. Zhang, Q.; Ding, J.J.; Shen, W.X.; Ma, J.H.; Li, G.L. Multiobjective Particle Swarm Optimization for Microgrids Pareto Optimization Dispatch. *Math. Probl. Eng.* **2020**, *2020*, 5695917. [CrossRef]
24. Abido, M.A. A Niched Pareto Genetic Algorithm for Multiobjective Environmental/Economic Dispatch. *Int. J. Electr. Power Energy Syst.* **2003**, *25*, 97–105. [CrossRef]
25. Jiang, S.Y.; Yang, S.X. A Strength Pareto Evolutionary Algorithm Based on Reference Direction for Multiobjective and Many-Objective Optimization. *IEEE Trans. Evol. Comput.* **2017**, *21*, 329–346. [CrossRef]

processes

Article

Reinforcement Learning and Stochastic Optimization with Deep Learning-Based Forecasting on Power Grid Scheduling

Cheng Yang [1,2,†], Jihai Zhang [2,*,†], Wei Jiang [2,†], Li Wang [2,†], Hanwei Zhang [2,†], Zhongkai Yi [3,†] and Fangquan Lin [2,†]

[1] Polytechnic Institute, Zhejiang University, Hangzhou 310027, China; 11821123@zju.edu.cn
[2] Alibaba Group, Hangzhou 311121, China; alice.jw@alibaba-inc.com (W.J.); feiyu.wl@alibaba-inc.com (L.W.); hanwei.zhanghw@alibaba-inc.com (H.Z.); fangquan.linfq@alibaba-inc.com (F.L.)
[3] School of Electrical Engineering and Automation, Harbin Institute of Technology, Harbin 150006, China; yzk_article@163.com
* Correspondence: jihai.zjh@alibaba-inc.com
† These authors contributed equally to this work.

Abstract: The emission of greenhouse gases is a major contributor to global warming. Carbon emissions from the electricity industry account for over 40% of the total carbon emissions. Researchers in the field of electric power are making efforts to mitigate this situation. Operating and maintaining the power grid in an economic, low-carbon, and stable environment is challenging. To address the issue, we propose a grid dispatching technique that combines deep learning-based forecasting technology, reinforcement learning, and optimization technology. Deep learning-based forecasting can forecast future power demand and solar power generation, while reinforcement learning and optimization technology can make charging and discharging decisions for energy storage devices based on current and future grid conditions. In the optimization method, we simplify the complex electricity environment to speed up the solution. The combination of proposed deep learning-based forecasting and stochastic optimization with online data augmentation is used to address the uncertainty of the dispatch system. A multi-agent reinforcement learning method is proposed to utilize team reward among energy storage devices. At last, we achieved the best results by combining reinforcement and optimization strategies. Comprehensive experiments demonstrate the effectiveness of our proposed framework.

Keywords: forecasting; reinforcement learning; power grid; planning and scheduling; uncertainty in artificial intelligence; agent-based systems; deep learning; stochastic optimization

Citation: Yang, C.; Zhang, J.; Jiang, W.; Wang, L.; Zhang, H.; Yi, Z.; Lin, F. Reinforcement Learning and Stochastic Optimization with Deep Learning-Based Forecasting on Power Grid Scheduling. *Processes* **2023**, *11*, 3188. https://doi.org/10.3390/pr11113188

Academic Editor: Jean-Louis Dirion

Received: 26 September 2023
Revised: 23 October 2023
Accepted: 31 October 2023
Published: 8 November 2023

1. Introduction

Nowadays, with the rapid development of artificial intelligence (AI), household appliances and equipment intelligence are gradually becoming popularized. More and more families are installing home solar power generation equipment and small-scale energy storage equipment, not only to meet their own electricity needs but also to sell excess power through the sharing network. If we can make home electricity use more efficient, then the community power grid will be more economical and low-carbon. Furthermore, the efficient and stable of community power grid can provide a guarantee for the stability of the national power grid.

Electricity research generally includes Large-scale Transmission Grids (LTG for short) and Small-scale Micro-Grids (SMG for short). LTG focuses on high-voltage and long-distance power transmission, while SMG focuses on electricity consumption in small areas such as schools, factories, or residential areas. We focus on smart scheduling techniques in SMG. For example, Figure 1 shows a case of SMG. Households can generate electricity from solar energy, store the excess power, and share with neighbors on the grid network (green arrows). When neither self-generated power nor a shared network can provide enough electricity, power is supplied by the national grid (orange lines). The national grid generates electricity through wind power, hydroelectric, and thermal. The cost of electricity and carbon emissions vary over time. In this paper, we use an AI-based approach to enable efficient scheduling of household storages. The AI-based scheduling method leads to economical and decarbonized electricity use.

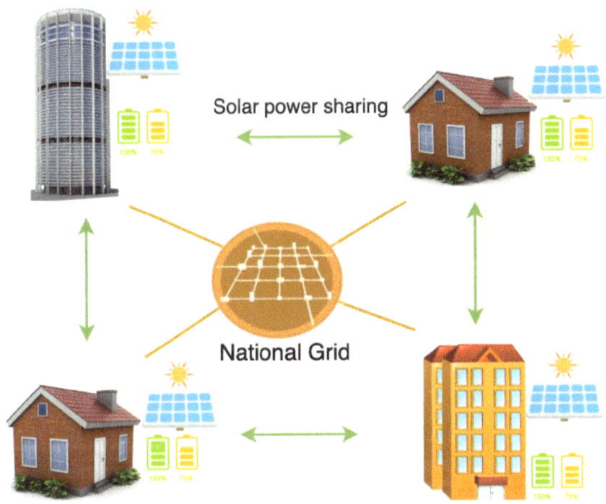

Figure 1. The micro-grid network framework. Green arrows denote solar power sharing among the micro-grid buildings and orange lines indicate how the micro-grid obtains power from the national grid.

In the power generation process, increasing the proportion of new energy sources is one of the most important methods to reduce carbon emissions. The use of new energy sources, such as wind power and solar power, reduces carbon emissions for the grid network but adds more uncertainty to the entire power network. For example, solar power generation is affected by the weather, and if future weather changes cannot be accurately predicted, then this will affect the scheduling program of other power generation methods in the power network. Uncertainty in new energy generation poses a great challenge to traditional dispatch systems. We categorize the uncertainty as data drift: the relation between input data and the target variables' changes over time [1]. For example, the sequential transition in a time series of renewable energy generation can be fluctuating (e.g., wind power and solar power).

The field of AI-based forecasting is continuously evolving. AI-based forecasting methods have been applied to predict the spread of contagious diseases such as COVID-19 [2], demonstrating their potential in public health applications. Deep learning techniques, including recurrent neural network (RNN) and long short-term memory (LSTM) networks [3], have been extensively studied for time series forecasting, showing promising results. Neural network architectures, such as feed-forward neural networks and convolutional neural network (CNN) [4], have been explored for time series forecasting, contributing to the advancement of AI-based forecasting models [5]. These studies provide insights into the use of advanced AI-based forecasting techniques and their applications in different domains, especially in the time series forecasting domain. Therefore, in the electricity power domain, we involve a deep learning-based method to predict future user demand and renewable generation (the task can be regarded as a sub-domain of time series forecasting domain).

For the problem of uncertainty, classical model predictive control (MPC)-based methods use rolling control to correct the parameters by realizing the feedback of rolling [6,7]. However, the effect is not up to expectations in practical applications. Taking industrial application as an example, the sequential MPC framework can usually be decomposed into point prediction of target variables (e.g., solar power generation), followed by deterministic optimization, which is unable to capture the uncertainty of probabilistic data distribution [8,9]. To solve the above problems, stochastic-based methods have been proposed, and they are able to eliminate the effects caused by some uncertainties.

Taking into account the uncertainty in forecasting, it is possible to improve energy efficiency by 13% to 30% [10,11]. Stochastic-based methods mainly include two types: one that requires prior knowledge of system uncertainty [12,13], and another is based on scenarios, generating values for multiple random variables [14,15]. Additionally, adaptive methods are also applied in the presence of uncertainty [16–18]. In this paper, enhanced generalization capability is achieved by combining stochastic optimization with online adaptive rolling updates.

Despite some recent progress, it is difficult for the existing system to meet the demand of real-time scheduling due to the huge number of SMGs and high model complexity. Under the requirement of real-time scheduling, the attempt of reinforcement learning in power grids is gradually emphasized.

Reinforcement learning has been proven to give real-time decisions in several domains and has the potential to be effectively applied in the power grid scenarios. In Large-scale Transmission Grids (LTG), reinforcement learning has not yet been successfully applied due to security concerns. In Small-scale Micro-Grids (SMG), where economy is more important (security can be guaranteed by the up-level grid network), reinforcement learning is gradually starting to be tried. In reinforcement learning, the model learns by trial and error through constant interaction with the environment [19] and ultimately obtains the best cumulative reward. Training for reinforcement learning usually relies on a simulation environment, which is assumed to be provided in this paper. Unlike the existing single agent approach, in this paper, we propose a multi-agent reinforcement learning method to adapt a grid scheduling task. Reinforcement learning in electricity power scheduling offers the potential to enhance the efficiency, reliability, and sustainability of power systems, leading to cost savings, reduced environmental impact, and improved overall performance. The main contributions of this paper are:

- To adapt to uncertainty, we propose two modules to achieve robust scheduling. One module combines deep learning-based prediction techniques with stochastic optimization methods, while the other module is an online data augmentation strategy, including stages of model pre-training and fine-tuning.
- In order to realize sharing rewards among buildings, we propose to use multi-agent PPO to simulate each building. Additionally, we provide the ensemble method between reinforcement learning and optimization methods.
- We conducted extensive experiments on a real-world scenario and the results demonstrate the effectiveness of our proposed framework.

2. Problem Statement

Generally, SMG contains various types of equipment, including solar generation machines (denoted as $\mathcal{G}$), storage devices (denoted as $\mathcal{S}$), and other user devices (denoted as $\mathcal{U}$). $\mathcal{M}$ denotes the markets, such as carbon and electricity. The total decision steps is set to T. We define the load demand of the user as: $L_{u,t}$, where step $t \in \mathcal{T} = \{1, \ldots, T\}$ and $u \in \mathcal{U}$. p_t is the market price as time t per unit or the average price among $\mathcal{M}$.

The variables in SMG include the electricity need from the national grid (denoted as $P_{\text{grid},t}$), the power generation of device $g \in \mathcal{G}$ (denoted as $P_{g,t}$), the charging or discharging of storage (denoted as $P^{+}_{s,t}$ or $P^{-}_{s,t}$), and the state of charge of device $s \in \mathcal{S}$ (denoted as $E_{s,t}$). We define the decision variables as: $X = \{P_{\text{grid},t}, P_{g,t}, P^{+}_{s,t}, P^{-}_{s,t}, E_{s,t}\}$, where $t \in \mathcal{T}$, $s \in \mathcal{S}$, $g \in \mathcal{G}$, and then the objective is to minimize the total cost of all markets, which is defined [20]:

$$\underset{X}{\text{minimize}} \quad \sum_{t=1}^{T} p_t \cdot P_{\text{grid},t} \tag{1}$$

s.t.:

$$P_{\text{grid},t} \geq 0 \qquad t \in \mathcal{T} \tag{2}$$

$$P_{g,t}^{\min} \leq P_{g,t} \leq P_{g,t}^{\max} \qquad g \in \mathcal{G},\, t \in \mathcal{T} \tag{3}$$

$$\left.\begin{array}{l} 0 \leq P_{s,t}^{+} \leq P_{s,t}^{+\max} \\ 0 \leq P_{s,t}^{-} \leq P_{s,t}^{-\max} \\ P_{s,t}^{+} \cdot P_{s,t}^{-} = 0 \end{array}\right\} s \in \mathcal{S},\, t \in \mathcal{T} \tag{4}$$

$$\begin{array}{ll} E_{s,t}^{\min} \leq E_{s,t} \leq E_{s,t}^{\max} & s \in \mathcal{S}, t \in \mathcal{T} \\ E_{s,t} = E_{s,t-1} + P_{s,t}^{+} - P_{s,t}^{-} & s \in \mathcal{S}, t \in \mathcal{T} \setminus \{1\} \end{array} \tag{5}$$

$$P_{\text{grid},t} + \sum_{g \in \mathcal{G}} P_{g,t} + \sum_{s \in \mathcal{S}} P_{s,t}^{-} = \sum_{s \in \mathcal{S}} P_{s,t}^{+} + \sum_{u \in \mathcal{U}} L_{u,t} \quad t \in \mathcal{T} \tag{6}$$

To facilitate the understanding of the above constraints, we explain each formula with details:

(2) Electricity need bounds from national grid: larger than zero and without upper bounds.
(3) ($P_{g,t}^{\min}$) denotes the lower bound of each electricity generation device, such as solar generation, while ($P_{g,t}^{\max}$) denotes the upper bound.
(4) ($P_{s,t}^{+\max}$) represents the upper limit for battery/storage charging at timestamp t, while ($P_{s,t}^{-\max}$) represents the upper limit for discharging.
(5) $E_{s,t}^{\min}$ represents the lower value of soc (state of charge), while $E_{s,t}^{\max}$ denotes the upper value, and the second equation denotes the updating of the soc.
(6) This equation makes sure the power grid is stable (the sum of power generation is equal to the sum of power consumption).

In practical application scenarios, it is not possible to obtain exact data on market prices, new energy generation, and user loads in advance when conducting power scheduling. Therefore, it is necessary to predict these values before making decisions. In the following, we will provide a detailed introduction to our solution.

3. Framework

3.1. Feature Engineering

Feature engineering provides input for the subsequent modules, including the forecasting module, reinforcement learning module, and optimization method module. We extract features for each building (the detailed building information will be introduced in the subsequent dataset section). Due to the different scales of features, we normalize all features X as follows:

$$x^{new} = \frac{x^{old}}{max(X) - min(X) + \epsilon} \tag{7}$$

where x^{new} is the normalized output, $max(X)$ denotes the max value of each domain, while $min(X)$ represents the minimum, and ϵ is a value that prevents the denominator from being zero.

Moreover, to eliminate the influence of some outliers, we also performed data denoising processes as:

$$x^{new} = \begin{cases} (1+\alpha) * avg(X), & \text{if } x^{old} \geq (1+\alpha) * avg(X) \\ (1-\alpha) * avg(X), & \text{if } x^{old} \leq (1-\alpha) * avg(X) \\ x^{old}, & \text{else} \end{cases} \tag{8}$$

where α is a pre-set adjustable parameter, and $avg(X)$ represents the average value of the feature. We truncate the outliers that exceed a certain percentage of the average value.

We show the key feature components of continuous modules. For the forecasting module:

- The user loads of past months;
- The electricity generation of past months;
- The radiance of solar direct or diffuse;
- Detailed time including the hour of the day, the day of the week, and the day of the month;
- The forecasting weather information including the values of humidity, temperature, and so on;

For the reinforcement learning module and optimization method module:

- The key components detailed before;
- The predictions of user load and electricity generation;
- The number of solar generation units in each building;
- The efficiency and capacity of the storage in each building;
- Market prices including the values for electricity and carbon;

3.2. Deep Learning-Based Forecasting Model

The deep learning-based forecasting module generates the corresponding input data for the next modules, including the optimization method module (or reinforcement learning module). The target variables include user load (denoted as $L_{u,t}$), market prices (denoted as p_t), and capacity of solar generation (denoted as $P_{g,t}^{\max}$). The input features of the forecasting models are listed in the Feature Engineering part before.

In sequence prediction tasks, deep neural network methods have gradually become state-of-the-art (SOTA). Gated Recurrent Unit (GRU for short) is one of the most commonly applied types of recurrent neural network with a gating mechanism [21]. We employ recurrent neural network (RNN) with a GRU in our approach. Additionally, our framework can easily adapt to any other neural networks, including CNNs and transformers. Compared to other variants of recurrent networks, RNN shows good performance in small datasets with a gated mechanism [22]. Thus, when given the input sequence $x = (x_1, \ldots, x_T)$, the RNN we used is described as [23]:

$$h_t = \phi_1(h_{t-1}, x_t), \quad y_t = \phi_2(h_t), \quad t \in \mathcal{T},$$

where h_t denotes the hidden state of RNN at time t, y_t denotes the corresponding output, and ϕ_1 and ϕ_2 represent the non-linear functions (active function or the combination with affine transformation). Fitting maximum likelihood on the training data, the model is able to predict f_{L_u}, f_p, and f_{P_g}, corresponding to user load, market prices, and capacity of solar generation, respectively. Moreover, since each of our modules is decoupled, it is easy to incorporate the predictions of any other forecasting methods into the framework.

3.3. Reinforcement Learning

In most scenarios, reinforcement learning can provide real-time decision-making, but the safety of these decisions cannot be guaranteed. Therefore, reinforcement learning has not been practically applied in LTG. However, SMG serves as a good testing ground for reinforcement learning. Due to the fact that SMG does not require the calculation of power flow in the network, in the training process, the interaction between the agent and the simulation environment can be conducted within a limited time. Since its proposal, Proximal Policy Optimization (PPO) [19] has been validated to achieve good results in various fields. Therefore, here, we model and adapt the power grid environment based on the PPO method.

The reinforcement learning framework we principally used for SGM, as shown in Figure 2, includes several parts: simulation environment module, external data input module, data preprocessor module, model module, and result postprocessor module.

The simulation environment simulates and models the microgrid, mainly using past years' real data for practice simulations. External input data includes real-time climate information obtained from websites. The data preprocessor filters and normalizes the observed data. The model module consists of multi-agent PPO (MAPPO), which includes multiple neural network modules and loss function design. The final result postprocessor module handles the boundaries of the model's output, such as checking whether the output of the generator exceeds the physical limits.

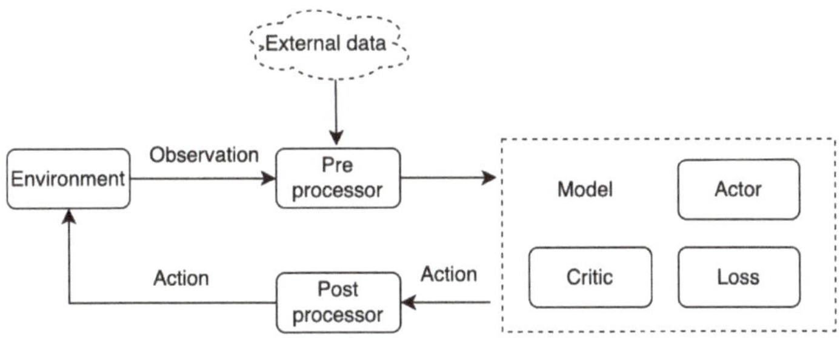

Figure 2. Reinforcement learning framework.

Most existing applications of reinforcement learning focus on single-agent methods, including centralized PPO (CPPO) and individual PPO (IPPO) [24]. As shown in Figure 3, CPPO learns the model by consolidating all inputs and interacting with the SMG. On the other hand, IPPO involves independent inputs for multiple learning instances. In the case of an SMG, each input represents a generation or consumption unit, such as a building.

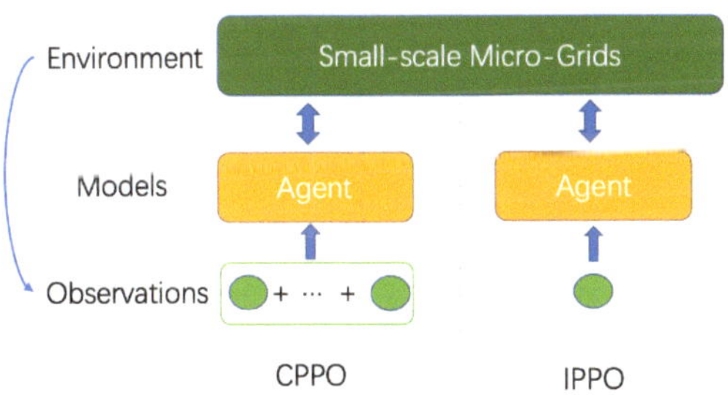

Figure 3. CPPO and IPPO framework.

In practical scenarios, there are various types of SMG, including factories, residential communities, schools, hospitals, etc. Therefore, the framework should be able to adapt to different types of SMG. The CPPO method mentioned above concatenates all inputs as one input each time, which cannot be applied to SMG with different inputs. For example, a model trained on a school SMG with 10 teaching buildings cannot be quickly adapted and applied to one with 20 teaching buildings. To address this issue, the IPPO method is introduced, which allows all teaching buildings to be inputted into the same agent in batches. However, in actual SMG, information sharing among teaching buildings is crucial. For example, the optimal power scheduling plan needs to be achieved through sharing solar energy between teaching buildings in the east and west. Since IPPO only has one agent, it cannot model the information sharing. Based on this, we propose a multi-agent PPO (MAPPO) model to address the information sharing problem in SMG.

As shown in the Figure 4, in the MAPPO framework, taking a school microgrid as an example, each agent represents a building, and each building has its own independent input. Additionally, the main model parameters are shared among all the buildings. If $\pi^i(a^i|\tau^i)$ is an agent model, the joint model is: $\pi(a|s) := \prod_{i=1}^{n} \pi^i(a^i|\tau^i)$, where n denotes the number of teaching buildings. The expected discounted accumulated reward is defined as [24]:

$$J(\pi) = \mathbb{E}_\pi[\sigma_{t=0}^{\infty}\gamma^t R(s_t, a_t, s_{t+1})] \quad (9)$$

where γ represents the discount ratio, R is the reward, and $s_t = [o_(t)^1, \ldots, o_t^n, a_t, \hat{r}_t]$ is the current state of the whole system.

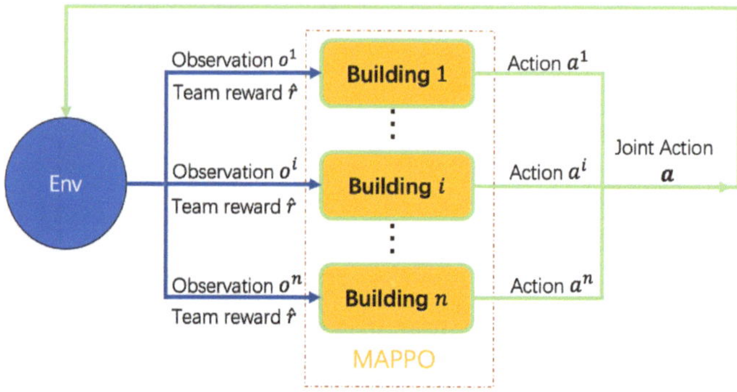

Figure 4. MAPPO framework.

3.4. Optimization

3.4.1. Stochastic Optimization

In the deep learning forecasting module, we have trained models that can predict user load ($\hat{L}_{u,t}$), market prices ($\hat{p}_t$), and the capacity of solar generation ($\hat{P}_{g,t}^{\max}$). In the validation dataset, we obtain the deviations of the models for these predictions, and their variances are denoted as $\hat{\Sigma}L_u$, $\hat{\Sigma}p$, and $\hat{\Sigma}_{P_g}$, respectively. These values represent the level of uncertainty. To mitigate the impact of uncertainty, we propose a stochastic optimization method as shown in Figure 5b. We use the predicted values as means and uncertainty as variances, for example, ($\hat{P}g, t^{\max}$, $\hat{\Sigma}P_g$), ($\hat{L}u, t$, $\hat{\Sigma}L_u$), and ($\hat{p}t$, $\hat{\Sigma}p$), to perform Gaussian sampling. Through Gaussian sampling, we can obtain multiple scenarios, which are considered as a multi-scenario optimization problem. Assuming we have N scenarios, the n-th scenario can be represented as ($n \in \mathcal{S}_{\mathcal{N}}$) [25]:

$$\begin{aligned}
(\tilde{P}_g^{\max})^n &= \left[(\tilde{P}_{g,1}^{\max})^n, (\tilde{P}_{g,2}^{\max})^n, \ldots, (\tilde{P}_{g,T}^{\max})^n\right], \\
(\tilde{L}_u)^n &= \left[(\tilde{L}_{u,1})^n, (\tilde{L}_{u,2})^n, \ldots, (\tilde{L}_{u,T})^n\right], \\
(\tilde{p})^n &= \left[(\tilde{p}_1)^n, (\tilde{p}_2)^n, \ldots, (\tilde{p}_T)^n\right].
\end{aligned}$$

Then, the objective function in our proposed stochastic optimization can be redefined as:

$$\underset{X}{\text{minimize}} \quad \sum_{t=1}^{T} \mathbb{E}_{n\in\mathcal{S}_{\mathcal{N}}}(\tilde{p}_t)^n \cdot P_{\text{grid},t}\,. \quad (10)$$

Constraint (3) is refined as:

$$P_{g,t}^{\min} \leq P_{g,t} \leq (\tilde{P}_{g,t}^{\max})^n \quad n \in \mathcal{S}_{\mathcal{N}}, g \in \mathcal{G}, t \in \mathcal{T}\,.$$

Constraint (6) is refined as:

$$P_{\text{grid},t} + \sum_{g\in\mathcal{G}} P_{g,t} + \sum_{s\in\mathcal{S}} P_{s,t}^{-} = \sum_{s\in\mathcal{S}} P_{s,t}^{+} + \sum_{u\in\mathcal{U}} (\tilde{L}_{u,t})^n \quad n \in \mathcal{S}_{\mathcal{N}}, t \in \mathcal{T}\,.$$

Through solving the stochastic optimization problem (10), we obtain the scheduling plan: $\dot{X} = \{\dot{P}_{\text{grid},t}, \dot{P}_{g,t}, \dot{P}_{s,t}^{+}, \dot{P}_{s,t}^{-}, \dot{E}_{s,t}\}$.

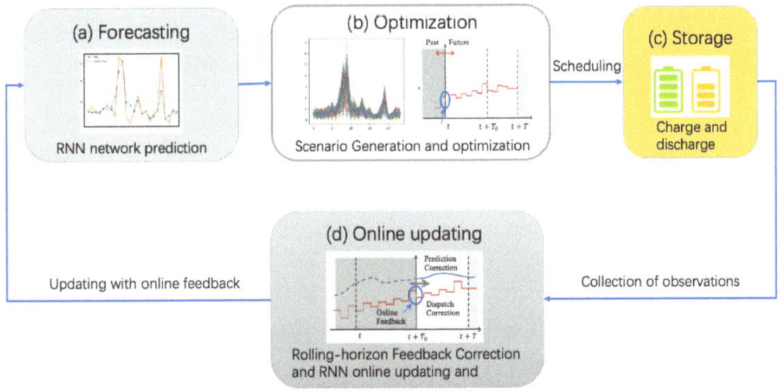

Figure 5. The whole optimization method framework. The subplot (**a**) represents the flowchart of prediction based on deep learning. The output of the prediction module serves as the input for the stochastic optimization module, as shown in (**b**). During the scheduling process, real-time data accumulates over time, and we update the predictions based on the real data, as demonstrated in (**d**), named the online data augmentation module. This framework enhances the robustness of scheduling under uncertain conditions.

3.4.2. Online Data Augmentation

In order to address the data drift problem, we propose the data augmentation method as shown in Figure 5c. The module contains two parts: pre-training/fine-tuning scheme and rolling-horizon feedback correction.

Pre-Training and Fine-Tuning

In practice, the real-time energy dispatch process is a periodic task (e.g., daily dispatch). Considering that the prediction models are trained based on historical data and future data and may not necessarily follow the same distribution as the past, we perform online data augmentation. Online data augmentation consists of two parts: pre-training and fine-tuning. Firstly, we pre-train the neural network model using historical data to obtain a model capable of predicting f_{L_u}, f_p, and f_{P_g}. Secondly, we fine-tune the neural network using the accumulated online data. Specifically, in the fine-tuning process, we employ partial parameter fine-tuning to obtain the refined network $\tilde{f}_{L_u}$, $\tilde{f}_p$, and $\tilde{f}_{P_g}$.

Rolling-Horizon Feedback Correction

In addition to updating the prediction models online, we also employ the rolling-horizon control technique. In the optimization process, we solve the optimization problem every horizon H (to incorporate the latest prediction models and trade-off computational time). This operation is repeated throughout the scheduling period.

4. Experiments

4.1. Experiment Setup

4.1.1. Dataset

We conducted experiments on building energy management using a real-world dataset from Fontana, California. The dataset includes one year of electricity scheduling for 17 buildings, including their electricity demand, solar power generation, and weather conditions. This dataset was also used for the NIPS 2022 Challenge. With our proposed framework, we achieved the global championship in the competition [20].

4.1.2. Metric

We follow the evaluation setup of the competition. The 17 buildings are divided into visible (5 buildings) and invisible data (12 buildings). The visible data are used as the training set, while the invisible data include the validation set and the testing set. Visible data contain all labels including user load demand and solar generation in a year. The labels of the invisible data can only be evaluated through limited interactions with the competition organizers' open API. The final leaderboard ranking is based on the overall performance of the model on all data sets. The evaluation metrics include carbon emissions, electricity cost, and grid stability. Specifically, the electricity consumption of each building i is calculated as $E_{i,t} = L_{i,t} - P_{i,t} + X_{i,t}$, where $L_{i,t}$ represents the load demand at timestamp t, $P_{i,t}$ represents the solar power generation of the building, and $X_{i,t}$ represents the electricity dispatch value provided by the model. The electricity consumption of the entire district is denoted as $E_t^{\text{dist}} = \sum_{i=1}^{I} E_{i,t}$.

Using the above notations, three metrics are defined as:

$$C_{\text{Emission}} = \sum_{t=1}^{T} \left(\sum_{i=1}^{I} \max(E_{i,t}, 0) \right) \cdot c_t, \quad C_{\text{Price}} = \sum_{t=1}^{T} \max\left(E_t^{\text{dist}}, 0\right) \cdot p_t,$$

$$\begin{aligned} C_{\text{Grid}} &= \frac{1}{2}\left(C_{\text{Ramping}} + C_{\text{Load Factor}}\right) \\ &= \frac{1}{2}\left(\sum_{t=1}^{T-1} \left| E_{t+1}^{\text{dist}} - E_t^{\text{dist}} \right| + \sum_{m=1}^{\#\text{months}} \frac{\text{avg}_{t \in [\text{month } m]} E_t^{\text{dist}}}{\text{max}_{t \in [\text{month } m]} E_t^{\text{dist}}} \right). \end{aligned}$$

4.1.3. Baseline

To evaluate the proposed MAPPO, Optimization, and their Ensemble method, we compare them with the following baseline methods:

- RBC: Rule-Based Control method. We tested several strategies and selected the best one: charging the battery by 10% of its capacity between 10 a.m. to 2 p.m., followed by discharging it by the same amount between 4 p.m. to 8 p.m.
- MPC [26]: A classical Model-Predictive-Control method. A GBDT-based model [27] is used to predict future features, and a deterministic optimization is used for daily scheduling.

Moreover, after the competition, we also compared the proposals of several top-ranked contestants:

- AMPC [26]: An adaptive Model-Predictive-Control method.
- SAC [28]: A Soft Actor-Critic method that uses all agents with decentralization.
- ES [29]: Evolution-Strategy method with adaptive covariance matrix.

4.1.4. Implementations

The environment simulator that employs reinforcement learning and an evaluation process is provided by the competition organizers [30]. The learning of deep learning networks is implemented using PyTorch. The optimization problem-solving utilizes our self-developed MindOpt [31]. All experiments are conducted on an Nvidia Tesla V100 GPU with eight cards.

4.2. Results

If only one metric is considered, any of the three metrics can perform very well. Therefore, the final effect needs to be seen in terms of the average value of the three metrics. In particular, as shown in Table 1, 'Emission', 'Price', and 'Grid' denote the metric $C_{Emimssion}$, C_{Price}, and C_{Grid}, respectively. Since the performance is compared with no use of storage, a lower value indicates a better performance. Our proposed MAPPO method and Optimization method both achieve better results than other competitors.

As shown in Table 1, the individual model has limited performance. By combining reinforcement learning and optimization, we can achieve the best results. Through observing the validation dataset, we found that reinforcement learning and optimization perform alternately in different months. By leveraging their advantages, we fuse their results based on the month to create a yearly schedule (named Ensemble), ultimately obtaining the best outcome. Besides, all calculations of the models above are completed within 30 min to generate the scheduling for the next year.

Table 1. Comparison of the performances of all methods in the entire building. All values are normalized against the simple baseline without strategy, i.e., not using the storage. Therefore, a lower value indicates a better performance.

Methods	Overall Performance			
	Average Cost	Emission	Price	Grid
RBC	0.921	0.964	0.817	0.982
MPC	0.861	0.921	0.746	0.916
AMPC	0.827	0.859	0.750	0.872
ES	0.812	0.863	0.748	0.827
SAC	0.834	0.859	0.737	0.905
MAPPO	0.810	0.877	0.726	0.826
Optimization	0.804	0.871	0.719	0.822
Ensemble	**0.801**	**0.864**	**0.718**	**0.821**

4.3. Ablation Studies

We conducted ablation studies on some modules to understand their contributions to the overall performance.

4.3.1. Analysis of Online Data Augmentation

We compare the performances of different online updating methods, as shown in Figure 6: **No-Ft**: no fine-tuning on online data; **Self-Adapt**: adaptive linear correction by minimizing the mean squared error between historical value and predicted value; **Scratch**: re-learning from scratch; **Small-LR**: continuous learning with a smaller learning rate; **Freeze**: continuous learning with online data but freezing the weights of the first few layers and only updating the last layer. To compare the efficiency of the models, we evaluate the average execution time of real-time scheduling within 24 h.

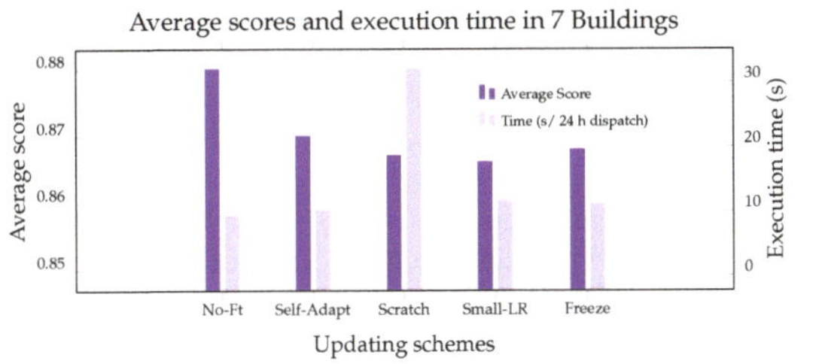

Figure 6. Analysis of online data augmentation, the evaluation about performance and execution time with various settings.

Results show that fine-tuning with a smaller learning rate has advantages in terms of efficiency and effectiveness.

4.3.2. Analysis of Forecasting Models

As shown in Table 2, we evaluated different forecasting models. The evaluation metrics include overall scheduling performance, execution time, and forecasting performance measured by the weighted mean absolute percentage error (WMAPE). The experimental results indicate that the RNN model with online fine-tuning achieves the best performance.

Table 2. Analysis of different forecasting models, including scheduling performance, forecasting performance, execution time, and updating methods.

Forecast Model	Online Update	Dispatch		Forecast (WMAPE)	
		Average	Time	Load	Solar
Linear	✗	0.878	8 s	42.1%	27.3%
GBDT		0.875	8 s	44.7%	10.7%
RNN		0.876	9 s	46.0%	10.7%
Transformer		0.879	11 s	45.3%	10.6%
Linear	✓ Self-Adaptive Linear Correction	0.871	8 s	39.4%	21.2%
GBDT		0.868	9 s	39.5%	9.4%
RNN		0.866	10 s	39.3%	9.3%
Transformer		0.869	11 s	39.9%	9.1%
RNN	✓ Online Fine-tuning	0.862	11 s	**39.0%**	**9.0%**
Transformer		0.864	12 s	39.3%	9.1%

4.3.3. Analysis of Stochastic Optimization

In stochastic optimization, the number of scenarios is a very important parameter. As shown in Figure 7, as the number of scenarios increases, the effectiveness of the model also gradually increases. This is in line with common sense, as a model that can cover more scenarios tends to have better performance.

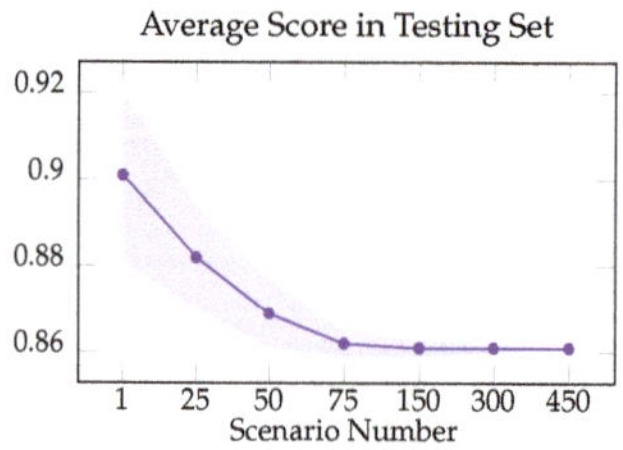

Figure 7. Effect of different number of scenarios N. The curve denotes the expected value, while the area is the standard deviation of the stochastic sample.

5. Conclusions

The challenge of power grid scheduling lies in the complexity of long-term decision-making. Through our research, we have learned that achieving end-to-end learning with a single strategy is difficult for such complex problems. We have identified that future load and solar energy generation are key information for decision-making. Our results show that using pre-trained auxiliary tasks to learn representation and prediction ahead of optimization and reinforcement learning outperforms directly feeding all the data into the decision model. By employing optimization and multi-agent reinforcement learning algorithms for decision-making, we have found that the optimization algorithm achieves better generalization on an unknown dataset through target approximation, data augmentation, and rolling-horizon correction. On the other hand, multi-agent reinforcement learning better models the problem and finds better solutions on a known dataset. The issue of data augmentation to improve generalization in energy management tasks warrants further research. We have also observed that the policies learned by the optimization algorithm and reinforcement learning perform differently in different months, which has motivated us to explore ensemble learning approaches. We left the ensemble of forecasting models as future work.

Author Contributions: Methodology, C.Y., J.Z., W.J., L.W., H.Z., Z.Y. and F.L.; software, L.W., H.Z., Z.Y. and F.L.; writing—original draft preparation, J.Z. and W.J.; writing—review and editing, C.Y., Z.Y. and F.L.; supervision, C.Y., J.Z., Z.Y. and F.L. All authors have read and agreed to the published version of the manuscript.

Funding: This research received no external funding

Data Availability Statement: Not applicable

Conflicts of Interest: The authors declare no conflict of interest.

References

1. Gama, J.; Žliobaitė, I.; Bifet, A.; Pechenizkiy, M.; Bouchachia, A. A survey on concept drift adaptation. *ACM Comput. Surv. (CSUR)* **2014**, *46*, 1–37. [CrossRef]
2. Elsheikh, A.H.; Saba, A.I.; Panchal, H.; Shanmugan, S.; Alsaleh, N.A.; Ahmadein, M. Artificial intelligence for forecasting the prevalence of COVID-19 pandemic: An overview. *Healthcare* **2021**, *9*, 1614. [CrossRef] [PubMed]
3. Torres, J.F.; Hadjout, D.; Sebaa, A.; Martínez-Álvarez, F.; Troncoso, A. Deep learning for time series forecasting: A survey. *Big Data* **2021**, *9*, 3–21. [CrossRef] [PubMed]
4. Lara-Benítez, P.; Carranza-García, M.; Riquelme, J.C. An experimental review on deep learning architectures for time series forecasting. *Int. J. Neural Syst.* **2021**, *31*, 2130001. [CrossRef] [PubMed]
5. Sina, L.B.; Secco, C.A.; Blazevic, M.; Nazemi, K. Hybrid Forecasting Methods—A Systematic Review. *Electronics* **2023**, *12*, 2019. [CrossRef]
6. Camacho, E.F.; Alba, C.B. *Model Predictive Control*; Springer Science & Business Media: London, UK, 2013.
7. Hewing, L.; Wabersich, K.P.; Menner, M.; Zeilinger, M.N. Learning-based model predictive control: Toward safe learning in control. *Annu. Rev. Control Robot. Auton. Syst.* **2020**, *3*, 269–296. [CrossRef]
8. Muralitharan, K.; Sakthivel, R.; Vishnuvarthan, R. Neural network based optimization approach for energy demand prediction in smart grid. *Neurocomputing* **2018**, *273*, 199–208. [CrossRef]
9. Elmachtoub, A.N.; Grigas, P. Smart "predict, then optimize". *Manag. Sci.* **2022**, *68*, 9–26. [CrossRef]
10. Lauro, F.; Longobardi, L.; Panzieri, S. An adaptive distributed predictive control strategy for temperature regulation in a multizone office building. In Proceedings of the 2014 IEEE International Workshop on Intelligent Energy Systems (IWIES), San Diego, CA, USA, 8 October 2014; pp. 32–37.
11. Heirung, T.A.N.; Paulson, J.A.; O'Leary, J.; Mesbah, A. Stochastic model predictive control—How does it work? *Comput. Chem. Eng.* **2018**, *114*, 158–170. [CrossRef]
12. Yan, S.; Goulart, P.; Cannon, M. Stochastic model predictive control with discounted probabilistic constraints. In Proceedings of the 2018 European Control Conference (ECC), IEEE, Limassol, Cyprus, 12–15 June 2018; pp. 1003–1008.
13. Paulson, J.A.; Buehler, E.A.; Braatz, R.D.; Mesbah, A. Stochastic model predictive control with joint chance constraints. *Int. J. Control* **2020**, *93*, 126–139. [CrossRef]
14. Shang, C.; You, F. A data-driven robust optimization approach to scenario-based stochastic model predictive control. *J. Process Control* **2019**, *75*, 24–39. [CrossRef]
15. Bradford, E.; Imsland, L.; Zhang, D.; del Rio Chanona, E.A. Stochastic data-driven model predictive control using gaussian processes. *Comput. Chem. Eng.* **2020**, *139*, 106844. [CrossRef]
16. Ioannou, P.A.; Sun, J. *Robust Adaptive Control*; Courier Corporation: Chelmsford, MA, USA, 2012.
17. Åström, K.J.; Wittenmark, B. *Adaptive Control*; Courier Corporation: Chelmsford, MA, USA, 2013.
18. Liu, X.; Paritosh, P.; Awalgaonkar, N.M.; Bilionis, I.; Karava, P. Model predictive control under forecast uncertainty for optimal operation of buildings with integrated solar systems. *Sol. Energy* **2018**, *171*, 953–970. [CrossRef]
19. Yu, C.; Velu, A.; Vinitsky, E.; Gao, J.; Wang, Y.; Bayen, A.; Wu, Y. The Surprising Effectiveness of PPO in Cooperative Multi-Agent Games. *Adv. Neural Inf. Process. Syst.* **2022**, *35*, 24611–24624.
20. Aicrowd. Neurips 2022 Citylearn Challenge. Available online: https://www.aicrowd.com/challenges/neurips-2022-citylearn-challenge (accessed on 18 July 2022).
21. Cho, K.; van Merriënboer, B.; Bahdanau, D.; Bengio, Y. On the Properties of Neural Machine Translation: Encoder–Decoder Approaches. In Proceedings of the SSST-8, Eighth Workshop on Syntax, Semantics and Structure in Statistical Translation, Doha, Qatar, 25 October 2014; pp. 103–111.
22. Chung, J.; Gulcehre, C.; Cho, K.; Bengio, Y. Empirical evaluation of gated recurrent neural networks on sequence modeling. *arXiv* **2014**, arXiv:1412.3555.
23. Graves, A. Generating sequences with recurrent neural networks. *arXiv* **2013**, arXiv:1308.0850.
24. Schulman, J.; Wolski, F.; Dhariwal, P.; Radford, A.; Klimov, O. Proximal policy optimization algorithms. *arXiv* **2017**, arXiv:1707.06347.
25. Wurdemann, H.A.; Stilli, A.; Althoefer, K. Lecture notes in computer science: An antagonistic actuation technique for simultaneous stiffness and position control. In Proceedings of the Intelligent Robotics and Applications: 9th International Conference, ICIRA 2015, Portsmouth, UK, 24–27 August 2015; Proceedings, Part III; Springer: Cham, Switzerland, 2015; pp. 164–174.

26. Sultana, W.R.; Sahoo, S.K.; Sukchai, S.; Yamuna, S.; Venkatesh, D. A review on state of art development of model predictive control for renewable energy applications. *Renew. Sustain. Energy Rev.* **2017**, *76*, 391–406. [CrossRef]
27. Ke, G.; Meng, Q.; Finley, T.; Wang, T.; Chen, W.; Ma, W.; Ye, Q.; Liu, T.Y. LightGBM: A Highly Efficient Gradient Boosting Decision Tree. In Proceedings of the 31st International Conference on Neural Information Processing Systems, NIPS'17, Red Hook, NY, USA, 4–9 December 2017; pp. 3149–3157.
28. Kathirgamanathan, A.; Twardowski, K.; Mangina, E.; Finn, D.P. A Centralised Soft Actor Critic Deep Reinforcement Learning Approach to District Demand Side Management through CityLearn. In Proceedings of the 1st International Workshop on Reinforcement Learning for Energy Management in Buildings & Cities, RLEM'20, New York, NY, USA, 17 November 2020; pp. 11–14.
29. Varelas, K.; Auger, A.; Brockhoff, D.; Hansen, N.; ElHara, O.A.; Semet, Y.; Kassab, R.; Barbaresco, F. A comparative study of large-scale variants of CMA-ES. In Proceedings of the Parallel Problem Solving from Nature—PPSN XV: 15th International Conference, Coimbra, Portugal, 8–12 September 2018; Proceedings, Part I 15; Springer: Cham, Switzerland, 2018; pp. 3–15.
30. Vázquez-Canteli, J.R.; Kämpf, J.; Henze, G.; Nagy, Z. CityLearn v1.0: An OpenAI gym environment for demand response with deep reinforcement learning. In Proceedings of the 6th ACM International Conference on Systems for Energy-Efficient Buildings, Cities, and Transportation, New York, NY, USA, 13–14 November 2019; pp. 356–357.
31. MindOpt. MindOpt Studio. Available online: https://opt.aliyun.com/platform/overview (accessed on 20 October 2022).

processes MDPI

Article

The Value of Energy Storage in Facilitating Renewables: A Northeast Area Analysis

Meng Zhu [1], Yong Sun [1], Yu Lu [1], Linwei Sang [2,*], Zhongkai Yi [2], Ying Xu [2] and Kerui Ma [1]

[1] Economic and Technical Research Institute, State Grid Jilin Electric Power Co., Ltd., Changchun 130022, China
[2] School of Electrical Engineering & Automation, Harbin Institute of Technology, Harbin 150006, China
* Correspondence: sanglinwei@gmail.com

Abstract: The cross-regional and large-scale transmission of new energy power is an inevitable requirement to address the counter-distributed characteristics of wind and solar resources and load centers, as well as to achieve carbon neutrality. However, the inherent stochastic, intermittent, and fluctuating nature of wind and solar power poses challenges for the stable bundled dispatch of new energy. Leveraging the regulation flexibility of energy storage offers a potential solution to mitigate new energy fluctuations, enhance the flexibility of the hybrid energy systems, and promote bundled dispatch of new energy for external transmission. This paper takes energy storage as an example and proposes a capacity configuration optimization method for a hybrid energy system. The system is composed of wind power, solar power, and energy storage, denoted by the wind–solar–energy storage hybrid energy systems. The objective is to quantify the support provided by energy storage to bundled dispatch of new energy, namely determining the new energy transmission capacity that can be sustained per unit of energy storage. The results demonstrate that the proposed method effectively improves the bundled dispatch capacity of new energy. Moreover, the obtained configuration results can be tailored based on different wind–solar ratios, allowable fluctuation rates, and transmission channel capacities, rendering the approach highly valuable for engineering practicality.

Keywords: energy storage system; cold scenario; machine learning

Citation: Zhu, M.; Sun, Y.; Lu, Y.; Sang, L.; Yi, Z.; Xu, Y.; Ma, K. The Value of Energy Storage in Facilitating Renewables: A Northeast Area Analysis. *Processes* **2023**, *11*, 3449. https://doi.org/10.3390/pr11123449

Academic Editors: Davide Papurello and Gang Wang

Received: 13 November 2023
Revised: 8 December 2023
Accepted: 14 December 2023
Published: 18 December 2023

1. Introduction

In the past two years, countries around the world have outlined blueprints for achieving carbon neutrality by 2050 or 2060 [1,2]. To effectively promote the low-carbon transformation of the energy system, there is a need to vigorously develop new energy sources to gradually replace traditional fossil-based generators [3,4]. It is anticipated that by 2050, renewable energy, especially solar and wind power, will take up the largest part of the energy supply both in China and worldwide. However, as the scale and penetration rate of new energy continue to increase, more upcoming issues such as power system fluctuations and instability become more prominent, posing significant challenges to the secure operation of the grid and the integration of new energy [5]. Compared to the conventional controllable fossil-based generators, the power outputs of renewables mainly rely on the corresponding weather conditions, with solar radiation for solar power and wind speed for wind power, which feature high uncertainty and variability. And the output may not satisfy the demand side. The supply–demand mode will change from the conventional "generation follow load" to a new pattern "load follow generation" or "generation follow load with storage". The new pattern relies on energy storage as the enabler to achieve the balance between energy supply and demand, which is the one of the keys for a future sustainable-energy society.

So, energy storage, due to its advantages of flexibility, rapid response, and clean and non-polluting nature, is expected to play a crucial role in the future energy system [6].

Moreover, its dual identity as both a power source and energy storage makes it an important approach to address the stochastic, intermittent, and fluctuating characteristics of renewable energy, facilitating the concentrated integration of new energy sources [5], reducing the difficulty of peak shaving in the system, increasing the utilization of cross-regional transmission channels [7–9], and enhancing the security and reliability of the power grid operation.

Most scholars have historically utilized historical annual wind and solar power outputs as input for forecasting models, which increases the complexity of the models. However, in actual grid planning, the uncertainty of new energy outputs significantly impacts capacity configuration decisions [10–12], making it a critical issue to effectively characterize uncertainty in the optimization of hybrid energy systems. Currently, common methods for dealing with uncertainty in optimization models include stochastic optimization and robust optimization [13–16]. Stochastic optimization uses scenario sets to represent the uncertainty of new energy outputs, aiming to simulate various operational scenarios based on typical scenarios derived from historical data. On the other hand, robust optimization employs parameter intervals to characterize the uncertainty of new energy outputs, encompassing all possible operational scenarios within the interval, ensuring that the optimization results satisfy operational constraints, even in the worst-case scenario [17,18].

Based on the above, considering the counter-distributed characteristics of domestic wind and solar resources and power load centers, large-scale cross-regional transmission of renewable energy is necessary [19]. However, the direct transmission of bundled renewable energy will lead to pressure on the receiving side and transmission line capacity enhancement, due to its variability and uncertainty. We study a hybrid energy system, composed of wind power, solar power, and energy storage, denoted by the wind–solar–energy storage hybrid energy systems. To address the above problems, it is essential to study wind–solar—storage resource configuration strategies that consider constraints on the renewable energy external transmission and match them with storage.

Using energy storage as an example, this study first conducts clustering analysis based on the actual wind and solar outputs and load data of a certain region in the northeast over one year to obtain typical scenarios of solar, wind, and load curve for the northeast region. The typical scenario can help analyze the complementary features of solar and wind power output. Then, based on the typical scenario, a wind–solar–storage ratio planning strategy that considers the value of storage support for new energy external transmission capacity is proposed, and the impacts of different photovoltaic ratios, allowable fluctuation rates, and transmission channel capacities on the optimization results are analyzed. To summarize, the key innovation of this paper lies in two aspects: (1) the total planning strategy with the clustering analysis and optimization analysis; (2) the comprehensive analysis of the value of energy storage under various scenarios.

The rest of this paper is organized as follows. Section 2 leverages the KMEANS++ clustering algorithm to conduct the scenario analysis on the solar–wind–load for obtaining the typical scenarios for later planning analysis. Section 3 proposes the wind–solar–storage ratio planning strategy that considers the value of storage support for the renewable energy external transmission capacity. Section 4 conducts comprehensive analysis to verify the effectiveness of the proposed models from the perspectives of clustering analysis, reduction analysis, sensitivity analysis, and numerical analysis. Section 5 concludes the paper and provides several advices for the system operation.

2. Scenario Analysis

In this study, the first step involves investigating the regional resource endowment to model photovoltaic power plants and wind farms, generating one year of wind and solar output data. These data are then combined with regional load data for joint analysis. Next, the KMEANS++ algorithm [20] is employed to conduct clustering analysis on the wind–solar–load data, extracting typical scenarios of wind–solar–load for subsequent planning research on the optimal configuration. Intuitively, the output of solar and wind power features the complementary feature. This feature means that solar power generates more in the daytime and less in the night, and, in comparison, the wind power usually generates more in the night and less in the daytime. The clustering analysis can help us to investigate this feature. During the clustering analysis, the effectiveness of clustering parameters is validated using both the average distance to cluster centers and dimensionality reduction visualization techniques.

2.1. Clustering Algorithm

In this study, the first step involves normalizing the wind–solar–load data for the entire year of 8760 time periods using the maximum value normalization method. And the multiple sources of data with different units are unified into the scale of 0–1, which can be seen as the p.u. values. After that, the KMEANS++ algorithm is applied with different numbers of clusters to perform clustering analysis. However, the number of clusters will have great impacts on the system. More clusters lead to complex computational power, and fewer clusters lead to less representative features. The effectiveness of clustering is evaluated by computing the average distance to the cluster centers and selecting the appropriate number of clusters. Additionally, dimensionality reduction visualization techniques are used to analyze the clustering results. If the clustering results do not meet the required criteria, the number of clusters is adjusted accordingly. Finally, the typical operational scenarios obtained after clustering are output and used for subsequent planning and analysis.

To enhance the computational efficiency, typical daily wind–solar–load data are extracted using clustering, as analyzing the full-year data of 8760 time periods directly would be computationally intensive. Currently, mainstream clustering algorithms include prototype-based clustering (KMEANS), hierarchical clustering, density-based clustering (DBSCAN), and graph-based clustering (AP). This study primarily adopts the KMEANS++ algorithm, a variant of the prototype-based KMEANS algorithm, due to its fast convergence and robustness, making it suitable for clustering large-scale high-dimensional real-world data.

The first step of the study involves normalizing the wind–solar–load output data for the entire year to the range of 0–1 based on their respective maximum output values. Subsequently, the KMEANS++ algorithm is applied to the normalized wind–solar–load data to extract typical operational scenarios. The core idea behind the KMEANS++ algorithm is to maximize the distance between initial cluster centers through iterative steps. The specific implementation steps are detailed in Algorithm 1:

Algorithm 1 The KEAMS++ algorithm for the solar–wind–load typical scenario extraction.

Input: The number of clusters N; the processed dataset of wind–solar–load data under the 0-1 normalization;

Step 1: Randomly select one wind–solar–load scenario as the initial cluster center C_0.

Step 2: Assign the remaining $N = C - 1$ cluster centers using a distance-based approach. For each of the N cluster centers, the algorithm calculates the distance between the wind–solar–load scenario and the existing cluster centers, including C_0.

Step 3: Continue the process of selecting $N - 1$ cluster centers by repeating step 2 until all N cluster centers have been chosen.

Step 4: Perform the KMEANS++ clustering algorithm using the N cluster centers obtained above to cluster the wind–solar–load data into C clusters.

Output: Typical wind–solar–load data pair scenarios with different corresponding clustering number.

2.2. Number of Clusters Analysis

Considering the diverse operational characteristics of regional wind and solar data, it is challenging to manually determine the number of cluster centers (C) for the KMEANS++ algorithm based on experience. Instead, this study selects the number of cluster centers using the sum of the squared error (SSE) of the clustering centroids for the wind–solar–load scenarios throughout the entire year. To determine the optimal number of clusters, the KMEANS++ algorithm is applied iteratively with different values of C. For each value of C, the clustering is performed, and the SSE is computed, representing the sum of the squared distances between data points and their respective cluster centroids. The value of C that results in the lowest SSE is selected as the optimal number of cluster centers, as it indicates a more compact and well-separated clustering, capturing the distinctive patterns of wind–solar–load scenarios effectively, which is calculated in Equation (1) .

$$\text{SSE} = \frac{1}{C}\sum_{i=1}^{C}\left(\frac{1}{|\Phi_i|}\sum_{m_j\in\Phi_i}||\ m_j - c_i\ ||\right) \tag{1}$$

By using the SSE as a criterion for selecting the number of cluster centers, this approach avoids relying on subjective judgment and ensures a data-driven determination, better reflecting the inherent structure and characteristics of the normalized wind–solar–load data for subsequent planning and analysis.

2.3. Dimension Reduction Analysis

Given that the typical daily wind–solar–load data are of high dimensionality (96 dimensions) with weak correlations, understanding the specific distribution of the typical days can be challenging. To address this, the study employs dimensionality reduction visualization techniques to extract the main features of the wind–solar–load data and project it onto a two-dimensional space, facilitating the visualization of the typical days.

The t-SNE (t-Distributed Stochastic Neighbor Embedding) algorithm is utilized in this study for dimensionality reduction visualization of the clustering results. The core idea behind t-SNE is to transform the high-dimensional distances in the original space into probability distributions using Gaussian distributions and then map similar probability distributions to distances in a lower-dimensional space. The t-SNE core principle is to perform dimensionality reduction on high-dimensional vectors by minimizing the Kullback–Leibler (KL) divergence between the probability distributions of the data points in the high-dimensional space and the low-dimensional space, as shown in Equation (2).

$$D_{KL} = \sum_{m\neq n} P_{mn}\frac{\mathcal{P}_{mn}}{S_{mn}} \tag{2}$$

By applying t-SNE, the typical daily wind–solar–load data are represented in a two-dimensional space, allowing for a more intuitive and interpretable visualization of the distinctive patterns and structures among the typical days. This visualization technique aids in better understanding the characteristics and relationships among the typical days, providing valuable insights for subsequent planning and operations analysis.

3. Ratio Planning Model

Based on the typical scenario analysis model and typical scenario, this section proposes the wind–solar–storage ratio planning strategy. It considers the value of storage to support the renewable energy transmission. Then, multiple parametric analysis is leveraged for various scenarios' analysis. Based on multiple parametric analysis, we can obtain the relationship among the needed capacity of the energy storage, different combinations of solar and wind power, and the different limits of transmission capacity.

3.1. Ratio Planning Framework

To achieve the goals of "carbon peaking" and "carbon neutrality," in the coming decades, the installed capacity of wind and solar power, represented by wind turbines and photovoltaics (PV) generation, will continue to increase in the power system. Regions abundant in wind and solar resources often experience wind and solar actual outputs that exceed their local demand. Therefore, it becomes essential to bundle and transmit the surplus energy to regions with higher demand.

However, due to the strong variability, intermittency, and uncertainty of wind and solar power outputs, directly bundling wind and solar energy for external transmission often results in combined output characteristics that do not align with the load patterns of the external grid, thus affecting the optimization operation of the external grid. Therefore, this study utilizes the regulation capabilities of energy storage to smooth the fluctuations in wind and solar power outputs, enabling the bundling of wind, solar, and storage energy. The bundled energy is then transmitted to the external grid via ultra-high-voltage transmission lines, ensuring efficient and optimized operation of the external grid.

3.2. Energy Storage System Formulation

Energy storage systems (ESS) usually work at the charging state when the electricity price is low and at the discharging state when the electricity price is high. The feasible region Φ_{ESS} of decision variable P_{ESS} is subject to the following constraints:

$$P_{ESS} = P_{dis} - P_{ch} \tag{3}$$

Equation (3) illustrates that the operating power P_{ESS} of ESS is composed of charging and discharging parts; P_{ch} denotes the charging of ESS from the grid, and P_{dis} denotes the discharging of ESS to the grid.

$$\begin{aligned} & E_t = E_{t-1} + \eta_{ch} P_{ch,t} \Delta_t - \frac{P_{dis,t}}{\eta_{dis}} \Delta_t \quad t = 2, .., T \\ & E_1 = E_{init} + \eta_{ch} P_{ch,1} \Delta_t - \frac{P_{dis,1}}{\eta_{dis}} \Delta_t \\ & E_{min} \le E_t \le E_{max} \quad t = 1, .., T \end{aligned} \tag{4}$$

where E_t (kWh) refers to the stored energy in the ESS. Equation (4) ensures that E_t in the ESS at time t lies in an allowable range, and E_{min}, E_{max} refer to the minimum and maximum capacity of the battery system.

$$\begin{aligned} & 0 \le P_{dis,t} \le \min\{P_{dis}^{max}, \eta_{dis} \frac{E_{t-1} - E_{min}}{\Delta_t}\} \quad t = 1, ..., T \\ & 0 \le P_{ch,t} \le \min\{P_{ch}^{max}, \frac{E_{max} - E_{t-1}}{\eta_{ch} \Delta_t}\} \quad t = 1, ..., T \\ & 0 \le P_{dis,t} \le M \mu_{dis,t} \quad t = 1, ..., T \\ & 0 \le P_{ch,t} \le M \mu_{ch,t} \quad t = 1, ..., T \\ & \mu_{dis,t} + \mu_{ch,t} \le 1 \quad t = 1, ..., T \end{aligned} \tag{5}$$

where M is a large positive number. μ_{dis} and μ_{ch} are binary indicators of discharging and charging state, where 1 means in the state and 0 means the opposite. P_{dis}^{max} and P_{ch}^{max} are the maximum values of charging and discharging power. Equation (5) prevents the simultaneous charging and discharging of ESS by utilizing the big-M relaxation method.

3.3. The Final Wind–Solar–Storage Ratio Planning Model

Based on the extracted typical scenarios, the ratio framework, and energy storage model, we formulate the wind–solar–storage ratio planning with the stochastic optimization to consider the uncertainty of the renewable power output. We note that most scenarios are similar, so we propose to leverage the KMEANS++ algorithm with SSE analysis to extract the typical scenarios for planning analysis with high efficiency. We leverage the above model to show the value of energy storage in facilitating renewables in the following (6):

$$\min_{E_{max}} C(E_{max}) \tag{6a}$$

$$\text{s.t.} P_{sw} = r_{pv} P_{PV} + (1 - r_{pv}) P_{WP} \tag{6b}$$

$$\sum_{g\in\mathcal{G}_n} P_{g,t} + \sum_{w\in\mathcal{W}_n, pv\in\mathcal{PV}_n} P_{WP} + P_{ES} + P_{trans}$$

$$+ \sum_{m:(n,m)} B_{nm}(\delta_{nt} - \delta_{mt}) = P_{n,t}^{D} - P_{n,t}^{cur} \quad \forall n, t \tag{6c}$$

$$|B_{nm}(\delta_{nt} - \delta_{mt})| \leq F_{nm}^{max} \quad \forall (n,m) \in \mathcal{L}, \forall t \tag{6d}$$

$$P_{trans} \leq C_{trans} \tag{6e}$$

$$\text{Constraints (3)–(5)} \tag{6f}$$

where $C(E_{max})$ is the cost of energy storage; r_{pv} is the ratio of PV to total renewables, which is composed of solar power P_{PV} and wind power P_{WP}; P_{trans} is the bundled transmission power of renewables; C_{trans} is the transmission capacity. Equation (6e) limits the transmission capacity of solar–wind–storage.

4. Case Study

In this study, the proposed algorithm includes two main components: the typical day clustering method based on KMEANS++ using photovoltaic wind turbine output, and load data from the simulation platform of the China Electric Power Research Institute in Northeast Province. The proportional planning scheme for energy storage considers stable external power transmission capacity. Building on the clustering analysis and the planning model for external output, the focus of this study is on the installation capacity of energy storage required per unit of wind power, i.e., the planning ratio between energy storage and wind-solar energy. The research investigates the enhancing role of energy storage for new energy, analyzing the effects of different wind–solar ratios, wind–solar operational volatility, and the allowed capacity of transmission channels on the system. For the proposed planning model, the power capacity is standardized by using 1 MW as the reference value. This approach allows for a comprehensive investigation into the optimal configuration of energy storage in relation to wind and solar energy, enabling a better understanding of the impacts of various factors on the system's performance.

All the clustering and optimization models are coded by python 3.6. The KMEANS++ algorithm and reduction analysis are based on the SKlearn package. The optimization models are based on the CVXPY package with the commercial solvers of GUROBI for effective solution analysis.

4.1. Clustering Analysis

According to the clustering analysis strategy for wind–solar–load data mentioned in the first section, this study starts by normalizing the wind–solar–load data for 365 days in a certain region of northeast province based on their respective maximum values. The wind power and solar power data are collected from the open-source website in [21], which can generate the proper renewables and load profiles given the proper location. The normalized data are then input into the KMEANS++ clustering model, and the change in SSE for different numbers of clusters is analyzed to determine the optimal number of clusters for the system. After identifying the optimal number of clusters, the clustering results are visualized and analyzed to assess the effectiveness of the clustering algorithm. Finally, the clustering results are output for further analysis and interpretation.

4.1.1. SSE Analysis

Figure 1 shows the change trend of SSE with the increase in the clustering number. The value of SSE decreases shapely at first and flattens latter, which means the typical scenarios can represent the features of the total scenarios. More clusters mean less SSE but a high computational requirement.

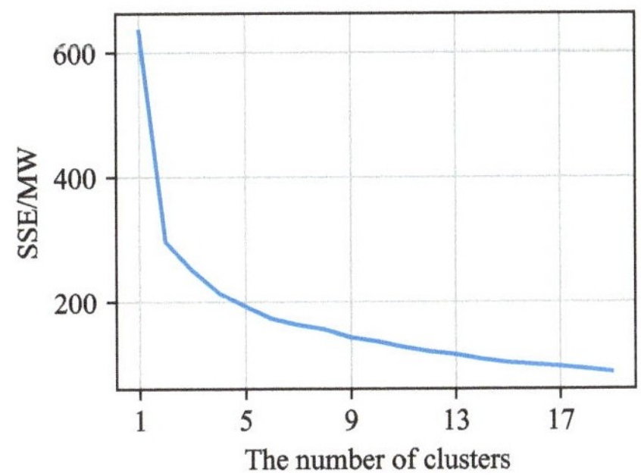

Figure 1. SSE change with the cluster.

4.1.2. tSNE Analysis

After determining the number of cluster centers as 9, this study employs the t-SNE algorithm to reduce the 96-dimensional wind–solar–load data of a typical day to a two-dimensional space for visualization. The scatter plot of the reduced data is shown in Figure 2. Different colors refer to different clusters, and different scenarios with the same color belong to the same cluster.

From the scatter plot, it is observed that the different operational scenarios are well-separated and effectively distinguished by the proposed clustering algorithm, demonstrating the effectiveness of the selection of clustering numbers. This demonstrates that the clustering algorithm can successfully identify and extract typical operational scenarios, allowing for a clear visualization of the distribution of different scenarios.

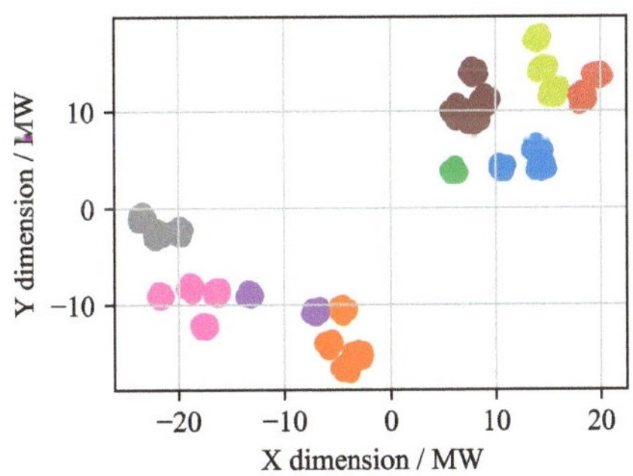

Figure 2. tSNE results analysis.

4.1.3. Clustered Results Analysis

The clustering results of typical solar–wind–load are shown in Figure 3, demonstrating that the ESS should adapt to different scenarios of solar–wind–load. Figure 3 shows the typical scenarios, where energy storage will have different impacts. For the row 1 and column 1 part of Figure 3, the solar and wind power show complementary feature, and energy storage will absorb the excess power when load is low. For the row 2 and column 1 part of Figure 3, the solar and wind power show similar features, and the energy storage will release the stored power when the load is high. The key analysis is based on the different change trends of solar, wind, and load profiles, where energy storage can bridge the gap.

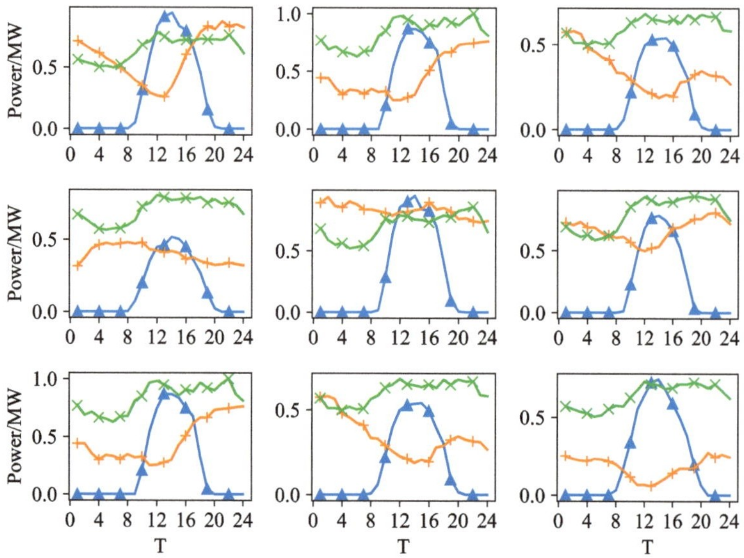

Figure 3. Typical normalized power scenarios of solar (blue), wind (green), and load (orange).

The above typical scenarios imply the different operation mismatching of solar, wind, and load. Energy storage takes the role of enabler to achieve the balance between sustainable energy supply and demand.

4.2. Optimization Results Analysis

4.2.1. Deterministic Optimization

Based on the typical scenarios obtained from the clustering analysis of wind–solar–load data, this study investigates the impact of different wind–solar ratios, allowable wind–solar operational volatility, and transmission channel capacities on the installation capacity of energy storage. Specifically, the study analyzes how varying the proportion of wind and solar energy in the mix affects the required capacity of energy storage. Additionally, different levels of wind and solar operational volatility are considered to assess their influence on the energy storage capacity requirements. Moreover, the study examines the effect of different transmission channel capacities on the installation capacity of energy storage. By exploring these different factors, the research aims to optimize the configuration of energy storage in relation to wind and solar energy, taking into account various operational conditions and transmission constraints, which will ultimately contribute to enhancing the integration and utilization of renewable energy in the power system in Figure 4.

As shown in the above Figure 4, with the increase in the transmission capacity, the ESS requirement reduces correspondingly. Because the transmission capacity can bear more renewables' variability. The cost of the transmission capacity expansion and energy storage installation should be a trade-off for the stakeholders and the policy makers.

Table 1 further presents the different ESS ratios under different transmission and PV ratios for numerical analysis, where ESS ratio refers to the ratio of ESS capacity to renewable capacity. Under the same transmission capacity, with the increase in PV ratio from 0.1 to 0.9, the requirement of ESS reduces from 0.186343 to 0.180739 and then increases from 0.1807390 to 0.232541, where a lower ESS ratio implies more solar–wind–demand coordination. It demonstrates the requirement of ESS relies on the PV ratio. And with the increase in transmission capacity from 0.9 to 1.1, the ESS ratio reduces from 0.280739 to 0.160909, implying the benefits of transmission capacity expansion. It means that more installation of the energy storage will not need a high transmission capacity to reduce the corresponding construction cost.

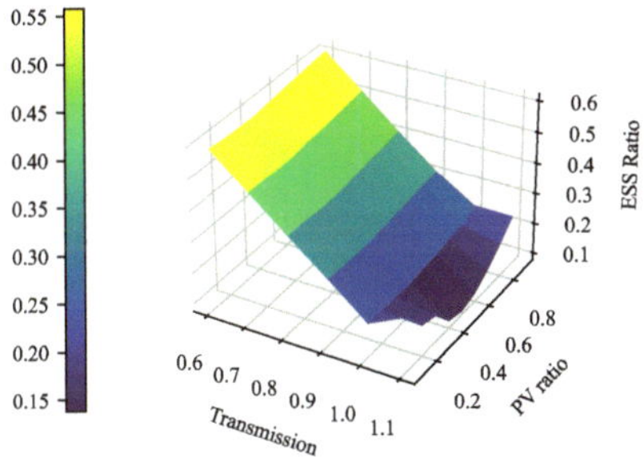

Figure 4. ESS storage capacity analysis under different transmission and PV ratios.

Table 1. ESS storage ratio under different transmission and PV ratio.

Transmission Capacity	PV Ratio	ESS Ratio
1.0	0.1	0.186343
1.0	0.3	0.180739
1.0	0.5	0.186674
1.0	0.7	0.198972
1.0	0.9	0.232541
0.9	0.3	0.280739
1.1	0.3	0.160909

When fixing the transmission capacity as 1.0, the change of ESS ratio with the PV ratio is shown in Figure 5, demonstrating the complementary impact of solar and wind on the energy storage.

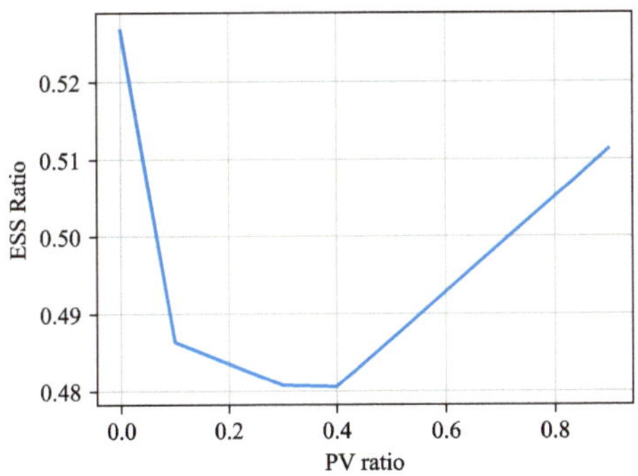

Figure 5. ESS storage capacity analysis under PV ratios.

The key intuition behind the above analysis is as follows. Due to the fluctuation in solar and wind power, the bundled power output of solar and wind could be high at some times and low at other times, but the transmission capacity will limit the high value. Without the energy storage, the excess power is wasted. Energy can absorb the excess power when high to satisfy the transmission capacity and release the stored power when low to flatten the bundled power out, which is the key value of energy storage.

4.2.2. Stochastic Optimization

Based on the above, we further consider the uncertainty of solar–wind–load by the stochastic optimization in Figure 6.

As shown in Figure 6, with the increase in uncertainty, the corresponding ESS requirement increases correspondingly.

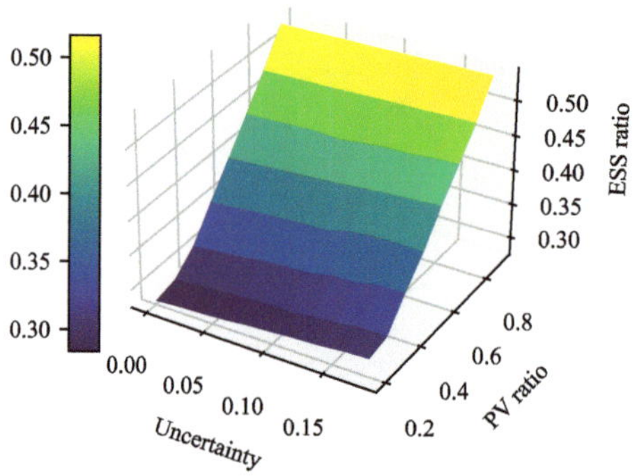

Figure 6. ESS storage capacity analysis under different uncertainty and PV ratios.

Table 2 shows the different ESS ratios under different uncertainty and PV ratios under the stochastic optimization approach. Intuitively, more uncertainty renewables will lead to more ESS requirements, which are evaluated numerically in the following Table 2. With the increase in uncertainty from 0 to 20%, the corresponding ESS requirement increases from 0.311330 to 0.324882 under the same PV ratio.

Table 2. ESS storage ratio under different transmission and PV ratios.

Uncertainty	PV Ratio	ESS Ratio
0	0.3	0.311330
2%	0.3	0.315273
4%	0.3	0.318878
10%	0.3	0.323113
20%	0.3	0.324882
4%	0.2	0.290579
4%	0.5	0.402065
4%	0.6	0.443731

So, the requirement of ESS for facilitating renewables mainly relies on PV ratio, transmission capacity, and renewable uncertainty, and its numerical requirements are analyzed based on the above experiments.

In summary, this section provides comprehensive analysis to verify the effectiveness of the proposed models from the perspectives of clustering analysis, reduction analysis, sensitivity analysis, and numerical analysis. These results can be leveraged by the corresponding stakeholders and policy makers to support the low-carbon energy society transformation of the northeast area by leveraging the enabler of the energy storage values.

5. Conclusions

As wind and solar power gradually take on a higher proportion in the power supply system, their strong intermittency poses challenges in directly bundling and supplying stable electricity to the external grid. Therefore, this study focuses on investigating the optimal configuration strategy of using energy storage to support the bundled output of wind and solar energy, aiming to achieve stable external supply. By leveraging the regulation capabilities of energy storage, the study aims to achieve a stable bundled output of wind, solar, and storage energy to the external grid. The study first employs the KMEANS++ algorithm to extract typical days from the wind–solar–load data. The optimal number of clusters is determined based on the SSE and t-SNE algorithms. Using the extracted typical days, the study formulates a planning model for the external output of wind, solar, and storage energy, with a focus on the energy storage capacity required per unit of wind and solar power. The research analyzes the impact of different photovoltaic ratios, allowable operational volatility, and transmission channel capacities on the required energy storage capacity. Through experimentation, the study plans the energy storage capacity for different scenarios.

The findings highlight that energy storage is a crucial means to achieve stable bundled output of wind and solar energy to the external grid, making it a vital pathway to achieving "carbon neutrality" and "carbon peaking". By optimizing the configuration of energy storage in relation to wind and solar energy, the study aims to contribute to the effective integration and utilization of renewable energy, supporting the broader goals of carbon reduction and sustainable energy development.

Author Contributions: Conceptualization, Z.Y. and Y.X.; methodology, M.Z., Y.S. and Y.L.; software, L.S. and K.M.; investigation, M.Z.; writing—review and editing, Y.S. and Y.L.; visualization, Y.L.; supervision, Z.Y. and Y.X.; project administration, Y.X. All authors have read and agreed to the published version of the manuscript.

Funding: This research was funded by the Science and Technology Project of State Grid Jilin Electric Power Company Limited, grant number 2022-44,SGJLIY00GPJS2200079.

Data Availability Statement: https://github.com/sanglinwei/ValueofESS (accessed on 20 September 2023).

Conflicts of Interest: Authors Meng Zhu, Yong Sun, Kerui Ma were employed by the company Economic and Technical Research Institute, State Grid Jilin Electric Power Co., Ltd. The remaining authors declare that the research was conducted in the absence of any commercial or financial relationships that could be construed as a potential conflict of interest.

Abbreviations

The main notation below is provided for unit commitment problem formulation. And other additional symbols are defined in the paper when needed.

SSE	The sum of the squared error
WP	Wind power
PV	Photovoltaic power
t-SNE	t-Distributed Stochastic Neighbor Embedding
ESS	Energy storage systems
$P_{ES}/P_{dis}/P_{ch}$	The operating/discharging/charging power of energy storage (p.u.)
Φ_{ES}	The feasible region of energy storage
E_t	The stored energy of time t in the energy storage (p.u.)
E_{min}/E_{max}	The minimum/maximum capacity of energy storage (p.u.)
t/T	The time period/ the maximum time (hour)
μ_{dis}/μ_{ch}	The binary indicators of discharging and charging state
$P_{dis}^{max}/P_{ch}^{max}$	The maximum values of charging and discharging power (p.u.)
M	The large positive number
$C(E_{max})$	The cost of energy storage (p.u.)
r_{pv}	The ratio of PV to total renewables (p.u.)
P_{PV}/P_{WP}	The solar/wind power output (p.u.)
P_{trans}	The bundled transmission power of renewables (p.u.)
C_{trans}	The transmission capacity (p.u.)

References

1. Chen, X.; Liu, Y.; Wang, Q.; Lv, J.; Wen, J.; Chen, X.; Kang, C.; Cheng, S.; McElroy, M. Pathway toward carbon-neutral electrical systems in China by mid-century with negative CO2 abatement costs informed by high-resolution modeling. *Joule* **2021**, *5*, 2715–2741. [CrossRef]
2. Kang, C.; Zhou, T.; Chen, Q.; Wang, J.; Sun, Y.; Xia, Q.; Yan, H. Carbon emission flow from generation to demand: A network-based model. *IEEE Trans. Smart Grid* **2015**, *6*, 2386–2394. [CrossRef]
3. Sang, L.; Xu, Y.; Long, H.; Wu, W. Safety-aware Semi-end-to-end Coordinated Decision Model for Voltage Regulation in Active Distribution Network. *IEEE Trans. Smart Grid* 2022, *early access*.
4. Sang, L.; Xu, Y.; Sun, H. Encoding Carbon Emission Flow in Energy Management: A Compact Constraint Learning Approach. *IEEE Trans. Sustain. Energy* 2023, *early access*.
5. Kaack, L.H.; Donti, P.L.; Strubell, E.; Kamiya, G.; Creutzig, F.; Rolnick, D. Aligning artificial intelligence with climate change mitigation. *Nat. Clim. Chang.* **2022**, *12*, 518–527. [CrossRef]
6. Sang, L.; Xu, Y.; Long, H.; Hu, Q.; Sun, H. Electricity Price Prediction for Energy Storage System Arbitrage: A Decision-Focused Approach. *IEEE Trans. on Smart Grid* **2022**, *13*, 2822–2832. [CrossRef]
7. Delille, G.; Francois, B.; Malarange, G. Dynamic frequency control support by energy storage to reduce the impact of wind and solar generation on isolated power system's inertia. *IEEE Trans. Sustain. Energy* **2012**, *3*, 931–939. [CrossRef]
8. Chitsaz, H.; Zamani-Dehkordi, P.; Zareipour, H.; Parikh, P.P. Electricity price forecasting for operational scheduling of behind-the-Meter storage systems. *IEEE Trans. Smart Grid* **2018**, *9*, 6612–6622. [CrossRef]
9. Arteaga, J.; Zareipour, H. A price-maker/price-taker model for the operation of battery storage systems in electricity markets. *IEEE Trans. Smart Grid* **2019**, *10*, 6912–6920. [CrossRef]
10. Comello, S.; Reichelstein, S. The emergence of cost-effective battery storage. *Nat. Commun.* **2019**, *10*, 2038. [CrossRef] [PubMed]
11. Khani, H.; Farag, H.E.Z. Joint arbitrage and operating reserve scheduling of energy storage through optimal adaptive allocation of the state of charge. *IEEE Trans. Sustain. Energy* **2019**, *10*, 1705–1717. [CrossRef]
12. Shen, Z.; Wei, W.; Wu, D.; Ding, T.; Mei, S. Modeling arbitrage of an energy storage unit without binary variables. *CSEE J. Power Energy Syst.* **2021**, *7*, 156–161.
13. Teng, F.; Trovato, V.; Strbac, G. Stochastic scheduling with inertia-dependent fast frequency response requirements. *IEEE Trans. Power Syst.* **2016**, *31*, 1557–1566. [CrossRef]
14. Badesa, L.; Teng, F.; Strbac, G. Simultaneous scheduling of multiple frequency services in stochastic unit commitment. *IEEE Trans. Power Syst.* **2019**, *34*, 3858–3868. [CrossRef]
15. Paturet, M.; Markovic, U.; Delikaraoglou, S.; Vrettos, E.; Aristidou, P.; Hug, G. Stochastic unit commitment in low-inertia grids. *IEEE Trans. Power Syst.* **2020**, *35*, 3448–3458. [CrossRef]
16. Agalgaonkar, Y.P.; Pal, B.C.; Jabr, R.A. Stochastic distribution system operation considering voltage regulation risks in the presence of PV generation. *IEEE Trans. Sustain. Energy* **2015**, *6*, 1315–1324. [CrossRef]
17. Li, S.; Wu, W.; Lin, Y. Robust data-driven and fully distributed volt/var control for active distribution networks with multiple virtual power plants. *IEEE Trans. Smart Grid* **2022**, *13*, 1210–1220. [CrossRef]
18. Chen, X.; Hu, Q.; Shi, Q.; Quan, X.; Wu, Z.; Li, F. Residential HVAC Aggregation Based on Risk-averse Multi-armed Bandit Learning for Secondary Frequency Regulation. *J. Mod. Power Syst. Clean Energy* **2020**, *8*, 1160–1167. [CrossRef]
19. Zhou, M.; Wang, M.; Li, J.; Li, G. Multi-area generation-reserve joint dispatch approach considering wind power cross-regional accommodation. *CSEE J. Power Energy Syst.* **2017**, *3*, 74–83. [CrossRef]
20. Goodfellow, I.; Bengio, Y.; Courville, A. *Deep Learning*; MIT Press: Cambridge, MA, USA, 2016.
21. Renewables. renewables.ninja. 2023. Available online: https://www.renewables.ninja/ (accessed on 20 September 2023).

processes MDPI

Article

Distributed Shared Energy Storage Double-Layer Optimal Configuration for Source-Grid Co-Optimization

Meng Yang [1], Yihan Zhang [1], Junhui Liu [1], Shuo Yin [1], Xing Chen [1], Lihui She [2,*], Zhixin Fu [2] and Haoming Liu [2]

[1] State Grid Henan Economic Research Institute, Zhengzhou 450052, China; yangmeng7@ha.sgcc.com.cn (M.Y.); zhangyihan2@ha.sgcc.com.cn (Y.Z.); liujunhui4@ha.sgcc.com.cn (J.L.); yinshuo@ha.sgcc.com.cn (S.Y.); chenxing13@ha.sgcc.com.cn (X.C.)

[2] College of Energy and Electrical Engineering, Hohai University, Nanjing 211100, China; zhixinfu@hhu.edu.cn (Z.F.); liuhaom@hhu.edu.cn (H.L.)

* Correspondence: 211606010138@hhu.edu.cn

Abstract: Shared energy storage is an energy storage business application model that integrates traditional energy storage technology with the sharing economy model. Under the moderate scale of investment in energy storage, every effort should be made to maximize the benefits of each main body. In this regard, this paper proposes a distributed shared energy storage double-layer optimal allocation method oriented to source-grid cooperative optimization. First, considering the regulation needs of the power side and the grid side, a distributed shared energy storage operation model is proposed. Second, a distributed shared energy storage double-layer planning model is constructed, with the lowest cost of the distributed shared energy storage system as the upper-layer objective, and the lowest daily integrated operation cost of the distribution grid-distributed new energy stations as the lower-layer objective. Third, a double-layer iterative particle swarm algorithm combined with tide calculation is used to solve the distributed shared energy storage configuration and distribution grid-distributed new energy stations' economic operation problem. Finally, a comparative analysis of four scenarios verifies that configuring distributed shared energy storage can increase the new energy consumption rate to 100% and reduce the net load peak-valley difference by 61%. Meanwhile, distributed shared energy storage operators have realized positive returns.

Keywords: distributed shared energy storage; double-layer optimal; new energy consumption; net load peak-to-valley difference; particle swarm algorithm

Citation: Yang, M.; Zhang, Y.; Liu, J.; Yin, S.; Chen, X.; She, L.; Fu, Z.; Liu, H. Distributed Shared Energy Storage Double-Layer Optimal Configuration for Source-Grid Co-Optimization. *Processes* **2023**, *11*, 2194. https://doi.org/10.3390/pr11072194

Academic Editors: Chenyu Wu, Zhongkai Yi and Chenhui Lin

Received: 17 June 2023
Revised: 14 July 2023
Accepted: 19 July 2023
Published: 21 July 2023

1. Introduction

In order to cope with the environmental problems caused by global warming, new energy power generation is attracting great attention from all over the world [1]. However, with the increasing scale of new energy access, the problem of imbalance between the intermittent power output and the spatial and temporal matching of the load has become more and more prominent, resulting in the phenomenon of wind and light curtailment and peak-to-valley differences increasing year by year [2,3]. As a flexible power regulation resource, energy storage can achieve energy leveling at the spatial and temporal levels, promote local consumption of new energy, and reduce peak-to-valley differences [4,5].

Reasonable selection of the location and capacity of energy storage is important to improve the safety and economy of power system operation [6,7]. There has been a lot of research on the optimal configuration of distributed energy storage. Ding et al. [8] established a double-layer coordinated siting and capacity optimization model for distributed PV and energy storage, where the upper layer optimizes the capacity and power of energy storage to minimize the annual integrated system cost, and the lower layer optimizes the grid connection location of energy storage with the objective of minimizing the system network loss. Gong et al. [9] used the dynamic planning method to solve for the distributed energy storage capacity and location to meet the operational needs of the active distribution network, with the whole life cycle cost of energy storage as the optimization objective. Li et al. [10] proposed a two-stage robust optimization model for the capacity configuration of integrated biogas–solar–wind energy systems applicable to rural areas, solving the configuration problem of energy storage in the first stage and the optimal operation of the system in the second stage. Guo et al. [11] first constructed a multi-attribute integrated index assessment model to determine the location of energy storage, and then a two-layer planning model to determine the storage capacity.

However, the current distributed energy storage investment costs are high [12,13], and the utilization efficiency is low [14]. To address this issue, some scholars have conducted research on shared energy storage models. Shuai et al. [15] constructed an optimal allocation model for shared energy storage under multi-regional integrated energy system interconnection. Xie et al. [16] constructed a multi-micro grid shared energy storage two-layer planning model that takes into account the economic consumption of new energy sources. Yang et al. [17] selected three types of industrial users with different peak types as research objects and established a shared energy storage optimal allocation model to maximize the overall net benefit of multiple users. Liu et al. [18] proposed a producer–consumer energy-sharing mechanism and verified that the new energy consumption rate can be improved by sharing energy storage.

In summary, research on shared energy storage configurations is still in its infancy. Existing research mainly focuses on centralized shared energy storage, a single type of shared energy storage user, with less analysis on the cost settlement between shared energy storage users and shared energy storage. Based on the above problems, this paper proposes a distributed shared energy storage double-layer optimal allocation method for source-grid co-optimization. First, a distributed shared energy storage operation model for source-grid co-optimization is proposed. Secondly, a distributed shared energy storage two-layer planning model is constructed, and a two-layer iterative particle swarm algorithm combined with tide calculation is used to solve the distributed shared energy storage configuration and distribution grid-distributed new energy stations' economic operation problem. Finally, the effectiveness and economy of the proposed configuration method are verified by simulation analysis of arithmetic cases.

In this paper, the main innovations of this paper are as follows:

1. There are limitations on storage power ratings, line transmission capacity, etc., so centralized shared storage no longer meets demand in actual use. Therefore, this paper investigates the optimal allocation of distributed shared energy storage;
2. This paper proposes a distributed shared energy storage operation model oriented to source-network co-optimization, and analyzes the operation mode of each subject and the profit mechanism of the shared energy storage operator;
3. Most of the existing literature considers energy storage sharing in the context of multiple single subjects, e.g., multi-industrial users, multi-microgrids, etc. In this paper, we consider the energy storage contribution of distributed new energy stations and distribution grids. In addition, this paper constructs a double-layer planning model for distributed shared energy storage, which comprehensively considers the operating costs of distributed shared energy storage operator and distribution grid-distributed new energy stations and realizes the maximization of the interests of each subject.

2. Distributed Shared Energy Storage Operation Model for Source-Grid Co-Optimization

Shared energy storage is an energy storage business application model that integrates traditional energy storage technology with the sharing economy model, which is an energy storage power plant invested by a third party to provide charging and discharging services for multiple subjects. The distributed shared energy storage studied in this paper takes into account the regulation needs of both the power side and the grid side, and the schematic diagram of the distributed shared energy storage operation model for source-grid co-optimization is shown in Figure 1.

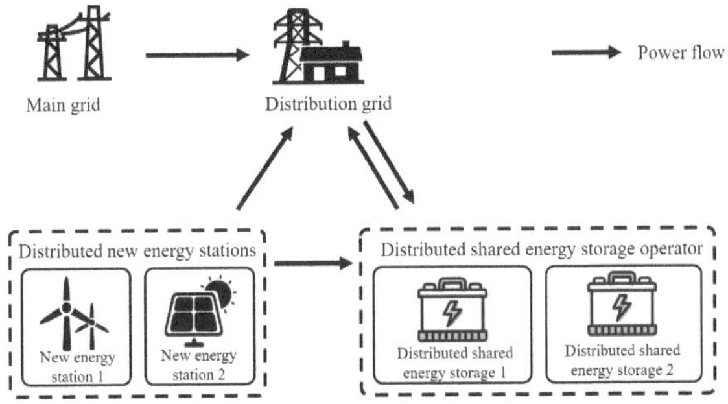

Figure 1. The schematic diagram of the distributed shared energy storage operation model for source-grid co-optimization.

Distributed shared energy storage operators are responsible for the operation and management of multiple energy storage plants. Each energy storage is connected to the distribution grid. A distributed new energy station is responsible for the operation and management of multiple new energy power stations. Each new energy station is connected to the distribution grid. The operational objectives of each of the distributed shared energy storage operators, distributed new energy stations, and distribution grid are described below:

Shared energy storage operators aim to provide charging and discharging services for distributed new energy sites and distribution grids to achieve the lowest capacity allocation and operating costs for distributed shared energy storage systems. At the optimal allocation level, energy storage operators will aggregate the charging and discharging needs of distributed new energy sites and active distribution grids to centralize and optimize the allocation of distributed shared energy storage system capacity. At the optimized operation level, shared energy storage operators provide charging and discharging services, and charge service fees while trading power through "low storage and high discharge" to achieve price arbitrage.

Distributed new energy stations aim to maximize the utilization of distributed new energy power generation and reduce the rate of wind and light curtailment by utilizing the charging and discharging services of distributed shared energy storage plants. The distributed new energy stations will give priority to the distribution grid to support its load, and if the new energy output exceeds the demand of the distribution grid, the excess new energy output will be charged to the distributed shared energy storage system in the form of electricity sales.

The distribution grid aims to reduce the net load peak-to-valley differential by utilizing the charging and discharging services of distributed shared energy storage plants. In the distribution grid, priority will be given to the consumption of new energy output to meet the load demand, and the power imbalance will be the net load of the distribution grid. During peak load periods, the distribution grid will shave peaks by discharging power from distributed shared energy storage systems or purchasing power from the main grid. During low load periods, the distribution grid will be filled by charging from a distributed shared energy storage system to further reduce the net load peak-to-valley difference.

3. Double-Layer Planning Model for Optimal Allocation of Distributed Shared Energy Storage

Double-layer planning divides the problem into two layers: upper-layer optimization and lower-layer optimization [19]. The upper- and lower-layer optimization models have their own optimization objectives, constraints, and decision variables. A schematic diagram of the distributed shared energy storage double-layer planning model in this paper is shown in Figure 2. From the figure, it can be seen that the upper and lower optimization problems are coupled with each other through the parameter transfer between layers. The upper-layer model passes the decision variables, i.e., the rated capacity and rated power of the distributed shared energy storage, to the lower-layer model as the constraints of the lower-layer model. The lower-layer model seeks the optimization of the charging and discharging power and position of each energy storage on this basis and feeds the optimization results of the power exchange between each body to the upper layer. The optimal values of the upper and lower layers are obtained through continuous iteration. Double-layer planning is used to find the lower layer optimum under the condition of the upper layer optimum, thus maximizing the interests of each subject.

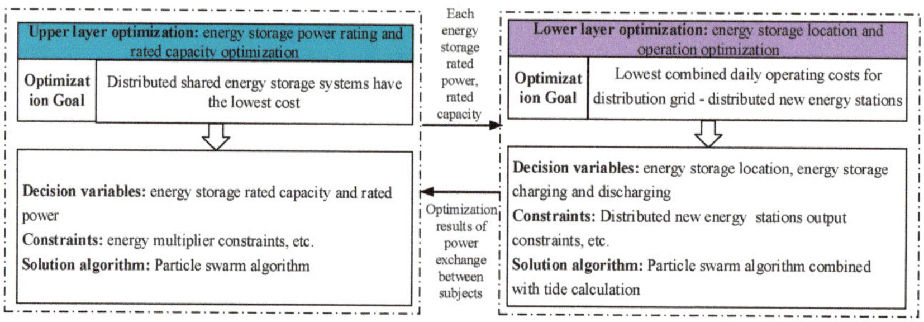

Figure 2. The schematic diagram of the distributed shared energy storage double-layer optimization model.

3.1. Upper-Layer Model

The upper-layer model is used to solve the distributed shared energy storage plant-rated capacity problem. The lowest cost of the distributed shared energy storage system is used as the objective function to plan the rated capacity and rated power of distributed shared energy storage.

3.1.1. Objective Function

The upper-layer optimization objective is the lowest cost of a distributed shared energy storage system, which can be expressed as

$$\min C_1 = C_{sto} - C_{ser} + C_{adn} - C_{new} \tag{1}$$

where C_1 is the cost of a distributed shared energy storage system; C_{sto} is the average daily investment and maintenance cost of distributed shared energy storage; C_{new} is the cost of trading electricity between distributed shared energy storage and distributed new energy stations; C_{adn} is the electricity transaction cost between distributed shared energy storage and the distribution grid; and C_{ser} is the distributed shared energy storage capacity lease service fee.

The average daily investment and maintenance cost of distributed shared energy storage is expressed as

$$C_{sto} = \sum_{i=1}^{n} \left[\frac{r(1+r)^y}{365[(1+r)^y - 1]} (\delta_p P_{sto,i} + \delta_e E_{sto,i}) + \delta_m P_{sto,i} \right] \tag{2}$$

where n is the number of energy storage units; r is the discount rate; y is the life cycle of energy storage equipment; δ_p and δ_e are the investment cost per unit power and capacity of energy storage, respectively; $P_{sto,i}$ and $E_{sto,i}$ are the rated power and capacity of energy storage, respectively; and δ_m is the maintenance cost per unit power.

The cost of trading electricity between distributed shared energy storage and distributed new energy stations is expressed as

$$C_{new} = \sum_{t=1}^{T} \sum_{j=1}^{N} \delta_{new}^{t} P_{sto,new,j}^{t} \tag{3}$$

where T is 24; N is the number of distributed new energy stations; δ_{new}^{t} is the selling electricity price per unit electricity of distributed new energy stations at time t; and $P_{sto,new,j}^{t}$ is the power selling from new energy station j to distributed shared energy storage system at time t.

The electricity transaction cost between distributed shared energy storage and the distribution grid is expressed as

$$C_{adn} = \sum_{t=1}^{T} (\delta_{sto}^{t} P_{sto,adn,d}^{t} - \delta_{adn}^{t} P_{sto,adn,c}^{t}) \tag{4}$$

where δ_{sto}^{t} is the selling electricity price per unit electricity of distributed shared energy storage at time t; δ_{adn}^{t} is the selling electricity price per unit electricity of the distribution grid at time t; $P_{sto,adn,d}^{t}$ is the electricity sold by the distributed shared energy storage system to the distribution grid at time t; and $P_{sto,adn,c}^{t}$ is the electricity sold by the distribution grid to the distributed shared energy storage system at time t.

The distributed shared energy storage capacity lease service fee is expressed as

$$C_{ser} = \delta_{s} \sum_{t=1}^{T} (P_{sto,adn,c}^{t} + P_{sto,adn,d}^{t}) + \delta_{s} \sum_{t=1}^{T} \sum_{j=1}^{N} P_{sto,new,j}^{t} \tag{5}$$

where δ_s is a unit power service fee paid by the distribution grid and distributed new energy stations to the distributed shared energy storage system.

3.1.2. Constraint Condition

The energy multiplier constraint can be expressed as

$$E_{sto,i} = \beta P_{sto,i} \tag{6}$$

where β is the energy storage battery rate, which refers to the energy ratio constraint between the capacity of the energy storage battery and the rated power.

The distributed shared energy storage power constraint can be expressed as

$$P_{sto,i,\min} \le P_{sto,i} \le P_{sto,i,\max} \tag{7}$$

where $P_{sto,i,\min}$ and $P_{sto,i,\max}$ are the minimum and maximum power of distributed shared energy storage installed at each node, respectively.

The distributed shared energy storage charging and discharging power constraint can be expressed as

$$\begin{aligned} &\sum_{i=1}^{n} (P_{sto,i,d}^{t} - P_{sto,i,c}^{t}) = P_{sto,sdn,d}^{t} - P_{sto,sdn,c}^{t} - \sum_{j=1}^{N} P_{sto,new,j}^{t} \\ &0 \le P_{sto,i,c}^{t} \le A_{sto,i,c}^{t} P_{sto,i} \\ &0 \le P_{sto,i,d}^{t} \le A_{sto,i,d}^{t} P_{sto,i} \\ &A_{sto,c,i}^{t} A_{sto,d,i}^{t} = 0 \end{aligned} \tag{8}$$

where $P^t_{sto,i,c}$ and $P^t_{sto,i,d}$ are the charging and discharging power of energy storage i at time t, respectively, and $A^t_{sto,i,c}$ and $A^t_{sto,i,d}$ are the charge and discharge flags of energy storage i at time t, respectively.

The distributed shared energy storage charge constraint can be expressed as

$$\begin{aligned} &E^t_{sto,i} = E^{t-1}_{sto,i} + (\eta_{sto,c} P^t_{sto,c,i} - P^t_{sto,d,i} \big/ \eta_{sto,d})\Delta t \\ &0.1E_{sto,i} \le E^t_{sto,i} \le 0.9E_{sto,i} \\ &E^0_{sto,i} = E^T_{sto,i} = 0.2E_{sto,i} \end{aligned} \tag{9}$$

where $E^t_{sto,i}$ is the charge of energy storage i at time t and $\eta_{sto,c}$ and $\eta_{sto,d}$ are the charging and discharging efficiency of energy storage, respectively.

3.2. Lower Layer Model

The lower-layer model is used for solving distributed shared energy storage siting and distribution grid-distributed new energy stations' economic operation problems. The objective is to optimize each energy storage's location and charging and discharging power, achieving the lowest comprehensive daily operating cost of the distribution of grid-distributed new energy stations.

3.2.1. Objective Function

The lower-layer optimization objective is to achieve the lowest integrated daily operating cost of the distribution grid-distributed new energy stations, which can be expressed as

$$\min C_2 = C_{grid} + C_{ser} + C_{adn} - C_{new} + C_{peak-valley} \tag{10}$$

where C_{grid} is the cost of electricity purchased from the main grid by the distribution grid and $C_{peak-valley}$ is the penalty cost of the net load peak-to-valley difference.

The cost of electricity purchased from the main grid by the distribution grid is expressed as

$$C_{grid} = \sum_{t=1}^{T} \delta^t_p P^t_{grid} \tag{11}$$

where δ^t_p is the price of electricity sold by the main grid at time t and P^t_{grid} is the power sold by the main grid to the distribution grid at time t.

The penalty cost of the net load peak-to-valley difference is expressed as

$$\begin{aligned} &C_{peak-valley} = \delta_{peak-valley}(L^{\max}_{load} - L^{\min}_{load}) \\ &L^t_{load} = \sum_{k=1}^{M} P^t_{load,k} + P^t_{sto,adn,c} - P^t_{sto,adn,d} - \sum_{j=1}^{N} P^t_{adn,new,j} \end{aligned} \tag{12}$$

where $\delta_{peak-valley}$ is the net load peak–valley difference unit power penalty cost of 0.65 Yuan/kW [20]; $L^{\max}_{load}$ and $L^{\min}_{load}$ are the net load maximum and minimum values, respectively; L^t_{load} is the net distribution grid load at time t; $P^t_{load,k}$ is the load at the node k at time t; and $P^t_{adn,new,j}$ is the power sold by the new energy station j to the distribution grid at time t.

3.2.2. Constraint Condition

The capacity constraint of a distributed new energy site can be expressed as

$$0 \le P^t_{new,j} \le P^t_{new\_0,j} \tag{13}$$

where $P^t_{new,j}$ is the actual output of the new energy station j at time t and $P^t_{new\_0,j}$ is the ideal output of the new energy station j.

The distributed new energy stations and distributed shared energy storage purchase and sale constraint can be expressed as

$$0 \leq P^t_{sto,new,j} \leq P^{\max}_{sto,new} \tag{14}$$

where $P^{\max}_{sto,new}$ is the maximum interactive power between the new energy station and the distributed shared energy storage.

The power balance constraint of distributed new energy stations can be expressed as

$$\begin{aligned} \sum_{j=1}^{N} P^t_{new,j} &= \sum_{j=1}^{N} (P^t_{adn,new,j} + P^t_{sto,new,j}) \\ \sum_{j=1}^{N} P^t_{adn,new,j} &= \min(\sum_{k=1}^{M} P^t_{load,k}, \sum_{j=1}^{N} P^t_{new\_0,j}) \end{aligned} \tag{15}$$

The distribution grid and distributed shared energy storage purchase and sale constraint can be expressed as

$$\begin{aligned} 0 \leq P^t_{sto,adn,d} \leq B^t_{sto,adn,d} P^{\max}_{sto,adn} \\ 0 \leq P^t_{sto,adn,c} \leq B^t_{sto,adn,c} P^{\max}_{sto,adn} \end{aligned} \tag{16}$$

where $B^t_{sto,adn,d}$ and $B^t_{sto,adn,c}$ are the flag bits of the power interaction between the distribution grid and distributed shared energy storage and $P^{\max}_{sto,adn}$ is the maximum interaction power between the distribution grid and distributed shared energy storage.

The distribution grid power balance constraint can be expressed as

$$P^t_{grid} + \sum_{j=1}^{N} P^t_{adn,new,j} + P^t_{sto,adn,d} = P^t_{sto,adn,c} + \sum_{k=1}^{M} P^t_{load,k} + P_{loss,t} \tag{17}$$

where $P_{loss,t}$ is the net loss of the distribution network at time t.

The node power balance constraint can be expressed as

$$\begin{aligned} P^t_i &= U^t_i \sum_{j \in i} U^t_j (G_{ij} \cos\theta_{ij} + B_{ij} \sin\theta_{ij}) \\ Q^t_i &= U^t_i \sum_{j \in i} U^t_j (G_{ij} \sin\theta_{ij} - B_{ij} \cos\theta_{ij}) \end{aligned} \tag{18}$$

where P^t_i and Q^t_i are the active and reactive power injected at node i at time t, respectively; U^t_i and U^t_j are the voltage amplitudes at node i at time t, respectively; G_{ij} and B_{ij} are the conductance and susceptance between nodes i and j, respectively; and θ_{ij} is the phase angle difference between nodes i and j.

The node voltage constraint can be expressed as

$$U_{i,\min} \leq U^t_i \leq U_{i,\max} \tag{19}$$

where $U_{i,\min}$ and $U_{i,\max}$ are the minimum and maximum values of the voltage amplitude of node i, respectively.

The branch circuit capacity constraint can be expressed as

$$S^t_{ij} \leq S_{ij,\max} \tag{20}$$

where S^t_{ij} is the transmitted power between nodes i and j at time t, and $S_{ij,\max}$ is the maximum value of the transmittable power between nodes i and j.

There is a nomenclature table in the Nomenclature section to navigate each symbol used in the paper.

3.3. Double-Layer Planning Model Solving

In the process of siting and setting the capacity of distributed shared energy storage, it is necessary to first consider the optimal economy of the distributed shared energy storage system, and then on this basis, consider the lowest operating costs of the distribution grid and distributed new energy station. The double-layer planning model based on the siting and capacity determination of distributed shared energy storage is a mutually coupled nonlinear multi-objective problem and contains multiple variables of different types. Therefore, this paper uses a double-layer iterative particle swarm algorithm combined with tidal wave calculation for the solution, as shown in Figure 3.

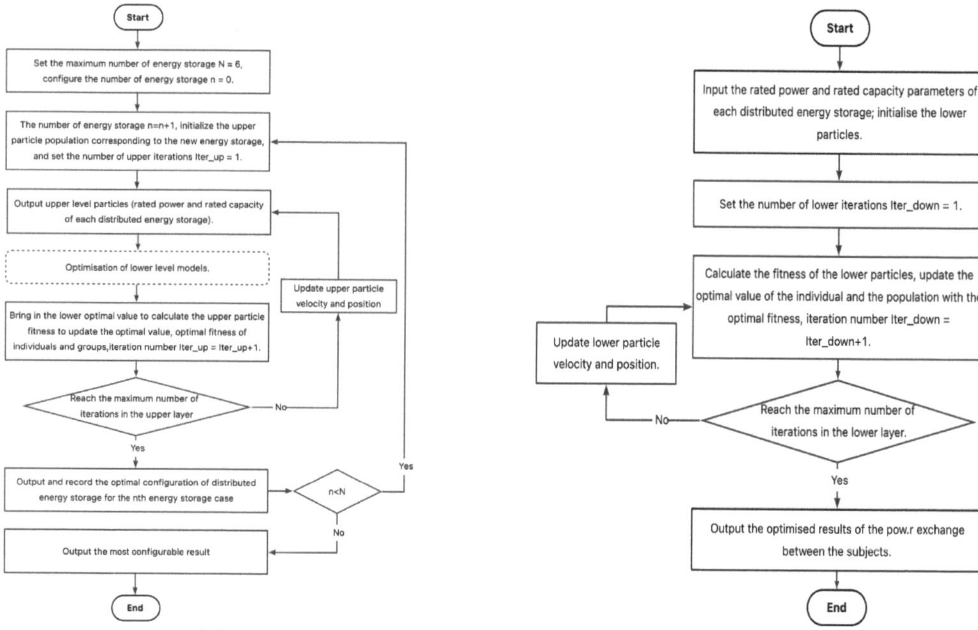

(a) (b)

Figure 3. The flowchart for solving a two-tier model for optimal distributed shared energy storage allocation. (**a**) Flowchart for solving the upper model; (**b**) flowchart for solving the lower model.

The upper model is solved by a particle swarm algorithm, where each particle consists of two parts: the rated power of each energy storage ($P_{sto,i}$) and the rated capacity of each energy storage ($E_{sto,i}$). The lower model is solved using a particle swarm algorithm combined with a tide calculation, where each particle also includes two parts: the location of each energy storage (x_i) and the charging and discharging power of each energy storage ($P^t_{sto,i,c}$ and $P^t_{sto,i,d}$, respectively). The upper model passes upper-level particles to the lower model as constraints for the lower model. The lower-layer model seeks the optimization of the charging and discharging power and position of each energy storage on this basis, and feeds the optimization results of the power exchange between the subjects to the upper layer. The optimal values of the upper and lower layers are obtained through continuous iteration.

4. Example Analysis

4.1. Case Setup

The algorithm uses a modified IEEE-33 node as the object of study. Among them, the IEEE-33 node is used as a distribution grid system [21]; 1000 kW PV is connected at node 9, and 1000 kW wind power is connected at node 20 as distributed new energy stations. For the distributed shared energy storage system, the allowed access nodes are 2–33, with a maximum of 6 energy storage accesses; the minimum rated power of energy storage access is 100 kW, the maximum rated power is 1000 kW, the discount rate of energy storage is 0.05 [20], the service life is 15 years [8], the unit power investment cost is 1173 Yuan/kW [20], the unit capacity investment cost is 1650 Yuan/(kW·h) [8], the unit power maintenance cost is 97 Yuan/(year·kW) [20], the energy storage unit power service fee is 0.05 Yuan/(kW·h) [20], the energy storage charging efficiency is 0.95 [20], and the energy storage discharging efficiency is 0.9 [8]. The modified IEEE-33 node is shown in Figure 4. The electricity sales tariffs between subjects [22] are shown in Table 1. The load power and the output of each new energy station for a typical day are shown in Appendix A.

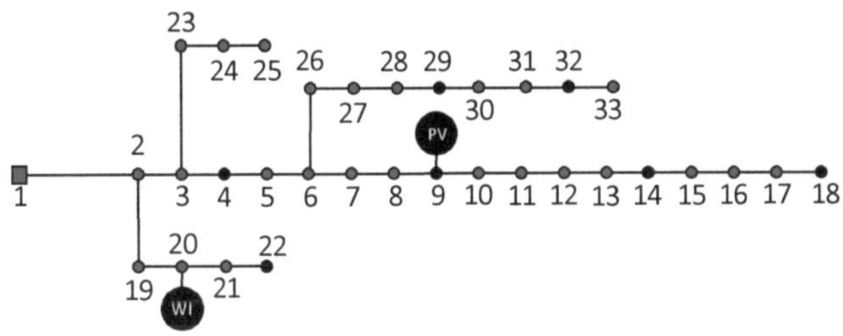

Figure 4. The modified IEEE-33 node.

Table 1. The electricity sales price between each subject.

Period		**Electricity Price/(Yuan [1]/(kW·h))**			
		Main Grid Electricity Sales Price	**Electricity Distribution Grid Sales Price**	**Distributed Shared Energy Storage Electricity Sales Price**	**Distributed New Energy Stations Electricity Sale Price**
peak	8:00–12:00 17:00–21:00	1.36	1.10	1.38	1.05
flat	12:00–17:00 21:00–24:00	0.82	0.8	0.82	0.65
valley	0:00–08:00	0.37	0.35	0.40	0.30

[1] 1 Yuan ≈ 0.1388 USD.

To analyze the rationality of distributed shared energy storage configuration, four scenarios are set up in this paper for comparative analysis.

Scenario 1: no energy storage is configured, the excess power from distributed new energy stations is directly curtailed, and the power imbalance of the distribution grid is directly purchased from the main grid.

Scenario 2: the distribution grid, new energy station 1 (node 20 access wind power station), and new energy station 2 (node 9 access PV station) invest in the construction of energy storage on their own to achieve peak shaving and fill the valley and improve the consumption rate of new energy. Parameters such as the energy storage discount rate are the same as for distributed shared energy storage.

Scenario 3: configuration of different numbers of shared energy storage. Discusses the economic impact of configuring shared energy storage on the system under the constraint of the number of shared energy storage.

Scenario 4: distributed shared energy storage is configured according to the method proposed in this paper, using distributed shared energy storage to cut peaks and fill valleys and improve the consumption rate of new energy.

4.2. Analysis of the Impact of Distributed Shared Energy Storage Systems on Peak Shaving and New Energy Consumption

The power balance of the distribution network for scenario 1 and scenario 4 is shown in Figure 5. In Figure 5, the positive power represents the power supplied to the distribution grid from outside, the negative power represents the network loss within the distribution grid and the power consumed by all electrical loads, and the difference between the maximum and minimum values of the net load curve is the peak-to-valley difference.

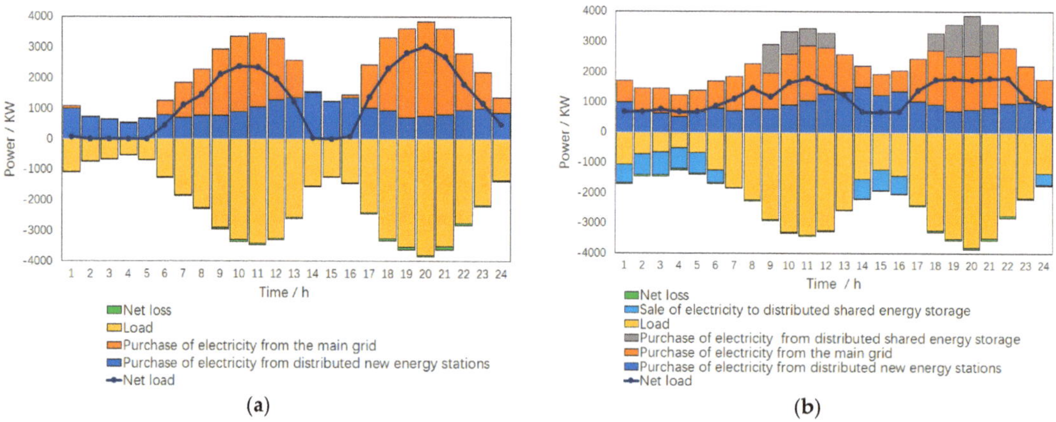

Figure 5. The distribution network power balance diagram. (**a**) Scenario 1; (**b**) scenario 4.

Analyzing the power balance diagram of the distribution network in scenario 1, we can see that the distribution grid gives priority to the power provided by the distributed new energy stations, and when the power provided by the distributed new energy stations is insufficient, the distribution grid purchases power directly from the main grid to meet the power demand of the load. The load has peak and valley characteristics, but the new energy output has anti-peak characteristics. From Figure 5a, we can see that in 1–5 h and 14–16 h, the load is less but the new energy output is larger, resulting in a net load curve close to 0. However, in 9–12 h and 18–21 h, the peak load increases but the new energy output decreases, and the distribution grid can only purchase a large amount of power from the main grid. Based on the net load curve, it can be seen that the peak-to-valley difference for scenario 1 is 3040 kW.

Analyzing the power balance diagram of the distribution grid in scenario 4, we can see that the distribution grid gives priority to consuming the power provided by distributed energy stations; during the low-load period of 1–8 h, the distribution grid fills the valley by selling power to distributed shared energy storage; during the peak load periods of 9–12 h and 18–21 h, the distribution grid cuts the peak by purchasing power from distributed shared energy storage, thus reducing the net load peak-to-valley difference of the distribution grid. Based on the net load curve, it can be seen that the peak-to-valley difference for scenario 4 is 1120 kW, which is 63% lower than that of scenario 1.

The power balance of distributed new energy sites for scenario 1 and scenario 4 is shown in Figure 6. In Figure 6, the positive power represents the power output of each new energy station, and the negative power represents the power sold by each new energy station to the distribution grid and the distributed shared energy storage system. The ideal power output of distributed new energy stations represents the sum of the maximum power available from all new energy stations in that period.

Analyzing the power balance diagram of distributed new energy stations in scenario 1, we can see that the distribution grid cannot consume all the new energy output at 2–6 h and 15 h, at which time there is power curtailment in distributed new energy stations, and the power curtailed by wind and light is 1455 kW. Scenario 4 is equipped with distributed shared energy storage. When the distribution grid cannot consume all the new energy output, the distributed new energy stations sell the excess power to distributed shared energy storage to improve the new energy consumption rate, and the new energy consumption rate of scenario 4 is 100%.

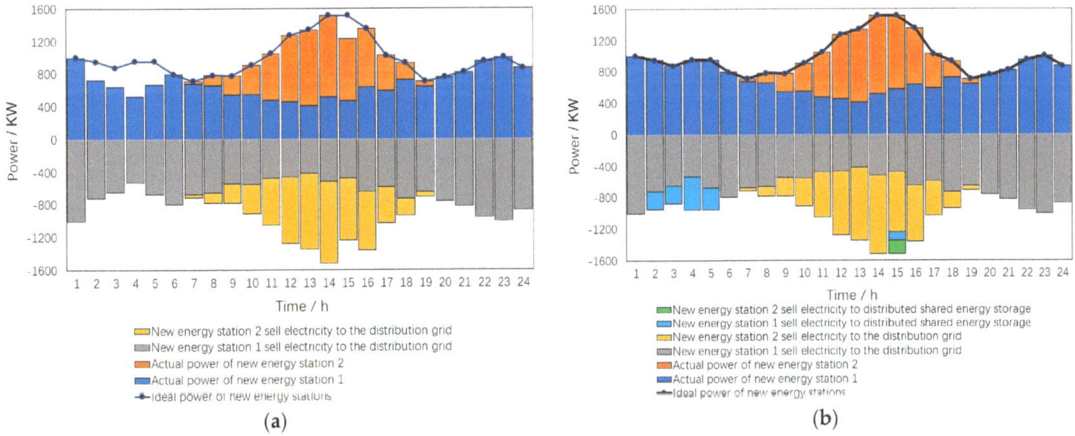

Figure 6. The power balance diagram of distributed new energy stations. (**a**) Scenario 1; (**b**) scenario 4.

The economic benefits of scenario 1 and scenario 4 are shown in Table 2. Scenario 1 does not configure energy storage, so the total cost of the distributed shared energy storage system is 0. The daily integrated operating cost of the distribution grid-distributed new energy stations is 35,873 Yuan, the net load peak-to-valley difference is 3040 kW, and the phenomenon of wind and light curtailment exists. Scenario 4 is configured with distributed shared energy storage; the cost of the distributed shared energy storage system is −84 Yuan, the energy storage is profitable, and the distribution grid-distributed new energy stations' daily integrated operation cost is reduced by 1786 Yuan compared to scenario 1; the net load peak-to-valley difference is reduced by 1920 kW compared to scenario 1, and the new energy consumption rate is 100%. The comparative analysis of scenario 1 and scenario 4 verifies that the configuration of distributed shared energy storage can effectively reduce the peak-to-valley difference and improve the consumption rate of new energy.

Table 2. Economic benefits of scenario 1 and scenario 4.

Scenario	Distributed Shared Energy Storage System Cost/Yuan [1]	Distribution Grid-Distributed New Energy Stations Comprehensive Daily Operating Cost/Yuan	Net Load Peak-to-Valley Difference/kW	New Energy Consumption Rate/%
1	-	35,873	3040	93
4	−84	34,087	1120	100

[1] 1 Yuan ≈ 0.1388 USD.

4.3. Analysis of Distributed Shared Energy Storage Optimal Allocation Results and Charging and Discharging Behavior

The results of scenario 2 and scenario 4 energy storage optimization configurations are shown in Table 3, where the distribution grid energy storage in scenario 2 is configured with the peak-to-valley difference derived from scenario 4 as the constraint. It can be seen that the total configured capacity in scenario 2 is 9580 kW·h and the total configured capacity of distributed shared energy storage in scenario 4 is 6870 kW·h, which is 28% less than the total configured capacity in scenario 2. It can be seen that by reasonably sharing distributed energy storage, realizing the time, sharing multiplexing of energy storage, and improving the utilization rate of energy storage resources, the configuration of smaller power and capacity of energy storage can meet the demand for energy storage in distributed new energy stations and distribution grid.

Table 3. The energy storage optimization configuration results of scenario 2 and scenario 4.

Scenario	Category	Access Node	Power Rating/kW	Rated Capacity/(kW·h)
2	Energy storage for new energy station 1	20	432	2160
	Energy storage for new energy station 2	9	175	875
	Energy storage for distribution grid	6	1000	5000
		3	309	1545
4	Distributed shared energy storage 1	6	987	4935
	Distributed shared energy storage 2	13	381	1935

To see the utilization of energy storage resources more intuitively, this paper will analyze the results of scenario 4 energy storage charging and discharging behavior and charge state optimization, as shown in Figure 7. Positive power represents energy storage charging and negative power represents energy storage discharging.

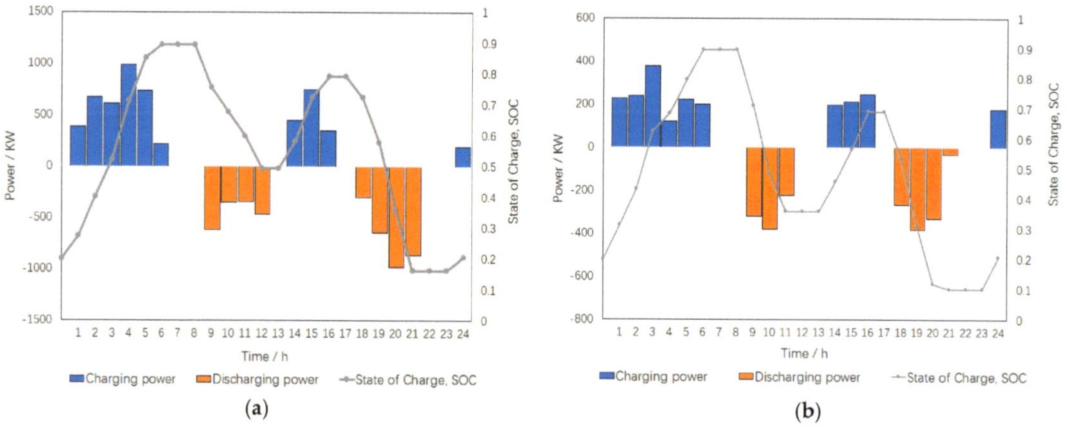

Figure 7. The distributed shared energy storage charge–discharge and charge state optimization results. (**a**) Distributed shared energy storage 1; (**b**) distributed shared energy storage 2.

From Figure 7, it can be seen that both distributed shared energy storage 1 and 2 reach the maximum charging power in the low valley period and the maximum discharging power in the peak load period, i.e., both distributed shared energy storage 1 and 2 have full charging and full discharging behaviors. In addition, the distributed shared energy storage 1 reaches a maximum charge state of 0.9 at 6 h and a minimum charge state of 0.16 at 21 h. Distributed shared energy storage 2 reaches a maximum charge state of 0.9 at 6 h and a minimum charge state of 0.1 at 21 h, indicating that all distributed shared energy storage power reaches the upper or lower capacity limit. Distributed shared energy storage makes full use of energy storage capacity resources by aggregating the energy demand of distribution grids and distributed new energy sites and reasonably allocating each energy storage charge and discharge.

The economic benefits of scenario 2 and scenario 4 are shown in Table 4. It can be seen that the distributed shared energy storage system in scenario 4 is profitable, with a total cost of −84 Yuan and a combined daily operating cost of 2409 Yuan less for scenario 4's distribution grid-distributed new energy field station compared to scenario 2. Through the comparative analysis of scenario 2 and scenario 4, it is verified that the configuration of distributed shared energy storage can reduce the operating cost of distribution grid-distributed new energy stations while taking into account the economics of shared energy storage investors to achieve a win–win situation for all parties.

Table 4. Economic benefits of scenario 2 and scenario 4.

Scenario	Distributed Shared Energy Storage System Cost/Yuan [1]	Distribution Grid-Distributed New Energy Stations Comprehensive Daily Operating Cost/Yuan	Net Load Peak-to-Valley Difference/kW	New Energy Consumption Rate/%
2	-	36,496	1120	100
4	−84	34,087	1120	100

[1] 1 Yuan ≈ 0.1388 USD.

4.4. Analysis of the Impact of Different Numbers of Energy Storage on the Economics of Distributed Shared System

To analyze the economic impact of configuring different numbers of energy storage on the distributed shared system, computational analysis was performed for scenario 3, and the cost of the distributed shared energy storage system with different numbers of energy storage as the constraint was obtained, as shown in Figure 8.

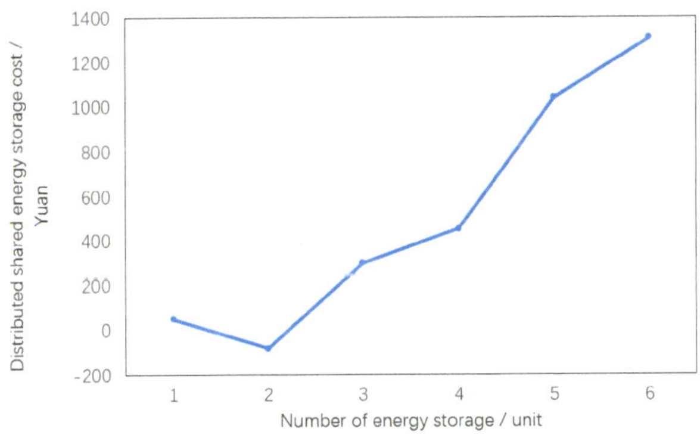

Figure 8. The distributed shared energy storage cost versus the number of energy storage.

As can be seen from Figure 8, the cost of distributed shared energy storage tends to decrease and then increase as the number of energy storage increases. Due to the constraints of energy storage rated power, line transmission capacity, etc., as the number and scale of energy storage increases, the ability of distributed shared energy storage systems to consume new energy and peak shaving is increasing (i.e., the revenue of distributed shared energy storage is increasing), so the cost of distributed shared energy storage is on a downward trend. However, as the number and scale of energy storage continue to increase, the effect of new energy consumption and peak shaving tends to saturate, but the investment cost of energy storage is increasing, so the cost of distributed shared energy storage is on the rise. In this calculation example, the cost of distributed shared energy storage is at least −84 Yuan when the number of energy storage sites is 2, and the distributed shared energy storage operator achieves profitability.

5. Conclusions

This paper proposes a distributed shared energy storage optimal allocation method that takes into account both power-side and grid-side regulation requirements, integrates the optimization problems at both planning and operation levels, constructs a double-layer model for distributed shared energy storage optimal allocation, and solves it using a double-layer iterative particle swarm algorithm combined with tide calculation, and draws the following main conclusions:

By deploying distributed shared energy storage, distribution grid and new energy stations receive energy storage charging and discharging services at a lower cost, increasing the new energy consumption rate to 100% and reducing the peak-to-valley difference by 61%.

Through the reasonable sharing of distributed energy storage, realize the time-sharing reuse of energy storage and improve the utilization rate of energy storage resources so that the configuration of smaller capacity energy storage can meet the demand for energy storage in distributed new energy stations and distribution grids. Distributed shared energy storage can reduce the allocated capacity by 28% compared to the standalone distribution storage scenario.

Through distributed shared energy storage system services and a reasonable number of energy storage configurations, distribution grids and distributed new energy stations can reduce their operating costs. At the same time, distributed shared energy storage operators realize positive returns, and there is potential for profitable investment in building distributed shared energy storage plants.

Author Contributions: Conceptualization, M.Y.; methodology, M.Y. and Y.Z.; software, J.L.; formal analysis, S.Y.; investigation, X.C.; resources, L.S.; writing—original draft preparation, H.L.; writing—review and editing, Z.F. All authors have read and agreed to the published version of the manuscript.

Funding: The research was supported by the R&D project "Research on the Optimized Configuration and Operation Model of Distributed Shared Energy Storage to Promote Local Consumption of Renewable Energy- Research on multi-objective optimization model and operational strategy for shared energy storage (5217L0230003)" from the State Grid Henan Economic Research Institute.

Data Availability Statement: Research data have been provided in the manuscript.

Conflicts of Interest: The authors declare no conflict of interest.

Nomenclature

C_1	the cost of a distributed shared energy storage system	C_{sto}	the average daily investment and maintenance cost of distributed shared energy storage
C_{new}	the cost of trading electricity between distributed shared energy storage and distributed new energy stations	C_{adn}	the electricity transaction cost between distributed shared energy storage and the distribution grid
C_{ser}	the distributed shared energy storage capacity lease service fee	δ_p	the investment cost per unit of power of energy storage
δ_e	the investment cost per unit capacity of energy storage	$P_{sto,i}$	the rated power of energy storage
$E_{sto,i}$	the rated capacity of energy storage	δ_m	the maintenance cost per unit of power
δ^t_{new}	the selling electricity price per unit of electricity of distributed new energy stations at time t	$P^t_{sto,new,j}$	the power sold from new energy station
δ^t_{sto}	the selling electricity price per unit of electricity of distributed shared energy storage at time t	δ^t_{adn}	the selling electricity price per unit of electricity of distribution grid at time t
$P^t_{sto,adn,d}$	the electricity sold by the distributed shared energy storage system to the distribution grid at time t	$P^t_{sto,adn,c}$	the electricity sold by the distribution grid to the distributed shared energy storage system at time t
δ_s	a unit power service fee paid by the distribution grid and distributed new energy stations to distributed shared energy storage system	β	the energy storage battery rate

$P_{sto,i,\min}$	the minimum power of distributed shared energy storage installed at each node	$P_{sto,i,\max}$	the maximum power of distributed shared energy storage installed at each node
$P^t_{sto,i,c}$	the charging power of energy storage i at time t	$P^t_{sto,i,d}$	the discharging power of energy storage i at time t
$A^t_{sto,i,c}$	the charge flags of energy storage i at time t	$A^t_{sto,i,d}$	the discharge flags of energy storage i at time t
$E^t_{sto,i}$	the charge of energy storage i at time t	$\eta_{sto,c}$	the charging efficiency of energy storage
$\eta_{sto,d}$	the discharging efficiency of energy storage	C_{grid}	the cost of electricity purchased from the main grid by the distribution grid
$C_{peak-valley}$	the penalty cost of the net load peak-to-valley difference	δ^t_p	the price of electricity sold by the main grid at time t
P^t_{grid}	the power sold by the main grid to the distribution grid at time t	$\delta_{peak-valley}$	the net load peak–valley difference unit power penalty cost
$L^{\max}_{load}$	the net load maximum values	$L^{\min}_{load}$	the net load minimum values
L^t_{load}	the net distribution grid load at time t	$P^t_{load,k}$	the load at the node k at time t
$P^t_{adn,new,j}$	the power sold by the new energy station j to the distribution grid at time t	$P^t_{new,j}$	the actual output of the new energy station j at time t
$P^t_{new\_0,j}$	the ideal output of the new energy station j	$P^{\max}_{sto,new}$	the maximum interactive power between the new energy station and the distributed shared energy storage
$B^t_{sto,adn,d}$	the discharge flag bits of the power interaction between the distribution grid and distributed shared energy storage	$B^t_{sto,adn,c}$	the charge flag bits of the power interaction between the distribution grid and distributed shared energy storage
$P^{\max}_{sto,adn}$	the maximum interaction power between distribution grid and distributed shared energy storage	$P_{loss,t}$	the net loss of the distribution network at time t
P^t_i	the active power injected at node i at time t	Q^t_i	the reactive power injected at node i at time t
U^t_i	the voltage amplitudes at node i at time t	U^t_j	the voltage amplitudes at node j at time t
G_{ij}	the conductance between nodes i and j	B_{ij}	the susceptance between nodes i and j
θ_{ij}	the phase angle difference between nodes i and j	$U_{i,\min}$	the minimum values of the voltage amplitude of node i
$U_{i,\max}$	the maximum values of the voltage amplitude of node i	S^t_{ij}	the transmitted power between nodes i and j at time t
$S_{ij,\max}$	the maximum value of the transmittable power between nodes i and j		

Appendix A

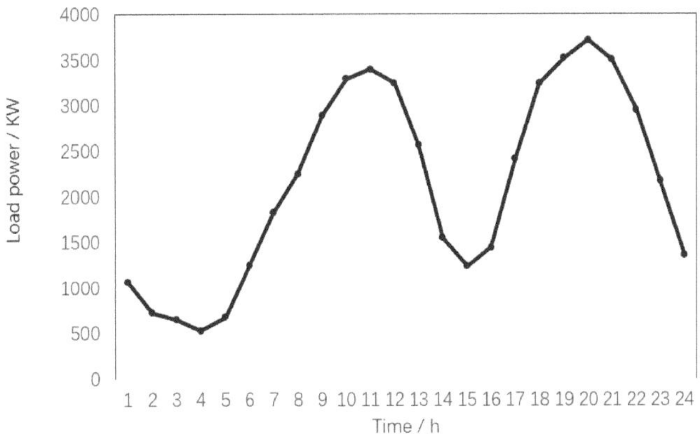

Figure A1. Load power graph for a typical day.

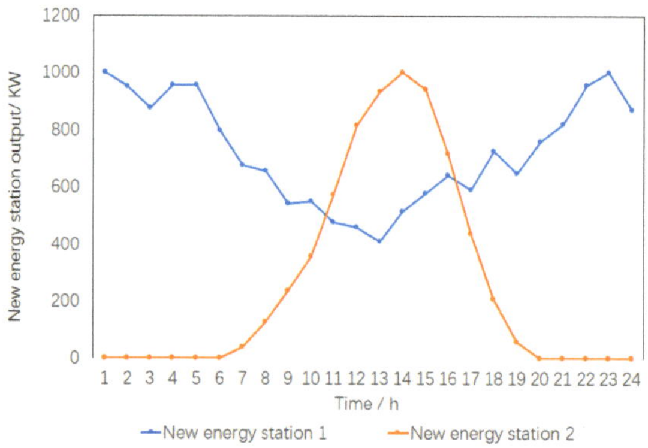

Figure A2. New energy power station output graph on a typical day.

References

1. Saranyaa, J.S.; Fathima, A.P. A Comprehensive Survey on the Current Trends in Improvising the Renewable Energy Incorporated Global Power System Market. *IEEE Access* **2023**, *11*, 24016–24038. [CrossRef]
2. Sun, L.; Zhang, Q.; Zhang, N.; Song, Z.; Liu, X.; Li, W. A Time-Sequence Simulation Method for Power Unit's Monthly Energy-Trade Scheduling with Multiple Energy Sources. *Processes* **2019**, *7*, 771. [CrossRef]
3. Yasuda, Y.; Bird, L.; Carlini, E.M. C-E (curtailment—Energy share) map: An objective and quantitative measure to evaluate wind and solar curtailment. *Renew. Sustain. Energy Rev.* **2022**, *160*, 112212. [CrossRef]
4. Tercan, S.M.; Demirci, A.; Gokalp, E.; Cali, U. Maximizing self-consumption rates and power quality towards two-stage evaluation for solar energy and shared energy storage empowered microgrids. *J. Energy Storage* **2022**, *51*, 104561. [CrossRef]
5. Peng, P.; Li, Y.; Li, D.; Guan, Y.; Yang, P.; Hu, Z.; Zhao, Z.; Liu, D. Optimized Economic Operation Strategy for Distributed Energy Storage With Multi-Profit Mode. *IEEE Access* **2021**, *9*, 8299–8311. [CrossRef]
6. Chen, Y.; Shi, Y.; Zhong, H.; Wang, X.; Lei, X.; Yin, H.; Liu, X. Hydrogen-Electric Hybrid Energy Storage System Configuration Method for Transmission Grid Containing High Percentage of Scenery Access. *Electr. Power Constr.* **2022**, *43*, 85–98.
7. Gong, Q.; Fang, J.; Qiao, H.; Liu, D.; Tan, S.; Zhang, H.; He, H. Optimal Allocation of Energy Storage System Considering Price-Based Demand Response and Dynamic Characteristics of VRB in Wind-PV-ES Hybrid Microgrid. *Processes* **2019**, *7*, 483. [CrossRef]
8. Ding, M.; Fang, H.; Bi, R.; Liu, X.; Pan, J.; Zhang, J. Distributed Photovoltaic and Energy Storage Siting and Capacity Planning for Distribution Networks Based on Cluster Delineation. *Proc. CSEE* **2019**, *39*, 2187–2201.
9. Gong, Q.; Wang, Y.; Fang, J.; Qiao, H.; Liu, D. Optimal Configuration of the Energy Storage System in ADN Considering Energy Storage Operation Strategy and Dynamic Characteristic. *IET Gener. Transm. Distrib.* **2020**, *14*, 1005–1011. [CrossRef]
10. Li, X.; Zhang, L.; Wang, R.; Sun, B.; Xie, W. Two-Stage Robust Optimization Model for Capacity Configuration of Biogas-Solar-Wind Integrated Energy System. *IEEE Trans. Ind. Appl.* **2023**, *59*, 662–675. [CrossRef]
11. Guo, W.; Xiu, X.; Li, W.; Li, J. A Method for Siting and Configuring Grid-Side Energy Storage Systems with Integrated Multi-attribute Metrics and Economics. *Electr. Power Constr.* **2020**, *41*, 53–62.
12. Gu, C.; Wang, J.; Li, Q.; Zhang, Y. A Review of Large-Scale Centralized Energy Storage Planning Research under New Energy Centralized Grid Integration. *Electr. Power* **2022**, *55*, 2–12+83.
13. Li, J.; Xing, Y.; Zhang, D. Planning Method and Principles of the Cloud Energy Storage Applied in the Power Grid Based on Charging and Discharging Load Model for Distributed Energy Storage Devices. *Processes* **2022**, *10*, 194. [CrossRef]
14. Du, X.; Li, X.; Chen, L.; Hao, Y.; Mei, S. Centralized Shared Energy Storage for Robust and Optimal Configuration of Multi-Scenario Regulation Requirements. *Trans. China Electrotech. Soc.* **2022**, *37*, 5911–5921.
15. Shuai, X.; Wang, X.; Huang, J. Optimal Allocation of Shared Energy Storage Capacity under Multi-region Integrated Energy System Interconnection. *J. Glob. Energy Interconnect.* **2021**, *4*, 382–392.
16. Xie, Y.; Luo, Y.; Li, Z.; Xu, Z.; Li, L.; Yang, K. Optimal Allocation of Shared Energy Storage Considering Economic Consumption of New Energy in Microgrid. *High Volt. Eng.* **2022**, *48*, 4403–4413.
17. Yang, S.; Hu, X.; Wang, H.; Ligao, J.; Meng, L.; Zhou, W.; Zhou, H. A Prosumer-Based Energy Sharing Mechanism of Active Distribution Network Considering Household Energy Storage. *IEEE Access* **2022**, *10*, 113839–113849. [CrossRef]
18. Liu, Y.; Dai, H.; Liu, Z.; Liu, R. Decentralized Shared Energy Storage Configuration and Investment Benefit Analysis for Multiple Types of Industrial Users. *Electr. Power Autom. Equip.* **2021**, *41*, 256–264.
19. Sun, T.; Zeng, L.; Zheng, F.; Zhang, P.; Xiang, X.; Chen, Y. Two-Layer Optimization Model for the Siting and Sizing of Energy Storage Systems in Distribution Networks. *Processes* **2020**, *8*, 559. [CrossRef]
20. Zhang, X.; Wang, Z.; Zhou, Z.Y.; Yin, X.Y.; Wang, Z.Y.; Liu, Y.Z. Optimization configuration of multi-agent shared energy storage considering photovoltaic integrated 5G base station energy consumption mode. *Electr. Meas. Instrum.* **2023**, *60*, 97–106.

21. Goswami, S.K.; Basu, S.K. A New Algorithm for the Reconfiguration of Distribution Feeders for Loss Minimization. *IEEE Trans. Power Deliv.* **1992**, *7*, 1484–1491. [CrossRef]
22. Wu, S.; Li, Q.; Liu, J.; Zhou, Q.; Wang, C. Two-layer Optimal Configuration of Cold, Heat and Power Multi-microgrid System Based on Energy Storage Plant Services. *Power Syst. Technol.* **2021**, *45*, 3822–3832. [CrossRef]

processes

Article

Optimal Allocation Method of Circuit Breakers and Switches in Distribution Networks Considering Load Level Variation

Guodong Huang [1], Yi Zhou [2], Chen Yang [1], Qiong Zhu [1], Li Zhou [1], Xiaofeng Dong [1], Junting Li [1] and Junpeng Zhu [2,*]

[1] Suzhou Power Supply Company, State Grid Jiangsu Electric Power Co., Ltd., Suzhou 215004, China; huangguodong0626@163.com (G.H.); xiaozhuzq@163.com (Q.Z.); fenfery@gmail.com (L.Z.); 18706200494@139.com (X.D.); lijunting0626@163.com (J.L.)

[2] College of Energy and Electrical Engineering, Hohai University, Nanjing 210098, China; yichow_99@163.com

* Correspondence: jzhu@hhu.edu.cn

Abstract: Reliability is a fundamental concept for power systems, and the optimal placement of switchable devices is a valuable tool for improvements in this area. The goal of this paper is to propose an optimal allocation method for circuit breakers and switches that can break the cost–reliability dilemma and simultaneously achieve reliability and economic improvement in terms of the distribution network. Moreover, in view of the fact that variations in the load level can affect the reliability of the distribution network, the variations of different load level scenarios are considered in this paper, where a mixed integer linear programming (MILP) model based on fictitious fault flows is established to derive the optimal allocation scheme that can adapt to the changes of multiple scenarios regarding the load. Meanwhile, due to the constraints of reliability indices, the post-fault reconfiguration scheme of a distribution network under different load level scenarios can also be obtained to enhance its overall reliability. Finally, the applicability and effectiveness of the proposed method are verified by numerical tests on a 54-node test system.

Keywords: distribution network; load level variation; circuit breakers; switches; allocation; reliability enhancement

Citation: Huang, G.; Zhou, Y.; Yang, C.; Zhu, Q.; Zhou, L.; Dong, X.; Li, J.; Zhu, J. Optimal Allocation Method of Circuit Breakers and Switches in Distribution Networks Considering Load Level Variation. *Processes* **2023**, *11*, 2235. https://doi.org/10.3390/pr11082235

Academic Editors: Chenyu Wu, Zhongkai Yi and Chenhui Lin

Received: 2 June 2023
Revised: 23 July 2023
Accepted: 24 July 2023
Published: 25 July 2023

1. Introduction

According to the definition of reliability, a power system must be capable of consistently providing end users with both the quantity and quality of electricity they require [1]. About 70% of total electric service interruptions are caused by contingencies in the distribution system [2]. In recent years, the stability of the distribution network has declined due to the large proportion of distributed generation (DG) usage with the continuous development of renewable energy power generation technology [3]. Attaining high reliability for a distribution system is not only important but is also crucial to ensure the uninterrupted supply of electricity to consumers. Reliable distribution systems minimize power interruptions and enhance customer satisfaction. This can be achieved by implementing robust infrastructure, such as redundant lines, automated switching devices, and protective measures against external disturbances.

At present, the main strategies to improve power supply reliability are reducing the equipment failure rate, shortening power restoration time, and improving fault isolation accuracy [4,5]. Among the various methods that could improve the reliability of distribution networks, the optimal placement of circuit breakers and switches has a significant impact on enhancing the reliability of the utility grids. It has been demonstrated in earlier studies that remote-controlled switches (RCSs) increase the distribution network reliability indices, such as the system average interruption duration index (SAIDI) and the expected energy not supplied (EENS), by reducing the duration that it takes to restore the power supply as well as speeding up the process of isolating the faulted area from the rest of the distribution network [6]. Given the fact that installing switchable devices is expensive, it is essential to select the location sites properly to balance benefits and costs [7]. Due to the relatively high investment costs of switchable devices and the budget limitations of utilities to improve the quality of customer services, it is necessary to study the optimal allocation of switchable devices in the distribution network to achieve the maximum improvement to the level of reliability with the lowest investment cost so that the cost–reliability dilemma could be broken.

Moreover, variations in the load level may lead to operating conditions beyond the design limits of the distribution network, which may cause voltage instability, transformer overloads, and other equipment failures leading to outages. Moreover, it can also significantly affect the planning and operation of the distribution network, making it difficult to optimize network utilization and ensure adequate reserve capacity. Overall, the impact of load level variation on the reliability of distribution networks cannot be ignored, and addressing it requires effective planning and operation strategies that take into account the complexity and variability of load levels. Hence, it is worthwhile to study the optimal allocation of circuit breakers and switches that can adapt to load level variations in order to weaken the impact of this on the reliability of the distribution network and maximize the effects of reliability improvements in the distribution network with lower economic investment costs.

The allocation problem of switchable devices has an underlying service restoration problem, consisting of choices about which switchable device must be opened or closed to minimize the unattended area after the isolation of a failure, which is categorized as a complex, combinatorial, and constrained optimization problem [8]. Plenty of pieces of literature have studied the optimal placement of these switchable devices in the distribution network.

The optimal allocation issue in the distribution network is classified as a combinatorial optimization problem that, especially for large cases, can be challenging to solve when optimally utilizing mathematical programming methods. Due to the complexity of the problem, heuristic algorithms are mainly used to solve this [9–11]. The reliability index of EENS was used in [12] to perform the optimal placement of remote-controlled switches, employing the differential search metaheuristic algorithm. In [13], the immune algorithm was used to determine the optimal location of switches by utilizing an objective function that minimizes the expense of investments in line switches and the cost of customer service outages while taking into account the failure rates of the load points concerned. The ant colony optimization algorithm is adopted in [14] to solve the fuzzy multi-objective problem of optimizing the location of sectionalizing switches, with the objectives of reliability improvements and the minimization of the cost of sectionalizing switches. Fuzzy logic and genetic algorithms were employed in a hybrid algorithm in [15] to improve the SAIDI index, which requires many network parameters for its application. A global combination criterion was proposed to simultaneously evaluate the combination performance of multiple switch positions in [16], which avoids the tedious traditional problem of adjusting only one switch position at a time and directly determining the optimal solution. However, some mathematical methods have also been proposed to solve the problem in recent years. In [17], the remote-controlled switch configuration problem is modeled as a mixed-integer linear programming model that divides loads into two categories according to the restoration time, and the configuration scheme is developed with the objective of minimizing the total cost and expected outage losses. A mixed-integer linear programming model for simultaneous switch and tie line placement in distribution systems with complex topologies is presented in the study in [18]. The studies on the allocation of switchable devices usually aim at improving the reliability and economy of the distribution network and mainly focus on the placement of switches with little consideration given to the deployment of circuit breakers. Meanwhile, the influence of load level variation fluctuations on the reliability of the distribution network is usually ignored.

In light of the progress in the above studies, this paper carries the analysis further. In order to break the cost–reliability dilemma and achieve a higher reliability level with a relatively low investment cost, this paper establishes an optimal allocation model for circuit breakers and switches in distribution networks with the objective function of summing the minimum outage loss and investment cost of circuit breakers and switches while taking the reliability index as the constraint. By considering the fact that changes in load level can have an impact on reliability, variations in load level were considered to enhance the adaptability of the derived optimal configuration scheme to different load level scenarios while improving the practicality of the method proposed in this paper.

The concept of fictitious fault flows is used to realize the linearized calculation of reliability indices and, at the same time, the reconfiguration scheme after the occurrence of faults in different scenarios, which can be solved according to the load level in different scenarios in a targeted manner to realize the effective improvement of distribution network reliability. The proposed model for the optimal allocation of circuit breakers and switches (considering load level variation) is a mixed-integer linear programming model that can achieve efficient solutions for the optimal allocation of switchable devices in distribution networks based on the goal of reliability enhancements and the result can be guaranteed to be the global optimal solution, with good practicality and engineering value.

2. Mathematical Model of Optimal Allocation of Circuit Breakers and Switches

The distribution network optimization allocation model for circuit breakers and switches in this study is based on the reliability assessment model put forward in [19], which utilizes the concept of fictitious fault flows to linearize the calculation of the reliability indices of the distribution system. Variations in load level are considered in this paper, and based on different load level scenarios, the optimal circuit breaker and switch allocation scheme and its initial operating state that can adapt to the corresponding load level variations are explored with the goal of improving the reliability and economy of the distribution network, and the reconfiguration scheme of the distribution network after a fault occurs under different load scenarios can be obtained at the same time.

2.1. Objective Function

In order to ensure the economic requirements of distribution network planning and operation, this paper minimizes the sum of the outage loss and investment cost of circuit breakers and switches in a distribution network as the objective function, as is shown in (1). For the calculation of outage loss, this is represented by the product of the unit price of outage loss per unit of power and EENS in the distribution network. The improvement of the reliability of the distribution network can be achieved simultaneously by taking the EENS in the system as one of the optimization objectives.

$$\begin{aligned} f &= minimize\ (F_{CB}\sum_{ij}(x_{ij}^{i,CB} + x_{ij}^{j,CB}) + F_{SW}\sum_{ij}(x_{ij}^{i,SW} + x_{ij}^{j,SW}) + \alpha EENS) = \\ &= minimize\ (\text{Cost} + \alpha EENS) \end{aligned} \quad (1)$$

where F_{CB} and F_{SW} are the unit prices of the circuit breaker and switch, respectively. $x_{ij}^{i,CB}$, $x_{ij}^{j,CB}$ and $x_{ij}^{i,SW}$, $x_{ij}^{j,SW}$ are the binary variables indicating the installation of circuit breakers and switches on side i and side j of the branch ij, respectively. If the value of the variable is 1, it means that a circuit breaker or switch is installed on that side; if the value of the variable is 0, there are no circuit breakers or switches installed on that side. α is the unit price of power outage loss per unit of electricity. The *EENS* is the expected amount of power not supplied in the distribution network after an outage has occurred.

2.2. Logical Constraints on the Installation of Circuit Breakers and Switches and the Corresponding Status

Since the operation status can only be switched between closed and open when a circuit breaker or switch is installed on the branch, it is essential to make constraints to ensure the reasonable operation of circuit breakers and switches. For normal operation scenarios, the logical restrictions between the installation status and the operating status of the circuit breakers (or switches) are depicted as (2), (3) (or (4), (5)).

$$b_{ij}^{i,NO} \geq 1 - x_{ij}^{i,CB}, \forall ij \in \Upsilon \quad (2)$$

$$b_{ij}^{j,NO} \geq 1 - x_{ij}^{j,CB}, \forall ij \in \Upsilon \quad (3)$$

$$s_{ij}^{i,NO} \geq 1 - x_{ij}^{i,SW}, \forall ij \in \Upsilon \quad (4)$$

$$s_{ij}^{j,NO} \geq 1 - x_{ij}^{j,SW}, \forall ij \in \Upsilon \quad (5)$$

where $b_{ij}^{i,NO}$ and $b_{ij}^{j,NO}$ represent the original operation status of circuit breakers placed at end i and j of branch ij. If the variable equals 1, the circuit breaker or switch in this circuit is closed; if both variables equal 0, the circuit is open. Similarly, $s_{ij}^{i,NO}$ and $s_{ij}^{j,NO}$ represent the original operation status of switches at ends i and j of branch ij.

Under outage scenarios, taking into account the variation in load levels, the above logical constraints on the installation of the circuit breakers and switches and their corresponding changeable operating status are rewritten as (6)–(9).

$$b_{ij}^{i,xy,SC} \geq 1 - x_{ij}^{i,CB}, \forall ij \in \Upsilon \quad (6)$$

$$b_{ij}^{j,xy,SC} \geq 1 - x_{ij}^{j,CB}, \forall ij \in \Upsilon \quad (7)$$

$$s_{ij}^{i,xy,SC} \geq 1 - x_{ij}^{i,SW}, \forall ij \in \Upsilon \quad (8)$$

$$s_{ij}^{j,xy,SC} \geq 1 - x_{ij}^{j,SW}, \forall ij \in \Upsilon \quad (9)$$

where xy represents the branch where the fault occurs, and SC represents different load level scenarios.

2.3. Constraints on Power Flow and the Capacity of Branches

Each normal operating scenario's load demand condition is constrained by (10) and (11). The load demand of a node under normal operating conditions is all the load power connected to that node under the corresponding load level scenario. Constraints (12)–(15) are derived from the linearized power flow equations in [20]. Equation (14) uses a method that combines binary variables with the large M method to constrain the power and voltage in the network to facilitate the linearization of the optimal allocation model. Nodal voltage constraints under normal scenarios are expressed as (16). Constraints (17)–(20) illustrate how the status of switches in the circuit restricts the power flow of branch ij; that is, the existence of power flow is possible when and only when the branch is connected.

$$P_i^{SC,NO} = P_i^{SC}, \forall i \in \Psi^{LN} \quad (10)$$

$$Q_i^{SC,NO} = Q_i^{SC}, \forall i \in \Psi^{LN} \quad (11)$$

$$P_{ki}^{SC,NO} = \sum_{j \in \Psi_i} P_{ij}^{SC,NO} + P_i^{SC,NO}, \forall ki \in \Upsilon \quad (12)$$

$$Q_{ki}^{SC,NO} = \sum_{j \in \Psi_i} Q_{ij}^{SC,NO} + Q_i^{SC,NO}, \forall ki \in \Upsilon \quad (13)$$

$$-\left(2-s_{ij}^{i,NO}-s_{ij}^{j,NO}\right)M+2\left(r_{ij}P_{ij}^{SC,NO}+x_{ij}Q_{ij}^{SC,NO}\right)\leq U_i^{SC,NO}-U_j^{SC,NO}\leq$$
$$\leq\left(2-s_{ij}^{i,NO}-s_{ij}^{j,NO}\right)M+2\left(r_{ij}P_{ij}^{SC,NO}+x_{ij}Q_{ij}^{SC,NO}\right),\forall ij\in\Upsilon \quad (14)$$

$$U_i^{SC,NO}=\left(V^S\right)^2,\forall i\in\Psi^{SS} \quad (15)$$

$$\underline{U}\leq U_i^{SC,NO}\leq\overline{U},\forall i\in\Psi^{LN} \quad (16)$$

$$-Ms_{ij}^{i,NO}\leq P_{ij}^{SC,NO}\leq Ms_{ij}^{i,NO},\forall ij\in\Upsilon \quad (17)$$

$$-Ms_{ij}^{i,NO}\leq Q_{ij}^{SC,NO}\leq Ms_{ij}^{i,NO},\forall ij\in\Upsilon \quad (18)$$

$$-Ms_{ij}^{j,NO}\leq P_{ij}^{SC,NO}\leq Ms_{ij}^{j,NO},\forall ij\in\Upsilon \quad (19)$$

$$-Ms_{ij}^{j,NO}\leq Q_{ij}^{SC,NO}\leq Ms_{ij}^{j,NO},\forall ij\in\Upsilon \quad (20)$$

where $P_i^{SC,NO}$ and $Q_i^{SC,NO}$ describe the active and reactive demand under different load level scenarios at the node under normal operating conditions. The active and reactive power flows through branch ij under different load level scenarios are denoted by $P_{ij}^{SC,NO}$ and $Q_{ij}^{SC,NO}$, respectively. $U_i^{SC,NO}$ is the square voltage under different load levels at the node, while V^S is the source voltage at a feeder's head end.

In constraints (21)–(23), the branch capacity limitations that are linearized and rely on piecewise relations are offered by [21], which presents a quadratic circular constraint to facilitate dualization in the solving process. Constraints (24) and (25) denote the power of the feeder f provided by the transformer tr^f that connects to it, whereas constraints (26)–(28) indicate the capacity restrictions of the transformers.

$$-S_{ij}^C\leq P_{ij}^{SC,NO}\leq S_{ij}^C,\forall ij\in\Upsilon \quad (21)$$

$$-S_{ij}^C\leq Q_{ij}^{SC,NO}\leq S_{ij}^C,\forall ij\in\Upsilon \quad (22)$$

$$-\sqrt{2}S_{ij}^C\leq P_{ij}^{SC,NO}\pm Q_{ij}^{SC,NO}\leq\sqrt{2}S_{ij}^C,\forall ij\in\Upsilon \quad (23)$$

$$P_f^{SC,NO}=P_{tr^f}^{SC,NO},\forall f\in\Psi^F,tr^f\in\Upsilon \quad (24)$$

$$Q_f^{SC,NO}=Q_{tr^f}^{SC,NO},\forall f\in\Psi^F,tr^f\in\Upsilon \quad (25)$$

$$P_f^{SC,NO}\leq S_f^C,\forall f\in\Psi^F \quad (26)$$

$$Q_f^{SC,NO}\leq S_f^C,\forall f\in\Psi^F \quad (27)$$

$$\pm Q_f^{SC,NO}+P_f^{SC,NO}\leq\sqrt{2}S_f^C,\forall f\in\Psi^F \quad (28)$$

where S_{ij}^C is the peak transmission capacity of branch ij, while S_f^C represents the transformer's capability in connection to feeder f.

2.4. Radial Constraints on Distribution Network Topology

Nowadays, a mesh-constructed distribution network architecture is popular in urban areas to improve the reliability of the power supply [22–24]; However, it operates radially. Reconfiguring a distribution network requires changing the topology in order to boost performance while preserving network radiality. The operating status of a switch on either side of the branch, as indicated in (29) for normal operation circumstances, is used to determine the connection status of branch ij. Constraint (30) ensures that the distribution network operates in a radial structure.

$$s_{ij}^{i,NO}+s_{ij}^{j,NO}-1\leq l_{ij}^{NO}\leq\left(s_{ij}^{i,NO}+s_{ij}^{j,NO}\right)/2,\forall ij\in\Upsilon \tag{29}$$

$$n^{LN}=\sum_{ij\in\Upsilon}l_{ij}^{NO} \tag{30}$$

where l_{ij}^{NO} equals 1 when branch ij is connected under normal operation conditions, and this shows whether branch ij is linked under a specified assignment of switchable devices. The number of load nodes if represented by n^{LN}.

2.5. Constraints Related to Reliability Assessment

This paper classifies the fictitious fault flows in the network into "RA" and "PF" during the fault recovery after an outage occurs. Just after the outage, a fictitious fault flow identified as "RA" appears and can only be stopped by circuit breakers. Fictitious fault flow, denoted by the symbol "PF," takes place after the fault branch has been isolated and can only be eliminated via switches. The reliability of load points and the power system can be assessed and optimized while the network reconfiguration is being carried out via systematic reliability indices and nodal reliability indices, which are calculated in the distribution network based on the distribution of the two fictitious fault flows in each branch.

The first block of constraints is given in (31)–(38). Constraint (31) sets the location of the branch xy where the fault occurred. As shown in Equations (32) and (33), the spread of the fictitious fault flow "RA" between branches is influenced by the initial switch operating state in the network and can be prevented by tripping the circuit breakers. Constraint (34) guarantees that only one circuit breaker can trip to stop the spread of fault currents under every outage scenario. Constraints (35) and (36) specify the upper and lower limits for representing the fictitious fault flow "RA" variable. Constraint (37) makes sure that the outage does not occur on substation nodes. Equation (38) uses the variable $p_i^{xy,SC}$ to represent the state of the power supply after a fault at each node, which is based on the value of the fictitious fault flow "RA" variable at each node.

$$f_{xy}^{xy,SC,RA}=0 \tag{31}$$

$$-\left(2-b_{ij}^{i,xy,SC}-s_{ij}^{i,NO}\right)M+f_i^{xy,SC,RA}\leq f_{ij}^{xy,SC,RA}\leq\left(2-b_{ij}^{i,xy,SC}-s_{ij}^{i,NO}\right)M+f_i^{xy,SC,RA},\forall ij\in\Upsilon \tag{32}$$

$$-\left(2-b_{ij}^{j,xy,SC}-s_{ij}^{j,NO}\right)M+f_j^{xy,SC,RA}\leq f_{ij}^{xy,SC,RA}\leq\left(2-b_{ij}^{j,xy,SC}-s_{ij}^{j,NO}\right)M+f_j^{xy,SC,RA},\forall ij\in\Upsilon \tag{33}$$

$$\sum_{ij\in\Upsilon_I^B}b_{ij}^{i,NO}+\sum_{ij\in\Upsilon_J^B}b_{ij}^{j,NO}-1=\sum_{ij\in\Upsilon_I^B}b_{ij}^{i,xy,SC}+\sum_{ij\in\Upsilon_J^B}b_{ij}^{j,xy,SC} \tag{34}$$

$$0\leq f_i^{xy,SC,RA}\leq 1,\forall i\in\Psi^{LN} \tag{35}$$

$$0\leq f_{ij}^{xy,SC,RA}\leq 1,\forall ij\in\Upsilon \tag{36}$$

$$f_i^{xy,SC,RA} = 1, \forall i \in \Psi^{SS} \tag{37}$$

$$p_i^{xy,SC} = 1 - f_i^{xy,SC,RA}, \forall i \in \Psi^{LN} \tag{38}$$

The second part of the constraints is given by (39)–(67). Constraint (39) sets the outage branch where 'PF' stems from. Constraints (40) and (41) show that the spread of 'PF' is restricted by the operating status of the switches. Constraints (42) and (43) restrict the variation in the fictitious fault flow of each node and branch. Constraint (44) makes sure that the outage does not occur on substation nodes. If and only if the load at node i is supplied after post-fault reconfiguration (including unscathed nodes and restored nodes), this condition is designated as $q_i^{xy,SC} = 1$, which is explained in (45); thus, the demand of load nodes can be given by (46) and (47).

$$f_{xy}^{xy,SC,PF} = 0 \tag{39}$$

$$-\left(1 - s_{ij}^{i,xy,SC}\right)M + f_i^{xy,SC,PF} \le f_{ij}^{xy,SC,PF} \le \left(1 - s_{ij}^{i,xy,SC}\right)M + f_i^{xy,SC,PF}, \forall ij \in \Upsilon \tag{40}$$

$$-\left(1 - s_{ij}^{j,xy,SC}\right)M + f_j^{xy,SC,PF} \le f_{ij}^{xy,SC,PF} \le \left(1 - s_{ij}^{j,xy,SC}\right)M + f_j^{xy,SC,PF}, \forall ij \in \Upsilon \tag{41}$$

$$0 \le f_i^{xy,SC,PF} \le 1, \forall i \in \Psi^{LN} \tag{42}$$

$$0 \le f_{ij}^{xy,SC,PF} \le 1, \forall ij \in \Upsilon \tag{43}$$

$$f_i^{xy,SC,PF} = 1, \forall i \in \Psi^{SS} \tag{44}$$

$$q_i^{xy,SC} = f_i^{xy,SC,PF}, \forall i \in \Psi^{LN} \tag{45}$$

$$P_i^{xy,SC} = P_i q_i^{xy,SC}, \forall i \in \Psi^{LN} \tag{46}$$

$$Q_i^{xy,SC} = Q_i q_i^{xy,SC}, \forall i \in \Psi^{LN} \tag{47}$$

The demands of the nodes that were not impacted by the outage could not be altered following the post-fault network reconfiguration, as this is assured by constraint (48). The radial structure of the network under outage scenarios is ensured by (49) and (50).

$$1 - p_i^{xy,SC} \le q_i^{xy,SC}, \forall i \in \Psi^{LN} \tag{48}$$

$$s_{ij}^{i,xy,SC} + s_{ij}^{j,xy,SC} - 1 \le l_{ij}^{xy,SC} \le \left(s_{ij}^{i,xy,SC} + s_{ij}^{j,xy,SC}\right)/2, \forall ij \in \Upsilon \tag{49}$$

$$\sum_{i \in \Psi^{LN}} q_i^{xy,SC} = \sum_{ij \in \Upsilon} l_{ij}^{xy,SC} \tag{50}$$

Linearized power flow equations under outage scenarios are shown in (51)–(55). The principles of them are the same as that of (12)–(16) and will not be further elaborated here.

$$P_{ki}^{xy,SC} = \sum_{j \in \Psi_i} P_{ij}^{xy,SC} + P_i^{xy,SC}, \forall ki \in \Upsilon \tag{51}$$

$$Q_{ki}^{xy,SC} = \sum_{j \in \Psi_i} Q_{ij}^{xy,SC} + Q_i^{xy,SC}, \forall ki \in \Upsilon \tag{52}$$

$$\begin{aligned} -\left(2 - s_{ij}^{i,xy,SC} - s_{ij}^{j,xy,SC}\right)M + 2\left(r_{ij}P_{ij}^{xy,SC} + x_{ij}Q_{ij}^{xy,SC}\right) \le U_i^{xy,SC} - U_j^{xy,SC} \le \\ \le \left(2 - s_{ij}^{i,xy,SC} - s_{ij}^{j,xy,SC}\right)M + 2\left(r_{ij}P_{ij}^{xy,SC} + x_{ij}Q_{ij}^{xy,SC}\right), \forall ij \in \Upsilon \end{aligned} \tag{53}$$

$$U_i^{xy,SC} = \left(V^S\right)^2, \forall i \in \Psi^{SS} \tag{54}$$

$$\underline{U} \le U_i^{xy,SC} \le \overline{U}, \forall i \in \Psi^{LN} \tag{55}$$

The status of the switches limits the amount of power that can flow through branch ij, as shown in constraints (56)–(59). Similar to the constraints placed on branch power flow by the circuit breakers and switches under normal operation scenarios, the power flow of the branch ij exists under outage scenarios only when the branch is connected.

$$-Ms_{ij}^{i,xy,SC} \le P_{ij}^{xy,SC} \le Ms_{ij}^{i,xy,SC}, \forall ij \in \Upsilon \tag{56}$$

$$-Ms_{ij}^{i,xy,SC} \le Q_{ij}^{xy,SC} \le Ms_{ij}^{i,xy,SC}, \forall ij \in \Upsilon \tag{57}$$

$$-Ms_{ij}^{j,xy,SC} \le P_{ij}^{xy,SC} \le Ms_{ij}^{j,xy,SC}, \forall ij \in \Upsilon \tag{58}$$

$$-Ms_{ij}^{j,xy,SC} \le Q_{ij}^{xy,SC} \le Ms_{ij}^{j,xy,SC}, \forall ij \in \Upsilon \tag{59}$$

The linearized capacity restrictions of the branches, feeders, and transformers are provided by (60)–(67). The principles are the same as that of (21)–(28) and will not be further elaborated here.

$$-S_{ij}^C \le P_{ij}^{xy,SC} \le S_{ij}^C, \forall ij \in \Upsilon \tag{60}$$

$$-S_{ij}^C \le Q_{ij}^{xy,SC} \le S_{ij}^C, \forall ij \in \Upsilon \tag{61}$$

$$-\sqrt{2}S_{ij}^C \le P_{ij}^{xy,SC} \pm Q_{ij}^{xy,SC} \le \sqrt{2}S_{ij}^C, \forall ij \in \Upsilon \tag{62}$$

$$P_f^{xy,SC} = P_{tr^f}^{xy,SC}, \forall f \in \Psi^F, tr^f \in \Upsilon \tag{63}$$

$$Q_f^{xy,SC} = Q_{tr^f}^{xy,SC}, \forall f \in \Psi^F, tr^f \in \Upsilon \tag{64}$$

$$P_f^{xy,SC} \le S_f^C, \forall f \in \Psi^F \tag{65}$$

$$Q_f^{xy,SC} \le S_f^C, \forall f \in \Psi^F \tag{66}$$

$$\pm Q_f^{xy,SC} + P_f^{xy,SC} \le \sqrt{2}S_f^C, \forall f \in \Psi^F \tag{67}$$

2.6. Constraints on the Allocation of Circuit Breakers and Switches

In order to reduce the impact of switchable equipment errors on distribution network reliability and power supply continuity, the circuit breakers are generally placed in parallel with the switches in the distribution network, as is shown in (68) and (69).

$$x_{ij}^{i,SW} \geq 1 - x_{ij}^{i,CB} \tag{68}$$

$$x_{ij}^{j,SW} \geq 1 - x_{ij}^{j,CB} \tag{69}$$

2.7. Calculation of Reliability Indices

The indexes $p_i^{xy,SC}$ and $q_i^{xy,SC}$ concerning whether the load points receive power during different stages of the fault (obtained with the previous constraints) make it possible to linearize the representation of each of the reliability indices of the distribution network under different load level scenarios, as shown in (70)–(75). Among them, the calculation of the outage time index CID_i^{SC} of the load nodes in different scenarios consists of two parts in (70): one is the switching interruption fault time experienced by load points that do not get restored after the action of circuit breakers, and the other part is the time taken for the manual repair of the faults experienced by the load nodes that still cannot receive power after switch action for post-fault reconfigurations. However, Equations (71)–(75) are all common expressions of reliability indices.

$$CID_i^{SC} = \sum_{xy \in \Upsilon} \lambda_{xy} \tau_{xy}^{SW} p_i^{xy,SC} + \sum_{xy \in \Upsilon} \lambda_{xy} \left(\tau_{xy}^{RP} - \tau_{xy}^{SW} \right) \left(1 - q_i^{xy,SC} \right), \forall i \in \Psi^{LN} \tag{70}$$

$$CIF_i^{SC} = \sum_{xy \in \Upsilon} \lambda_{xy} p_i^{xy,SC}, \forall i \in \Psi^{LN} \tag{71}$$

$$SAIDI_{SC} = \frac{\sum_{i \in \Psi^{LN}} NC_i CID_i^{SC}}{\sum_{i \in \Psi^{LN}} NC_i} \tag{72}$$

$$SAIFI_{SC} = \frac{\sum_{i \in \Psi^{LN}} NC_i CIF_i^{SC}}{\sum_{i \in \Psi^{LN}} NC_i} \tag{73}$$

$$ASAI_{SC} = 1 - \frac{SAIDI_{SC}}{8760} \tag{74}$$

$$EENS_{SC} = \sum_{h \in H} \frac{\Delta h}{8760} \sum_{i \in \Psi^{LN}} CID_i^{SC} \mu_h L_i^{SC} \tag{75}$$

where λ_{xy} is the failure rate of branch xy, τ_{xy}^{SW} and τ_{xy}^{SW} represent the switching-only interruption duration and the repair and switch interruption duration of each branch, respectively.NC_i denotes the number of customers at each load node. Δh is the duration of load level h. μ_h represents the load factor of load level h. L_i^{SC} is the peak demand at node i under different load level scenarios.

After obtaining the reliability of the distribution network under each load level scenario, the final annual reliability indices of the distribution network need to be calculated based on the probability of occurrence of different scenarios, as is shown in (76)–(79).

$$SAIDI = \sum_{SC \in Scene} g_{SC} SAIDI_{SC} \tag{76}$$

$$SAIFI = \sum_{SC \in Scene} g_{SC} SAIFI_{SC} \tag{77}$$

$$ASAI = \sum_{SC \in Scene} g_{SC} ASAI_{SC} \tag{78}$$

$$EENS = \sum_{SC \in Scene} g_{SC} EENS_{SC} \tag{79}$$

where g_{SC} is the probability of various load level scenarios.

2.8. Constraints of Reliability Indices

The constraints on the reliability indices are necessary to balance the demand for reliability and economy regarding the distribution network and to ensure that the distribution network has a relatively high reliability level while improving the cost-effectiveness of planning. The reliability indices of the distribution network include *SAIDI, SAIFI, ASAI,* and *EENS*. Any constraint on any index can ensure the corresponding reliability, so in practice, any constraint from constraints (80)–(83) can be selected.

$$SAIDI \leq \varepsilon_{SAIDI} \tag{80}$$

$$SAIFI \leq \varepsilon_{SAIFI} \tag{81}$$

$$ASAI \leq \varepsilon_{ASAI} \tag{82}$$

$$EENS \leq \varepsilon_{EENS} \tag{83}$$

where ε_{SAIDI}, ε_{SAIFI}, ε_{ASAI}, and ε_{EENS} are the preset requirements for the reliability indices, respectively.

The entire optimal allocation model of circuit breakers and switches in distribution networks can be described as follows after specifying the objective functions and constraints:

$$\begin{cases} f = minimize\ (F_{CB} \sum_{ij} (x_{ij}^{i,CB} + x_{ij}^{j,CB}) + F_{SW} \sum_{ij} (x_{ij}^{i,SW} + x_{ij}^{j,SW}) + \alpha EENS) = \\ = minimize\ (\mathrm{Cost} + \alpha EENS) \\ subject\ \mathrm{to{:}}\ (2)\text{–}(83) \end{cases} \tag{84}$$

3. Numerical Test

The proposed model is validated on a 54-node distribution test system. This test system is a 1 MVA, 13.5 kV radial network with four substations, 50 load nodes, eight feeders, and 61 branches, for which the corresponding specific data can be obtained from [25]. The switching-only interruption duration for each branch is 0.5 h, and the repair-and-switching interruption duration is 3 h. The three load levels, with loading factors equal to 70% (2000 h/year), 83% (5760 h/year), and 100% (1000 h/year), of the associated peak demand are used to depict the loading condition [26]. The topological structure of the 54-node test system utilized in this paper, as well as the selection of circuit breaker and switch candidate locations, is shown in Figure 1.

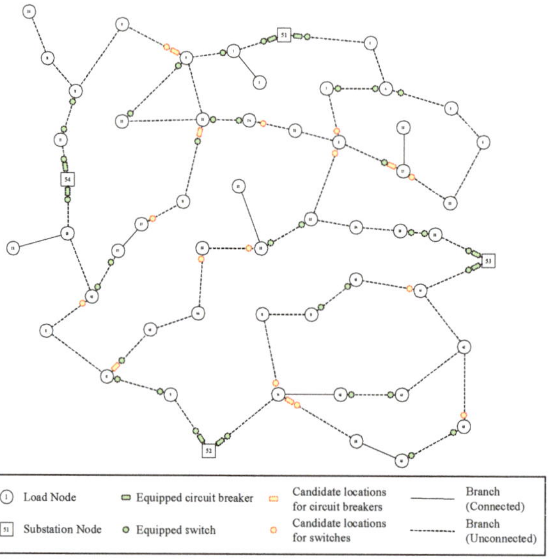

Figure 1. Topological structure of the 54-node distribution test system.

In order to validate the effectiveness of the proposed model in improving the reliability and economy of the distribution network by optimizing the allocation of circuit breakers and switches in the distribution network, this paper selects some of the branches of the system as the candidate locations for circuit breakers or switches and also sets a small number of already-equipped circuit breakers and switches in the system. In this paper, the reliability index *SAIDI* is constrained by the reliability requirement ε_{SAIDI}, set as 1.2 h/year. The outage loss per unit of power is set to RMB 30 yuan/MWh. The investment unit prices of the circuit breakers and switches are set at RMB 42,000 and RMB 15,000, respectively. The sum of outage loss and investment cost is to be minimized as the objective function, and the reliability of the distribution network under the corresponding allocation scheme is evaluated as well.

In a distribution network, the daily load level exhibits peak and off-peak variations, representing the fluctuations in electricity demand throughout the day. The peak and off-peak variations in load level have important implications for the management and operation of the power distribution network. Utilities are needed to ensure that they have enough capacity to meet the peak demand during the day while efficiently utilizing resources during the off-peak period. Therefore, four different load level scenarios, as shown in Figure 2, are set up to reflect the impact of peak and off-peak variations in load level, which are 0.5 times the load level, 1 times the load level, 2 times the load level, and 2.5 times the load level. Therefore, the adaptability of the optimal allocation of circuit breakers and switches to scenario changes, as well as the effectiveness of comprehensive optimization on the reliability and economy of the distribution system, could be optimized. The proposed method is modeled in MATLAB and was solved using the CPLEX solver regarding the optimal allocation scheme of circuit breakers and switches, which takes about 2 min to solve for the 54-node test system used in this paper.

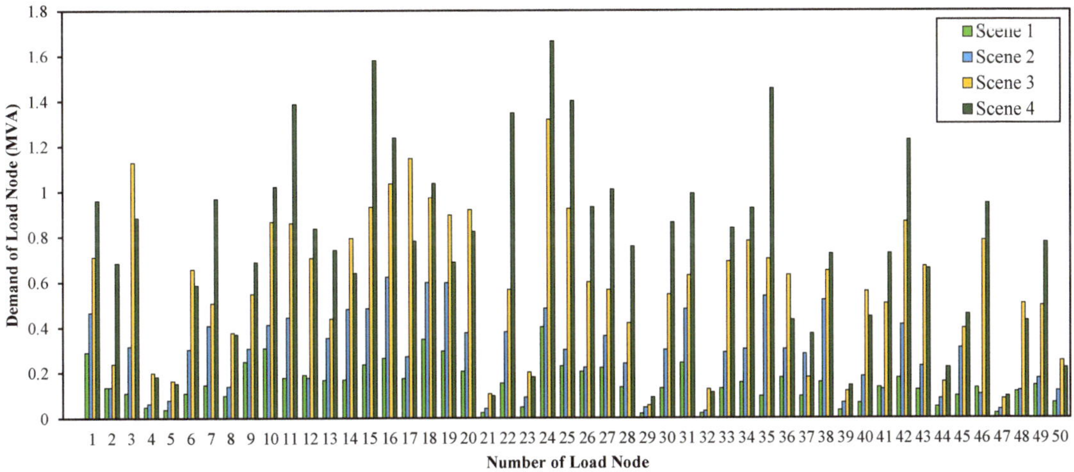

Figure 2. Demand of load nodes in 54-node test systems.

Figure 3 shows the results of the optimal allocation of circuit breakers and switches obtained by solving the model in this paper. Figure 4 illustrates the post-fault reconfiguration strategy of the distribution network under different load level scenarios when a fault occurs at branches 1–9. The reliability indices of each load node in the distribution network under different load level scenarios are shown in Table 1. Table 2 shows the reliability indices of the system under each load level scenario, as well as the overall reliability indices, the investment costs of circuit breakers and switches, and the overall outage losses of the system under the comprehensive consideration of load level variation.

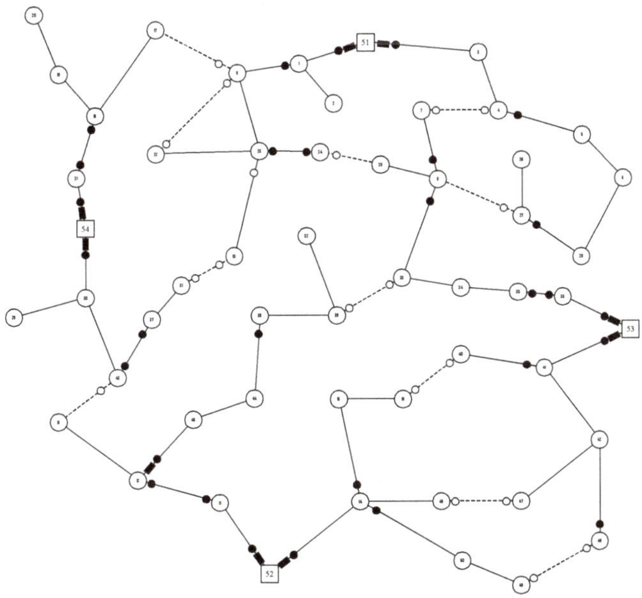

Figure 3. Results of the optimal allocation of circuit breakers and switches.

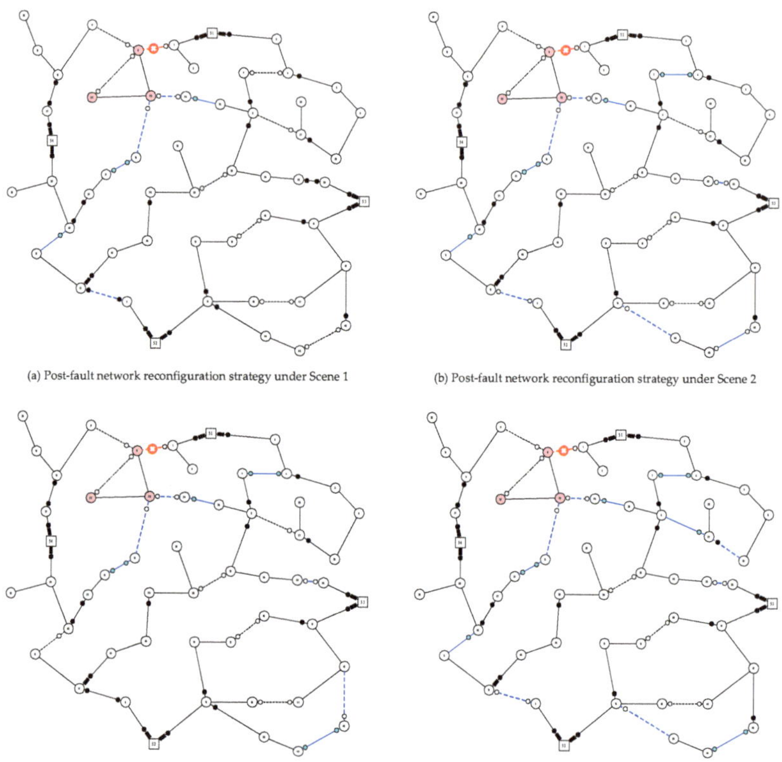

(a) Post-fault network reconfiguration strategy under Scene 1

(b) Post-fault network reconfiguration strategy under Scene 2

(c) Post-fault network reconfiguration strategy under Scene 3

(d) Post-fault network reconfiguration strategy under Scene 4

Figure 4. Post-fault reconfiguration strategies for distribution network when a fault occurred at branches 1-9 under different load level scenarios.

Table 1. Nodal CID and CIF for the 54-node test system.

Node	CID (h/Year)					CIF (Interruption/Year)				
	Scene 1	Scene 2	Scene 3	Scene 4	Integrated Index	Scene 1	Scene 2	Scene 3	Scene 4	Integrated Index
1	0.077	0.348	0.302	0.206	0.933	0.077	0.342	0.291	0.203	0.913
2	0.077	0.348	0.302	0.206	0.933	0.077	0.342	0.291	0.203	0.913
3	0.111	0.468	0.376	0.284	1.238	0.082	0.344	0.283	0.208	0.916
4	0.111	0.468	0.376	0.284	1.238	0.082	0.344	0.283	0.208	0.916
5	0.158	0.668	0.561	0.402	1.789	0.082	0.344	0.283	0.208	0.916
6	0.158	0.668	0.561	0.402	1.789	0.082	0.344	0.283	0.208	0.916
7	0.068	0.290	0.231	0.170	0.759	0.087	0.380	0.306	0.225	0.998
8	0.119	0.518	0.417	0.307	1.361	0.087	0.380	0.306	0.225	0.998
9	0.132	0.578	0.486	0.342	1.537	0.077	0.342	0.291	0.203	0.913
10	0.080	0.360	0.309	0.214	0.962	0.077	0.342	0.291	0.203	0.913
11	0.058	0.264	0.214	0.155	0.691	0.051	0.228	0.190	0.133	0.602
12	0.090	0.391	0.327	0.227	1.035	0.051	0.228	0.190	0.133	0.602

Table 1. *Cont.*

Node	CID (h/Year)					CIF (Interruption/Year)				
	Scene 1	Scene 2	Scene 3	Scene 4	Integrated Index	Scene 1	Scene 2	Scene 3	Scene 4	Integrated Index
13	0.090	0.391	0.327	0.227	1.035	0.051	0.228	0.190	0.133	0.602
14	0.123	0.540	0.420	0.318	1.402	0.087	0.376	0.300	0.223	0.986
15	0.107	0.460	0.383	0.275	1.224	0.087	0.376	0.300	0.223	0.986
16	0.107	0.460	0.383	0.275	1.224	0.087	0.376	0.300	0.223	0.986
17	0.145	0.641	0.544	0.382	1.712	0.061	0.266	0.226	0.158	0.711
18	0.145	0.641	0.544	0.382	1.712	0.061	0.266	0.226	0.158	0.711
19	0.145	0.641	0.544	0.382	1.712	0.061	0.266	0.226	0.158	0.711
20	0.145	0.641	0.544	0.382	1.712	0.061	0.266	0.226	0.158	0.711
21	0.045	0.191	0.164	0.114	0.513	0.061	0.266	0.226	0.158	0.711
22	0.132	0.578	0.486	0.342	1.537	0.077	0.342	0.291	0.203	0.913
23	0.132	0.578	0.486	0.342	1.537	0.077	0.342	0.291	0.203	0.913
24	0.039	0.171	0.146	0.101	0.457	0.077	0.342	0.291	0.203	0.913
25	0.119	0.518	0.417	0.307	1.361	0.087	0.380	0.306	0.225	0.998
26	0.058	0.240	0.194	0.146	0.638	0.082	0.344	0.283	0.208	0.916
27	0.058	0.240	0.194	0.146	0.638	0.082	0.344	0.283	0.208	0.916
28	0.158	0.668	0.561	0.402	1.789	0.082	0.344	0.283	0.208	0.916
29	0.108	0.468	0.391	0.284	1.251	0.047	0.205	0.170	0.124	0.546
30	0.108	0.468	0.391	0.284	1.251	0.047	0.205	0.170	0.124	0.546
31	0.033	0.147	0.124	0.089	0.393	0.047	0.205	0.170	0.124	0.546
32	0.060	0.271	0.221	0.158	0.711	0.057	0.251	0.204	0.148	0.660
33	0.122	0.539	0.443	0.320	1.424	0.087	0.380	0.306	0.225	0.998
34	0.122	0.539	0.443	0.320	1.424	0.087	0.380	0.306	0.225	0.998
35	0.122	0.539	0.443	0.320	1.424	0.087	0.380	0.306	0.225	0.998
36	0.072	0.318	0.251	0.189	0.830	0.087	0.380	0.306	0.225	0.998
37	0.033	0.147	0.124	0.089	0.393	0.047	0.205	0.170	0.124	0.546
38	0.057	0.244	0.195	0.145	0.641	0.057	0.251	0.204	0.148	0.660
39	0.060	0.271	0.221	0.158	0.711	0.057	0.251	0.204	0.148	0.660
40	0.070	0.304	0.243	0.186	0.803	0.077	0.329	0.274	0.201	0.881
41	0.199	0.849	0.716	0.517	2.280	0.077	0.329	0.274	0.201	0.881
42	0.199	0.849	0.716	0.517	2.280	0.077	0.329	0.274	0.201	0.881
43	0.108	0.468	0.391	0.284	1.251	0.047	0.205	0.170	0.124	0.546
44	0.112	0.487	0.400	0.289	1.288	0.057	0.251	0.204	0.148	0.660
45	0.112	0.487	0.400	0.289	1.288	0.057	0.251	0.204	0.148	0.660
46	0.123	0.540	0.420	0.318	1.402	0.087	0.376	0.300	0.223	0.986
47	0.199	0.849	0.716	0.517	2.280	0.077	0.329	0.274	0.201	0.881
48	0.038	0.165	0.137	0.100	0.440	0.077	0.329	0.274	0.201	0.881
49	0.118	0.505	0.397	0.297	1.317	0.087	0.376	0.300	0.223	0.986
50	0.118	0.505	0.397	0.297	1.317	0.087	0.376	0.300	0.223	0.986

Table 2. Reliability indices and economic indices of the 54-node test system.

	SAIFI (1/Year)	SAIDI (h/Year)	EENS (MWh/Year)	ASAI (%)	Investment (Yuan)	Cost (Yuan)
Scene 1	0.0714	0.1040	0.7640	-	-	-
Scene 2	0.3103	0.4527	6.3444	-	-	-
Scene 3	0.2553	0.3730	9.9622	-	-	-
Scene 4	0.1850	0.2697	9.8624	-	-	-
Integrated Indices	0.8219	1.1994	26.9330	99.9863%	1,113,000	1,113,807.989

Under different load level scenarios, the optimal allocation model proposed in this paper can optimize the post-fault distribution network reconfiguration strategy according to the requirements of the required reliability level, as can be seen in Figure 4. The reconfiguration strategy under different load levels is optimized to minimize the outage loss in order to maximize the reliability of the distribution system. For fault scenarios under different load levels, the proposed model can form an optimal post-fault reconfiguration scheme for that load level scenario, which has strong adaptability and practicality and can effectively ensure the level of reliability of the distribution network.

As can be seen from Table 1, different load levels affect the nodal reliability index. From Figure 2, it can be seen that the load levels of scenes 1–4 are distributed from low to high. The higher the load level, the higher the frequency; the longer the duration of the outages at the load nodes, the worse the reliability of the indices of the nodes in the distribution network is. When the load level is low, i.e., the demand for power supply is low, the balance of supply and demand in the distribution network is easier to maintain, and the possibility of outage is relatively low. When the load level is high, i.e., when the customer's demand for electricity is relatively high, the time of electricity consumption is greater, or the load is unstable, then the reliability indices *CID* and *CIF* of the distribution network are usually affected to a certain extent. This is because, at this time, the distribution network faces a power load that may exceed the capacity of the design, leading to problems such as overload, excessive current, and aging equipment, which, in turn, can lead to breakdowns and may also lead to problems such as insufficient power supply.

As can be seen from Table 2, the proposed method can effectively maximize the economic benefits while ensuring that the distribution network reliability level meets the requirements. The optimal allocation model of circuit breakers and switches established in this paper takes into account both the economic indices represented by the equipment investment costs and outage losses, and the reliability index represented by EENS of the distribution network. The proposed model can effectively ensure the required reliability level while optimizing the allocation of circuit breakers and switches in the distribution network at the lowest investment cost, effectively reducing the outage losses in the distribution network by optimizing the post-fault reconfiguration scheme under various load levels and successfully achieving a balance between the optimization of reliability and economy.

In summary, the method proposed in this paper can better solve the optimal placement of circuit breakers and switches in the actual distribution network. The model uses faster calculation speed and better calculation quality to obtain an optimal allocation scheme of circuit breakers and switches that ensures the reliability of the distribution network to meet demand at the lowest investment cost, which achieves a balance between the optimization of reliability and economy under the objective function. Furthermore, since changes in load are considered in the model, the model developed in this paper can adapt to the uncertain changes in loads in the distribution network in practical applications, meaning the obtained allocation scheme has stronger adaptability and practicality.

4. Conclusions

The number and location of circuit breakers and switches in a distribution network may have a significant impact on the reliability of the distribution network. When the investment budget is low, the number of switchable devices equipped in the network is usually small, which cannot guarantee satisfactory reliability levels in the distribution network. However, in practice, increasing the investment cost means the utility faces more financial risks, which, to some extent, limits the enhancement of the reliability of the distribution network. Simultaneously, the peak and valley fluctuations in the load levels can also have an impact on the level of reliability of the distribution network. Hence, it is essential to study optimal allocation methods for the placement of circuit breakers and switches in distribution networks and consider the relevant economic and reliability requirements by exploring the establishment of a solution to the cost–reliability dilemma and searching for the best allocation solution that can adapt to load level variations in the distribution network. This will reduce the negative impact of load level variations on the reliability of the distribution system.

This paper presents a linear programming model for the optimal allocation of circuit breakers and switches based on reliability and economic improvements to distribution networks with consideration given to load level variation. In this context, the sum of outage losses and the investment costs of switchable equipment is minimized as the objective function, and the reliability index of the distribution network is taken as the constraint to ensure that the distribution network meets reliability demands while minimizing the investment cost of equipment so as to achieve a balance between the optimization of reliability and the economic indices. In addition, the proposed model takes into account the impact of load level variation on the reliability of the distribution network and considers various load level scenarios to obtain the optimal allocation scheme of switchable equipment that can also adapt to different load fluctuation scenarios as well as target post-fault reconfiguration schemes under different scenarios. By using numerical examples, we verified that the optimal allocation scheme obtained by the proposed model could reduce investment costs and outage loss while ensuring the reliability index of the system meets the requirements. The method provides a practical reference for distribution network planners to carry out distribution network optimization design, and it has strong engineering research significance for optimizing distribution network structures. Last but not least, the linearization of the optimal allocation model was achieved through the concept of fictitious fault flow, which ensured the global optimality of the derived results and also enhanced the computational efficiency and speed of the model in this paper, improving the practicality and scalability of the model.

Author Contributions: Conceptualization, G.H., Y.Z., C.Y. and Q.Z.; methodology, G.H., Y.Z., C.Y. and Q.Z.; software, G.H., Y.Z., C.Y., Q.Z., L.Z., X.D., J.L. and J.Z.; validation, G.H., Y.Z., C.Y., Q.Z., L.Z., X.D., J.L. and J.Z.; formal analysis, G.H., Y.Z., C.Y., Q.Z. and L.Z.; investigation, C.Y., Q.Z., L.Z. and X.D.; resources, J.L. and J.Z.; data curation, G.H., Y.Z. and L.Z.; writing—original draft preparation, G.H. and Y.Z.; writing—review and editing, G.H., Y.Z. and C.Y.; project administration, G.H. and Q.Z; funding acquisition, G.H. and J.Z. All authors have read and agreed to the published version of the manuscript.

Funding: This research was funded by the Science and Technology Project of State Grid Jiangsu Electric Power Co., Ltd. (Grant no. J2022151).

Data Availability Statement: The data presented in this study are available upon request from the corresponding author. The data are not publicly available due to the project not being completed.

Conflicts of Interest: The authors declare no conflict of interest.

References

1. Willis, H.L. *Power Distribution Planning Reference Book*, 2nd ed.; Marcel Dekker: New York, NY, USA, 2004.
2. Billinton, R.; Allan, R.N. *Reliability Evaluation of Power Systems*, 2nd ed.; Plenum: New York, NY, USA, 1996.
3. Lu, Z.; Li, Z.; Guo, X.; Yang, B. Optimal Planning of Hybrid Electricity–Hydrogen Energy Storage System Considering Demand Response. *Processes* **2023**, *11*, 852. [CrossRef]
4. Zhou, C.; Gui, S.; Liu, Y.; Ma, J.; Wang, H. Fault Location of Distribution Network Based on Back Propagation Neural Network Optimization Algorithm. *Processes* **2023**, *11*, 1947. [CrossRef]

5. Tu, N.; Fan, Z. IMODBO for Optimal Dynamic Reconfiguration in Active Distribution Networks. *Processes* **2023**, *11*, 1827. [CrossRef]
6. Zare-Bahramabadi, M.; Ehsan, M.; Farzin, H. An MILP Model for Switch, DG, and Tie Line Placement to Improve Distribution Grid Reliability. *IEEE Syst. J.* **2022**, *17*, 1316–1327. [CrossRef]
7. Carvalho, P.; Ferreira, L.; CerejodaSilva, A. A Decomposition Approach to Optimal Remote Controlled Switch Allocation in Distribution Systems. *IEEE Trans. Power Deliv.* **2005**, *20*, 1031–1036. [CrossRef]
8. Abraham, A.; Das, S. *Computational Intelligence in Power Engineering*, 1st ed.; Springer Publishing Company, Incorporated: Berlin/Heidelberg, Germany, 2010.
9. Teng, J.-H.; Liu, Y.-H. A novel ACS-based optimum switch relocation method. *IEEE Trans. Power Syst.* **2003**, *18*, 113–120. [CrossRef]
10. Levitin, G.; Mazal-Tov, S.; Elmakis, D. Optimal sectionalizer allocation in electric distribution systems by genetic algorithm. *Electr. Power Syst. Res.* **1994**, *31*, 97–102. [CrossRef]
11. Bernardon, D.P.; Sperandio, M.; Garcia, V.J.; Canha, L.N.; Abaide, A.D.R.; Daza, E.F.B. AHP Decision-Making Algorithm to Allocate Remotely Controlled Switches in Distribution Networks. *IEEE Trans. Power Deliv.* **2011**, *26*, 1884–1892. [CrossRef]
12. Ray, S.; Bhattacharya, A.; Bhattacharjee, S. Optimal Placement of Switches in a Radial Distribution Network for Reliability Improvement. *Int. J. Electr. Power Energy Syst.* **2016**, *76*, 53–68. [CrossRef]
13. Chen, C.-S.; Lin, C.-H.; Chuang, H.-J.; Li, C.-S.; Huang, M.-Y.; Huang, C.-W. Optimal Placement of Line Switches for Distribution Automation Systems Using Immune Algorithm. *IEEE Trans. Power Syst.* **2006**, *21*, 1209–1217. [CrossRef]
14. Falaghi, H.; Haghifam, M.-R.; Singh, C. Ant Colony Optimization-Based Method for Placement of Sectionalizing Switches in Distribution Networks Using a Fuzzy Multiobjective Approach. *IEEE Trans. Power Deliv.* **2009**, *24*, 268–276. [CrossRef]
15. Alves, H.N. A hybrid algorithm for optimal placement of switches devices in electric distribution systems. *IEEE Lat. Am. Trans.* **2012**, *10*, 2218–2223. [CrossRef]
16. Cao, W.; Li, Z.; Li, Y. Optimisation Configuration of Overhead-Line Segmentation Switch of a Distribution Network Based on Global Combination Criterion. *Processes* **2022**, *10*, 1976. [CrossRef]
17. Abiri-Jahromi, A.; Fotuhi-Firuzabad, M.; Parvania, M.; Mosleh, M. Optimized Sectionalizing Switch Placement Strategy in Distribution Systems. *IEEE Trans. Power Deliv.* **2011**, *27*, 362–370. [CrossRef]
18. Jooshaki, M.; Karimi-Arpanahi, S.; Lehtonen, M.; Millar, R.J.; Fotuhi-Firuzabad, M. An MILP Model for Optimal Placement of Sectionalizing Switches and Tie Lines in Distribution Networks With Complex Topologies. *IEEE Trans. Smart Grid* **2021**, *12*, 4740–4751. [CrossRef]
19. Li, Z.; Wu, W.; Tai, X.; Zhang, B. Optimization Model-Based Reliability Assessment for Distribution Networks Considering Detailed Placement of Circuit Breakers and Switches. *IEEE Trans. Power Syst.* **2020**, *35*, 3991–4004. [CrossRef]
20. Gan, L.; Li, N.; Topcu, U.; Low, S.H. Exact Convex Relaxation of Optimal Power Flow in Radial Networks. *IEEE Trans. Autom. Control* **2014**, *60*, 72–87. [CrossRef]
21. Chen, X.; Wu, W.; Zhang, B.; Lin, C. Data-Driven DG Capacity Assessment Method for Active Distribution Networks. *IEEE Trans. Power Syst.* **2016**, *32*, 3946–3957. [CrossRef]
22. Islam, F.R.; Prakash, K.; Mamun, K.A.; Lallu, A.; Pota, H.R. Aromatic Network: A Novel Structure for Power Distribution System. *IEEE Access* **2017**, *5*, 25236–25257. [CrossRef]
23. Gandioli, C.; Alvarez-Hérault, M.-C.; Tixador, P.; Hadjsaid, N.; Medina, D.-M.R. Innovative Distribution Networks Planning Integrating Superconducting Fault Current Limiters. *IEEE Trans. Appl. Supercond.* **2013**, *23*, 5603904. [CrossRef]
24. Kumar, D.S.; Srinivasan, D.; Sharma, A.; Reindl, T. Adaptive directional overcurrent relaying scheme for meshed distribution networks. *IET Gener. Transm. Distrib.* **2018**, *12*, 3212–3220. [CrossRef]
25. Li, Z.; Wu, W.; Zhang, B.; Tai, X. Test Bench of Optimization Model-based Reliability Assessment for Distribution Networks Considering Detailed Placement of Circuit Breakers and Switches. 2019. Available online: https://drive.google.com/open?id=1HlsK9BThu5UIIMCV06VeVHK_zketBE-u (accessed on 4 September 2019).
26. Li, Z.; Wu, W.; Zhang, B.; Tai, X. Analytical Reliability Assessment Method for Complex Distribution Networks Considering Post-Fault Network Reconfiguration. *IEEE Trans. Power Syst.* **2019**, *35*, 1457–1467. [CrossRef]

processes

MDPI

Article

Thermoeconomic Modeling as a Tool for Internalizing Carbon Credits into Multiproduct System Analysis

José Joaquim C. S. Santos [1], Pedro Rosseto de Faria [1,2,*], Igor Chaves Belisario [3], Rodrigo Guedes dos Santos [1,3] and Marcelo Aiolfi Barone [1]

[1] Department of Mechanical Engineering, Federal University of Espírito Santo (UFES), Vitória 29075-910, Brazil; jjcssantos@yahoo.com.br (J.J.C.S.S.); rodrigo.guedes@ifes.edu.br (R.G.d.S.); mabacz@gmail.com (M.A.B.)
[2] Department of Railways, Federal Institute of Espírito Santo (IFES), Cariacica 29150-410, Brazil
[3] Department of Mechanical Engineering, Federal Institute of Espírito Santo (IFES), Vitória 29040-780, Brazil; igor.belisario@ifes.edu.br
* Correspondence: pedro.faria@ifes.edu.br

Abstract: In the context of emissions, carbon dioxide constitutes a predominant portion of greenhouse gases (GHGs), leading to the use of the term "carbon" interchangeably with these gases in climate-related discussions. The carbon market has emerged as a pivotal mechanism for emission regulation, allowing industries that struggle to meet emission reduction targets to acquire credits from those who have successfully curbed their emissions below stipulated levels. Thermoeconomics serves as a tool for analyzing multiproduct systems prevalent in diverse sectors, including sugarcane and alcohol mills, paper and pulp industries, steel mills, and cogeneration plants. These systems necessitate frameworks for equitable cost/emission allocation. This study is motivated by the need to expand the scope of thermoeconomic modeling to encompass expenses or revenues linked with the carbon market. By utilizing a cogeneration system as a representative case, this research aims to demonstrate how such modeling can facilitate the allocation of carbon market costs to final products. Moreover, it underscores the adaptability of this approach for internalizing other pertinent costs, encompassing expenses associated with environmental control devices, licenses, and permits. Although certain exergy disaggregation models depict an environmental component within diagrams, which is integral for addressing environmental burdens, even models without explicit environmental devices can effectively internalize carbon credits and allocate them to final products. The integration of carbon credits within thermoeconomic modeling introduces the capability to assess both the financial and environmental implications of emissions. This integration further incentivizes the reduction in GHGs and supports optimization endeavors concerning system design and operation. In summary, this study delves into the incorporation of carbon market dynamics into thermoeconomic modeling. It demonstrates the potential to allocate carbon-related costs, facilitates comprehensive cost analysis, encourages emission reduction, and provides a platform for enhancing system efficiency across industrial sectors.

Keywords: thermoeconomic modeling; carbon credit; carbon market; environmental cost; cost allocation; multiproduct system

Citation: Santos, J.J.C.S.; Faria, P.R.d.; Chaves Belisario, I.; dos Santos, R.G.; Barone, M.A. Thermoeconomic Modeling as a Tool for Internalizing Carbon Credits into Multiproduct System Analysis. *Processes* **2024**, *12*, 705. https://doi.org/10.3390/pr12040705

Academic Editors: Chenyu Wu, Zhongkai Yi and Chenhui Lin

Received: 7 August 2023
Revised: 23 August 2023
Accepted: 31 August 2023
Published: 30 March 2024

1. Introduction

According to the Intergovernmental Panel on Climate Change (IPCC) [1], greenhouse gas emissions (GHG) have shown an increase since 1990 worldwide. The largest contributors to these emissions in 2018 were the combination of electricity and heat generation (cogeneration) and transportation, accounting for over two-thirds of the total [2]. Given that CO_2 constitutes the predominant portion of GHGs and is linked to global warming, the term "carbon" has been adopted in climate discussions to encompass these gases.

To regulate emissions, the carbon market offers industries and sectors that are unable to meet emission reduction targets the opportunity to purchase credits from those who have successfully reduced their emissions below the required levels. Conventionally, one carbon credit is equivalent to one ton of carbon dioxide. Consequently, it can be considered a valuable asset, both financially and environmentally, representing the reduction or removal of one ton of CO_2 equivalent. These credits are recognized and issued within the carbon market, irrespective of whether they are obtained voluntarily or through regulation [3].

This market is subject to regulation in certain jurisdictions, such as the European Union, where well-defined credit values have been established [4]. However, in many other countries, such as Brazil, the market remains voluntary. According to the World Bank's report in 2022 [5] and the IPCC in 2023 [6], the carbon market, which is associated with environmental preservation measures, is experiencing global expansion. Nonetheless, it has yet to reach the necessary levels to effectively address environmental challenges and meet the objectives set forth in the Paris Agreement to combat climate change.

Thermoeconomics is an interdisciplinary field that combines principles from thermodynamics and economics to provide insights not available through conventional energetic and economic analyses. The information derived from thermoeconomics is essential for the design and operation of thermal systems [7]. Initially, the primary objective of thermoeconomics was to mathematically integrate the Second Law of Thermodynamics with economic principles. However, contemporary analyses must also incorporate environmental considerations [8]. In this context, exergy emerges as the most appropriate thermodynamic property to utilize, as it accounts for the quality of energy and allows for the identification and quantification of irreversibilities in processes [7]. Furthermore, exergy serves as a crucial link between the Second Law and the assessment of environmental impact, as it measures a system's deviation from its equilibrium state with the environment [9].

Illustrative cases, including sugarcane and alcohol mills [10], paper and pulp industries [11], steel mills [12], and cogeneration plants, exemplify instances of multiproduct systems that require established guidelines for the allocation of costs and emissions from the fuel source to the final products. In the case of cogeneration, which generates useful heat and power simultaneously from a single combustible source, rational criteria for allocating the cost/emission of the fuel among the various final products are required. In such scenarios, thermoeconomics enables a rational allocation (based on physical criteria) of monetary, exergetic, and environmental costs for these final products. Consequently, a comparison becomes feasible between the exergetic/monetary/environmental costs of each product and the production cost of each individual product in separate systems, as shown in papers assessing the exergetic unit cost (in a regenerative gas turbine cogeneration system [13]; in a cogeneration system with gas turbine, intercooler, and supplementary firing [14]; and in a combined cycle [15]), monetary unit cost (in a gas turbine cogeneration system [16] and in a power generation system of a steel mill plant [17]), and emissions pollutant allocation (in gas and steam cogeneration systems [18], in a gas cogeneration system with supplementary firing [19], in a combined cycle [20], and in a dual product heat pump [21]). Despite the practical relevance of this aspect, in the literature review, no studies were found that dealt with the incorporation and internalization of the monetary unit cost linked to the carbon market in the thermoeconomic evaluations of multi-product plants.

Thermoeconomic methodologies have been previously applied to incorporate environmental factors, such as specific CO_2 emissions. In this paper, thermoeconomic methods are applied to internalize and allocate monetary costs associated with environmental concerns, such as carbon credits. Thus, the novelty introduced in this study is to demonstrate how this internalization can be accomplished using thermoeconomic principles.

Moreover, thermoeconomics plays a fundamental role in the analysis of energy conversion systems. This study aims to elucidate the application of thermoeconomic modeling as a valuable tool for incorporating expenses or revenues related to the carbon market into thermal systems analyses and allocating them to the system's internal and final products. The conventional models employed to compute the monetary costs of internal flows and final products can be suitably adapted to account for these environmental costs. To achieve this, the adaptation is explicated using matrix notation and demonstrated through a case study of a gas turbine cogeneration system. The study also illustrates how this inclusion of environmental costs can impact the monetary evaluation of the system's final products. Furthermore, it highlights the potential of this modeling approach to internalize other costs, such as those associated with environmental control devices, environmental licenses, and permits.

It is crucial to underscore that a methodology utilized here to exemplify and expound the internalization of environmental costs, particularly carbon credits, introduces a definition of the environmental device in thermoeconomic diagrams, precisely allocating environmental costs to this environmental representation. The H&S Model is adopted as the methodological framework for this purpose; nonetheless, any other exergy-based thermoeconomic methodology that consistently defines such an environmental device can be employed to conduct similar analyses. Even models that do not explicitly designate this device to represent the environment can internalize carbon credits (as is the case with the E Model utilized in this paper). However, in this instance, environmental burdens are not internalized in the environmental device. In other words, this study aims to demonstrate how to adapt any thermoeconomic model to incorporate carbon credits and allocate them to the final products.

Certainly, an aspect that could be incorporated into this study to facilitate the development and execution of future research is the inclusion of real-world data from industries, which will permit greater accuracy in the simulations. However, according to a group of recognized experts in this field [22], the beauty of a theory is usually shown in the simplicity of its forms and the generality of its message, but its power resides in its capacity to solve practical cases. Thus, a simple gas turbine cogeneration plant was used for the illustration of the method application.

Future investigations should encompass various cogeneration systems. The range of industry types, each with its own operational complexities, provides a diverse array of insights into the adaptability and effectiveness of the modeling here proposed. Exploring different types of cogeneration systems can enhance the understanding of the pros and cons of incorporating carbon market factors into thermoeconomic analyses. These advancements hold the potential to enrich scholarly inquiry while driving practical industrial progress.

2. Thermoeconomic Modeling

Furthermore, apart from employing traditional modeling techniques to ascertain the monetary and exergetic unit costs of the system's internal flows and final products, this section demonstrates the general adaptations made to the modeling process for the allocation of specific pollutant emissions. Additionally, it elaborates on the integration of carbon credits in thermoeconomic modeling.

2.1. Conventional Modeling

The purpose of thermoeconomic modeling is to derive a system of cost equations that mathematically represents the cost formation process, i.e., the process of allocating external resources until the final cost of products is established.

Costs can be deemed satisfactory if they belong to a viable region of solutions for a given problem, and the procedure for cost validation must be founded on the plant's behavior and thermodynamics, as this irreversibility constitutes the cost-generating magnitude [7].

Equations (1) and (2) are utilized to ascertain the monetary (c) and exergetic (k^*) unit costs, respectively, of the internal flows and the final products within the systems. The allocation of specific pollutant emissions, such as CO_2, NOx, and SOx, can be performed using Equation (3). In these equations, the subscripts "out" and "in" denote the outputs and inputs of flows, respectively. The variable Y represents a general thermodynamic magnitude that can be assessed by power, heat, exergy flows, or its components. E_F denotes the exergy of the external fuel, while c_F and k_F^* represent its monetary and exergetic unit costs, respectively. Furthermore, λ_F signifies the amount of emission generated due to the combustion of one unit of exergy from the external fuel. It is customary for Z to represent the external hourly cost of the subsystem, accounting for capital and equipment operation and maintenance.

$$\sum(c_{out} \cdot Y_{out}) - \sum(c_{in} \cdot Y_{in}) = Z + c_F \cdot E_F \quad (1)$$

$$\sum(k^*_{out} \cdot Y_{out}) - \sum(k^*_{in} \cdot Y_{in}) = k^*_F \cdot E_F \quad (2)$$

$$\sum(\lambda_{out} \cdot Y_{out}) - \sum(\lambda_{in} \cdot Y_{in}) = \lambda_F \cdot E_F \quad (3)$$

Equation (2) is derived from Equation (1), where the Z term has to be zero. The exergy unit cost of the external fuel (k^*_F) is typically assumed to be equal to its exergy value, resulting in an exergy unit cost of 1 kW/kW [7]. Both the monetary and exergetic unit costs serve as measures of economic and thermodynamic efficiency, respectively, for a flow production process [7]. Conversely, the balance depicted by Equation (3) can be interpreted as a measure of environmental efficiency for this flow production process [19].

In all cases involving Equations (1)–(3), auxiliary equations are generally required to complete the modeling equation system. The formulation of these auxiliary equations is based on the chosen thermoeconomic diagram. In the context of productive diagrams, the equality criterion [23] is applied. According to this criterion, all products of a subsystem share the same unit cost as they are generated within the same productive process with identical irreversibilities.

2.2. Inclusion of Monetary Costs of Environmental Charges

The utilization of Equation (3) in the allocation of specific emissions to internal flows and thermal systems' final products represents an analysis that incorporates environmental considerations within thermoeconomic modeling. However, this approach lacks the inclusion of monetary expenses related to environmental factors, such as carbon credits and the cost of equipment for environmental treatment/control.

To address this limitation, variable Z (as defined in Equation (1)) plays a crucial role in allocating environmental costs. In a conventional monetary cost evaluation, Z denotes the subsystem's external hourly rate concerning capital, operation, and maintenance expenses. Nonetheless, it can also serve as a means to distribute environmental costs by introducing an environmental device into the thermoeconomic diagrams. An energy conversion system encompasses a collection of interconnected components that interact with each other and with the environment through flows of matter, work, or heat [24]. Consequently, the environment is considered an integral part of the system, and certain models propose its representation through an environmental device in thermoeconomic diagrams.

Equation (4) exemplifies how a conventional thermoeconomic model based on the monetary unit cost (Equation (1)) can be adapted to decompose the term Z into hourly costs associated with environmental charges (Z_{env}) and capital, operation, and maintenance expenses (O&M).

$$Z = Z_{cap} + Z_{O\&M} + Z_{env} \quad (4)$$

The environmental device presents no upfront acquisition cost; however, in some thermoeconomic methodologies, it serves as a mechanism for internalizing and redistributing environmental charges to other equipment and final products. For example, when waste control devices such as an electrostatic precipitator for ash disposal in flue gas or a bag filter for air pollution control (or any other equipment designed to mitigate environmental impacts by reducing GHG emissions into the atmosphere) are installed in a plant, its costs related to their capital and operation and maintenance (O&M) can be attributed to either the environmental device itself.

The same principle applies to devices used in carbon capture and storage, expenses related to environmental permits and licensing, fines incurred for emitting pollutants, and any other abatement costs (e.g., resources employed in waste treatment or disposal). Thus, it becomes possible to precisely allocate environmental charges to the device represented in the diagram as the environment, given that term Z is consistently associated with specific equipment. Consequently, a viable option is to link the environmental cost directly to the device symbolizing the environment in the diagrams.

Inclusion of Carbon Credits

In addition to the financial costs discussed in the preceding section, this paper proposes that the environmental aspect can be incorporated into the analysis by considering carbon pricing and internalizing the associated expenses or revenues from carbon credits. In this context, variable Z_{env} may exhibit positive or negative values. A negative value indicates a revenue stream generated by emission reduction or removal, resulting in the availability of credits for sale. Conversely, a positive value suggests an additional cost incurred by a facility that failed to achieve emission reduction targets, leading to the need to purchase carbon credits from those entities that have successfully reduced their emissions below the prescribed levels. An industry may also elect to procure carbon credits on account of environmental conscientiousness and as an investment strategy to align with sustainability concerns, with the aim of contributing to planetary decarbonization efforts.

A comprehensive exposition of thermoeconomic modeling, incorporating the dynamics of the carbon market, is presented in Section 3.

3. Case Study—Gas Turbine Cogeneration System

The cogeneration system with a simple gas turbine (Figure 1) is selected as an example to demonstrate how thermoeconomic modeling can effectively incorporate carbon credits. This system comprises four main components: an air compressor (AC), a combustion chamber (CC), a gas turbine (GT), and a recovery boiler (RB). The turbine generates power, part of which is utilized to drive the air compressor (W_{AC}). The system produces two final products: net power (W_N) and useful heat (Q_U). The fuel consumption is represented by (Q_F).

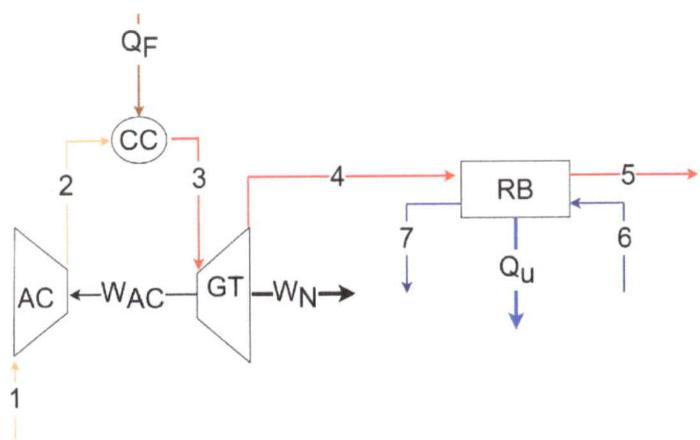

Figure 1. The physical structure: cogeneration system.

The main flow parameters of the physical structure, obtained through the Engineering Equation Solver—EES software [25], are presented in Table 1. Additionally, Table 2 provides the quantities of the primary productive flows. The reference conditions are specified as $T_0 = 25$ °C and $P_0 = 1.0132$ bar. Under these conditions, the mass flow of CO_2 from the exhaust gases is $\dot{m}_{CO_2} = 2228$ kg/h, considering natural gas as the fuel. Further details regarding this system can be found in [26]. The monetary unit cost of natural gas fuel is USD 24.04/MWh, which is based on the average value for the year 2022 in the international market [27].

Table 1. Main physical flow parameters of the system.

Physical Flow		$\dot{m}$ (kg/s)	T (°C)	P (bar)
No.	**Description**			
1	Air	14.72	25.00	1.0132
2	Air	14.72	230.20	5.1040
3	Gases	14.94	850.00	4.8480
4	Gases	14.94	537.30	1.0207
5	Gases	14.94	151.10	1.0132
6	Water	2.487	60.00	20.400
7	Steam	2.487	212.4	20.000

Table 2. Productive flows (exergetic basis).

Equipment	Flow	Quantity (kW)
Air compressor (AC)	W_{AC}	3113.03
Combustion chamber (CC)	Q_F	11,630.96
Gas turbine (GT)	W_{GT} W_N	5546.50 2433.47
Recovery boiler (RB)	Q_U	2246.32

In Table 3, the external monetary flows resulting from the cycle's equipment are presented. These values were sourced from [26] and updated using the Chemical Engineering Cost Index (CEPCI) up until the year 2022 [28]. The cost of the carbon credit utilized in the analysis amounts to USD 85/ton, representing the average for the year 2022 as reported in [4].

Table 3. Equipment external monetary cost.

Equipment	Z (USD/h)
Air compressor (AC)	25.33
Combustion chamber (CC)	9.04
Gas turbine (GT)	34.37
Recovery boiler (RB)	21.71

Thermoeconomic modeling employs various types of diagrams: physical, productive, and comprehensive. While the physical diagram alone may not suffice to identify the waste cost formation process [26], the productive diagram is commonly utilized in most methodologies. One distinguishing aspect of functional methodologies like Thermoeconomic Functional Analysis (TFA) [23] and Engineering Functional Analysis (EFA) [8] is their ability to describe the cost formation process of thermal systems based on productive flows. This original feature has been adopted not only by TFA and EFA but also by other thermoeconomic methodologies, including the H&S Model [26] used in this paper.

3.1. Thermoeconomic Models

Thermoeconomic modeling employs the widely recognized E Model, which utilizes the total exergy flows to represent the physical and/or productive flows in diagrams. However, there are scenarios where it becomes necessary to decompose the exergy into its components. This is particularly important to isolate dissipative equipment and allocate waste costs accurately in thermal systems. An exergy disaggregation model known as the H&S Model [26] is available for this purpose in some cases. The H&S Model analyzes the behavior of thermodynamic cycles on the h–s plane, considering the variations in enthalpy and entropy of the working fluid, as proposed by [29]. This model allows for the separation of physical exergy into its enthalpic component (E^H) and its entropic component (E^S), as described in Equation (5).

The total exergy (E) can be mathematically defined by Equation (6) as the sum of the physical (E^{PH}) and chemical (E^{CH}) components. Notably, this definition excludes nuclear, magnetic, electrical, surface tension, kinetic, and potential effects [30].

$$E^{PH} = E^H - E^S \quad (5)$$

$$E = E^H - E^S + E^{CH} \quad (6)$$

The H&S Model introduces the concept of the environmental device (ENV) within the framework of the productive diagram, facilitating its interaction with other plant subsystems. This device assumes a critical role, in this methodology, in analyzing thermal systems, particularly in the context of waste cost allocation and internalization of environmental costs. Within ENV, both the physical component and the chemical component of the waste are dissipated with the device receiving air from the compressor inlet. The chemical component originates in the CC as a result of the combustion reaction, wherein the air and fuel mixture is transformed into combustion gases. It should be noted that the E Model does not include an explicit representation of the environment in the diagram. Nevertheless, this study also presents an approach for the integration and allocation of carbon credits into this methodology, in order to establish a comparison with the one that defines the environmental device with the same purpose and to demonstrate that it is also possible to adapt any thermoeconomic model for this objective.

Moreover, the environmental device (as employed in the H&S Model) assumes the responsibility of closing the cycle (Figure 2), thereby redistributing the costs associated with waste management across other plant components and, subsequently, to the final products.

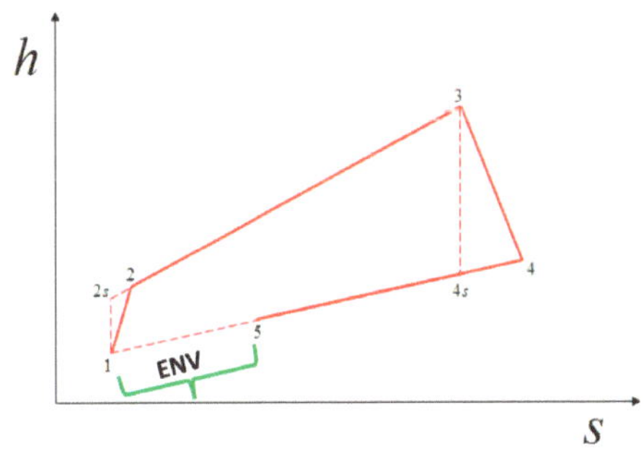

Figure 2. The environment device in open cycles.

Figure 2 illustrates the cogeneration cycle on the *h*–*s* diagram, with the various processes denoted by numerical labels representing the respective components involved:

- Process 1–2 corresponds to the compressor, with 1–2 s indicating isentropic compression.
- Process 2–3 represents the combustion chamber.
- Process 3–4 corresponds to the gas turbine, with 3–4 s denoting isentropic expansion.
- Process 4–5 corresponds to the recovery boiler.

Upon reaching the exit of the recovery boiler (at point 5), the exhaust gases possess exergy, rendering them waste products. Despite the slight reduction in entropy of the working fluid caused by the recovery boiler (RB), the cycle remains incomplete. In contrast, a Rankine cycle is capable of achieving full closure through the condenser, wherein the entropy of the turbine's output steam is reduced to that of saturated liquid at the pump inlet.

To complete the cycle in Figure 2, an environmental device (ENV) intervenes and facilitates process 5–1, effectively closing the loop. Within this device, flow 5 represents the exhaust gases, while flow 1 symbolizes the air drawn in by the compressor.

3.1.1. Productive Diagram

Figures 3 and 4 illustrate the productive diagrams of the gas turbine cogeneration system as per the E and H&S Models, respectively. In the E Model, the depicted flows represent the changes in exergy between two physical states, denoted as *i* and *j*, following the expression given in Equation (7). Conversely, in the H&S Model, the productive flows on the diagrams represent the alterations in the enthalpic, entropic, and chemical components of exergy between states *i* and *j*, according to Equations (8)–(10), respectively.

$$E_{i:j} = E_i - E_j \tag{7}$$

$$E^H_{i:j} = E^H_i - E^H_j \tag{8}$$

$$E^S_{i:j} = E^S_i - E^S_j \tag{9}$$

$$E^{CH}_{i:j} = E^{CH}_i - E^{CH}_j \tag{10}$$

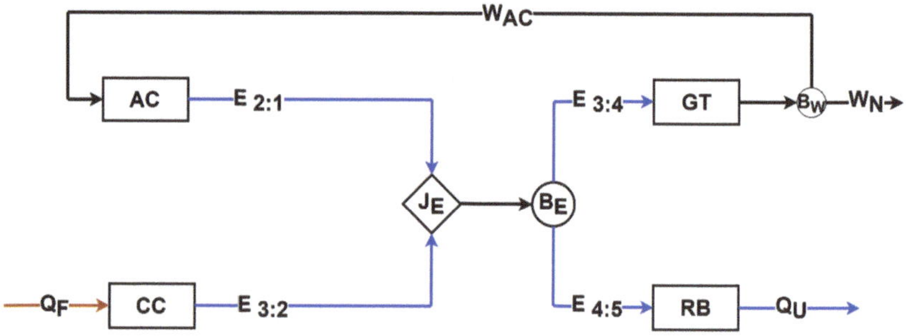

Figure 3. Productive diagram—E Model.

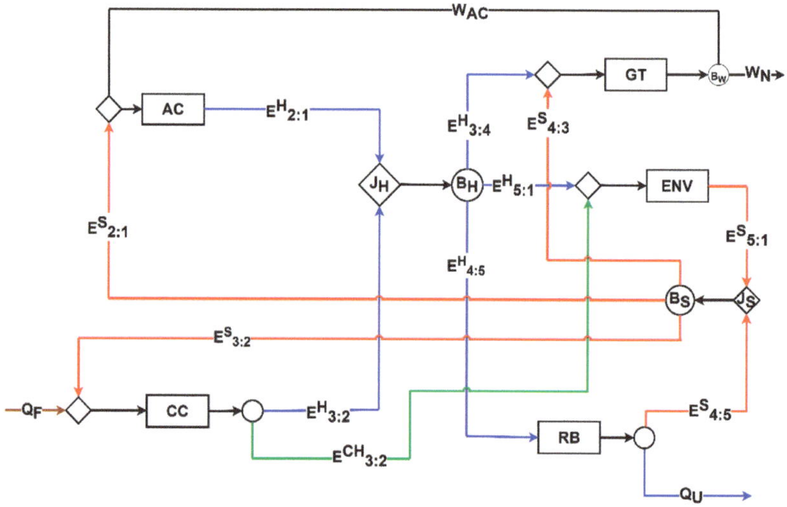

Figure 4. Productive diagram—H&S Model.

In Figures 3 and 4, the system components are depicted as real units or subsystems in the form of rectangles, while fictitious units called junctions (J) and bifurcations (B), represented by rhombuses and circles, respectively, are utilized to establish interconnections between these subsystems.

The characterization of fuel and product follows the SPECO approach [31] and is as follows: if the variation of specific exergy (or its components with a positive contribution to the exergy definition) exhibits a positive trend throughout the process; this variation, along with the exergy of energy flows generated within the component, determines the product. Conversely, if the variation of specific exergy (or its components with a positive contribution to the exergy) shows a negative trend throughout the process, this variation is combined with the exergy of energy flows supplied to the component in the input definition. The approach is reversed for components with a negative contribution to the exergy definition, such as the entropic component in the H&S Model. In this particular case, the H&S Model designates the entropic ($E^{H}_{5:1}$) and chemical ($E^{CH}_{3:2}$) components as inputs from the environment, while the entropic ($E^{S}_{5:1}$) component is identified as the product, as shown in Figure 4. On the other hand, E Model E (Figure 3) allocates the residue implicitly to the final products (W_L and Q_U), proportionally to the exergetic inputs of the equipment generating the respective final products GT and RB.

3.1.2. Monetary Cost Balance

Figure 5 presents the monetary cost balance for the H&S Model, represented in an expanded matrix format. This matrix is derived by applying the cost balance equation (Equation (1)) to each of the five subsystems (AC, CC, GT, RB, and ENV) as well as to the enthalpic (J_H-B_H) and entropic (J_S-B_S) junction–bifurcations within the productive diagram (Figure 4).

$$\underbrace{\begin{bmatrix} E^{H}_{2:1} & 0 & -W_{AC} & 0 & 0 & 0 & -E^{S}_{2:1} \\ 0 & E^{H}_{3:2}+E^{CH}_{3:2} & 0 & 0 & 0 & 0 & -E^{S}_{3:2} \\ 0 & 0 & W_{AC}+W_{N} & 0 & 0 & -E^{H}_{3:4} & -E^{S}_{4:3} \\ 0 & 0 & 0 & E^{S}_{4:5}+Q_{U} & 0 & -E^{H}_{4:5} & 0 \\ 0 & -E^{CH}_{3:2} & 0 & 0 & E^{S}_{5:1} & -E^{H}_{5:1} & 0 \\ -E^{H}_{2:1} & -E^{H}_{3:2} & 0 & 0 & 0 & E^{H}_{3:4}+E^{H}_{4:5}+E^{H}_{5:1} & 0 \\ 0 & 0 & 0 & -E^{S}_{4:5} & -E^{S}_{5:1} & 0 & E^{S}_{2:1}+E^{S}_{3:2}+E^{S}_{4:3} \end{bmatrix}}_{\text{Internal valuation}} \cdot \underbrace{\begin{bmatrix} c_{AC} \\ c_{CC} \\ c_{GT} \\ c_{RB} \\ c_{ENV} \\ c_{JB-H} \\ c_{JB-S} \end{bmatrix}}_{\text{Cost}} = \underbrace{\begin{bmatrix} Z_{AC} \\ c_{F}.Q_{F}+Z_{CC} \\ Z_{GT} \\ Z_{RB} \\ Z_{ENV} \\ 0 \\ 0 \end{bmatrix}}_{\text{External valuation}}$$

Figure 5. Monetary cost balance in matrix form.

The internal valuation matrix comprises flows of exergy components, power, and useful heat, representing the entire process from the distribution of external resources to the determination of the final product costs. The cost matrix (or vector) constitutes the unknown factor in the modeling, incorporating the monetary unit cost of the flows generated within each subsystem. For instance, c_{AC} denotes the monetary unit cost of the compressor (AC) product, which corresponds to the flow $E^{H}_{2:1}$.

Due to the application of the equality criterion, certain flows possess identical unit costs. Notable examples include $E^{H}_{3:2}$ and $E^{CH}_{3:2}$; $E^{S}_{4:5}$ and Q_U; and W_{AC} and W_N. Moreover, all entropic component flows departing from J_H-B_H share the same unit cost, as do all entropic component flows leaving J_S-B_S.

The external valuation matrix encompasses the exergy of the fuel along with its corresponding unit cost, in addition to the external hourly cost of each subsystem resulting from capital and equipment operation and maintenance (Z). The junction–bifurcation components are considered dummy elements and, therefore, have a Z-cost of zero, as illustrated in the external valuation matrix depicted in Figure 5.

The Z term and the environmental representation device depicted in the diagrams play a crucial role in incorporating environmental costs within the field of thermoeconomics. Figure 6 provides a comprehensive overview of this device, illustrating its input and output flows, which contribute to the overall monetary cost equilibrium. The environmental device itself does not entail any costs associated with its acquisition, operation, or maintenance, as it serves as a symbolic representation of the atmospheric environment.

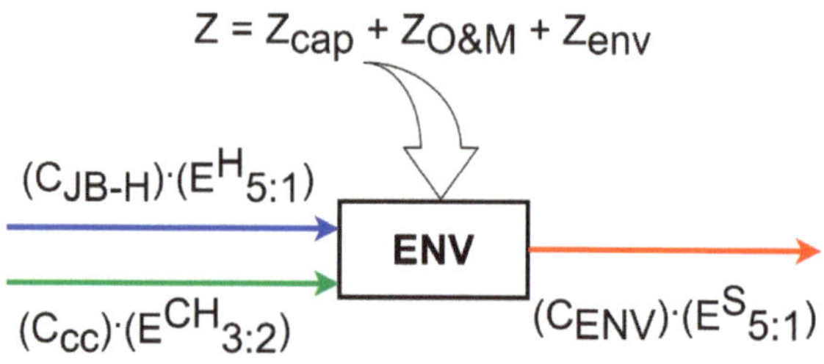

Figure 6. Cost balance in the environmental device.

However, when an environmental treatment component (such as a filter, electrostatic precipitator, etc.) is introduced into the thermal system, which is not typically represented in the physical diagram, its cost can be internalized in the environmental device through the initial two terms on the right-hand side of Equation (4). Consequently, this cost is distributed across the entire system.

Furthermore, expenses related to fines, environmental licenses, permits, and carbon pricing values should also be internalized within the Z_{env} factor. The exact treatment of carbon pricing values depends on whether they are considered revenue or expenses within the carbon market.

In the context of revenue generation, which can arise from a reduction in emissions below the specified level, resulting in the creation of tradable credits, a variable denoted as Z_{env} assumes a negative value on the balance sheet. The environmental device responsible for closing the loop reallocates the associated costs to other equipment and final products within the plant. Consequently, this credit serves to diminish the overall monetary expenses and can exert an influence on the plant's production decisions.

Conversely, expenses incurred in relation to carbon credits, such as the necessity to purchase credits due to the company's failure to meet the stipulated emission reduction targets (or simply due to environmental consciousness and concerns regarding climate change), lead to a positive value for the Z_{env} variable. This, in turn, results in an increase in the costs of other internal processes and final products within the plant.

To summarize, the equation depicted in Figure 6, referred to as Equation (11) in the text, plays a crucial role in understanding the aforementioned dynamics. An in-depth analysis of this equation can be conducted as follows:

- The environmental device (ENV) does not have any hourly costs related to capital and operations and maintenance (O&M). However, if environmental treatment equipment, which is not typically part of the physical structure of the system, is used, these costs can be considered within Z_{ENV};
- The expenses for licenses and permits associated with the environment are accounted for in the Z_{env} term;
- Similarly, the costs related to the carbon market are also accounted for through the Z_{env} term. When there is revenue, this term is represented as negative, and when there are expenditures, it is represented as positive.

In all three cases, since the ENV device closes the loop (Figure 2), the costs are systematically distributed to the other subsystems and, consequently, to the final products of the plant, as shown in the case of the H&S Model (Figure 4).

$$Z_{ENV} = Z_{cap} + Z_{O\&M} + Z_{env} \tag{11}$$

In the case of thermoeconomic models that do not define the environment as an explicit device in the diagram, as is the case with Model E (Figure 3) in this paper, the internalization of carbon credits must be performed through the equipment generating emissions, the CC in such case. Conversely, when considering the cost of acquiring environmental control/treatment equipment, this model would need to depict such equipment in the physical and productive diagrams to facilitate appropriate cost allocation. However, these devices are dissipative, and several studies [32–34] have already demonstrated that Model E is not adequate for isolating them within the productive structure and thus analyzing them in a separate and appropriate manner.

3.1.3. Results

Figure 7 illustrates a generic cogeneration system, also known as a combined heat and power (CHP) system. This system generates two products, W_N and Q_U, from a single fuel source, Q_F, similar to the gas turbine system depicted in Figure 1. By employing the cost balance equation (Equation (1)) to this generic cogeneration system, one obtains Equation (12), wherein c_{W_N} and c_{Q_U} represent the respective monetary unit costs of the final products, W_N and Q_U. It is important to note that Equation (12) takes the form of a linear equation, $y = A{\cdot}x + B$, and can be expressed in accordance with Equation (13).

$$c_{W_N} = -\frac{Q_U}{W_N} c_{Q_U} + \frac{c_F{\cdot}Q_F + Z}{W_N} \quad (12)$$

$$c_{W_N} = -A{\cdot}c_{Q_U} + B \quad (13)$$

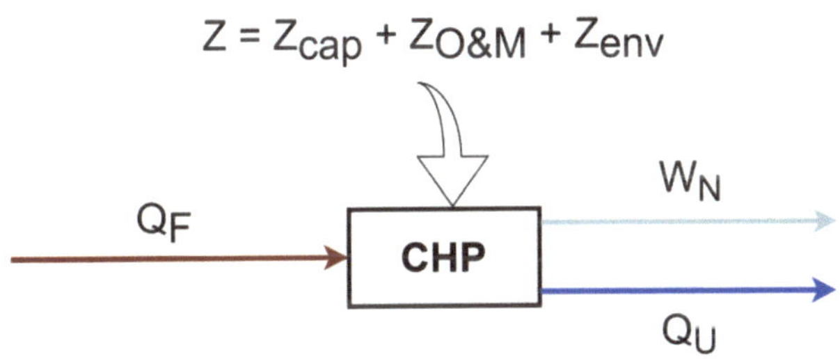

Figure 7. Accounting flows in cogeneration.

Irrespective of the applied thermoeconomic methodology, the solution to Equation (12) will consist of an ordered pair representing the monetary unit costs of the net power (c_{W_N}) and the useful heat (c_{Q_U}). Several studies [16,35–37] have previously compared various methodologies for such problems and verified that these ordered pairs yield identical straight-line solutions when the system's operational conditions are defined, including the net power-to-useful heat ratio and the global exergetic efficiency.

Figure 8 illustrates generic possibilities for this straight-line solution. In all cases, there is an inverse relationship between the unit cost of power and the unit cost of heat; that is, when the unit cost of power increases, the unit cost of heat decreases, and vice versa.

The central straight line, represented by the continuous line, denotes the specific condition for a cogeneration system. Changes in the thermodynamic model cause the straight line to shift to new positions parallel to the initial one [35,38], as depicted by the dashed lines in Figure 8.

Figure 9 illustrates the unit cost of the final products (represented as ordered pairs) in the cogeneration system under various scenarios. The cost values were determined using Equation (1) and applied to the diagrams in Figures 3 and 4 for the E and H&S Models, respectively. In the case of the H&S Model, the cost balance is presented in matrix form in Figure 5, with a specific focus on Z_{ENV}.

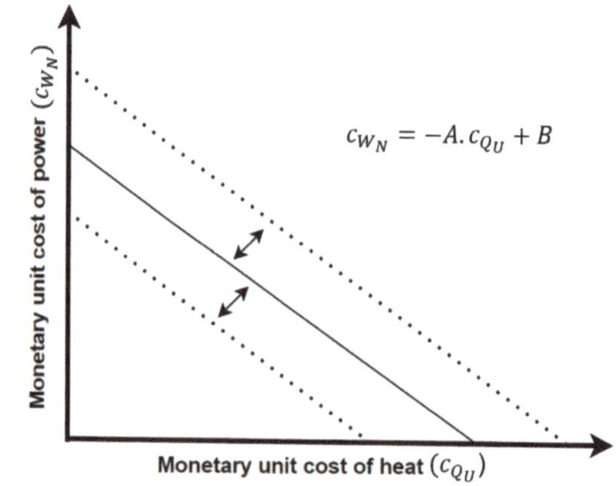

Figure 8. Unit cost solution line.

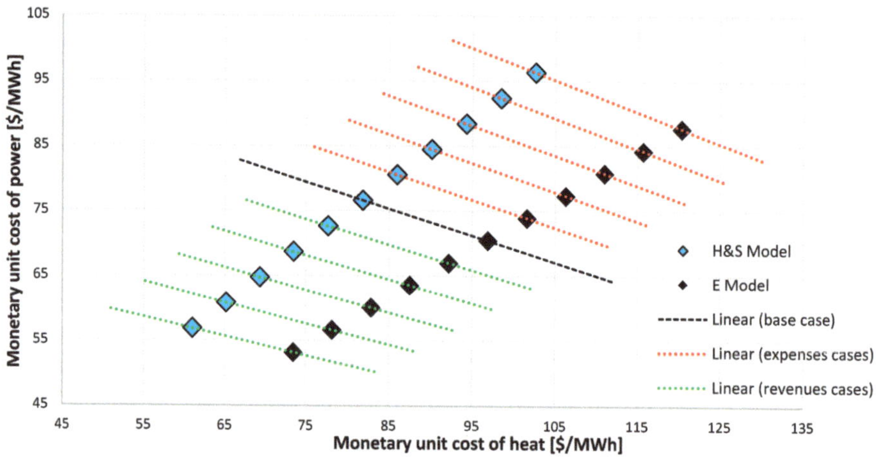

Figure 9. Monetary unit cost variation due to emissions.

The data points situated along the central line in Figure 9 correspond to the base case, as mentioned in the figure's caption. In this base case, carbon credit values are not taken into account or internalized.

It is noteworthy that an increase in emissions signifies a reduction in process efficiency, leading to higher production costs. Consequently, the solution line moves away from the origin, and the costs of the final products rise. Conversely, a reduction in emissions brings the solution line closer to the origin, indicating an improvement in process efficiency and a consequent decrease in the costs of the final products.

In order to conduct an analysis of the expenses and revenues within the carbon market, certain hypotheses were formulated and implemented. The initial scenario, referred to as the base case, considers that an increase in CO_2 emissions leads to the system emitting beyond the established limit, necessitating the purchase of carbon credits and resulting in an expense for the plant. Another plausible real scenario that could lead to an escalation in costs involves a rise in the unit price of carbon credits owing to stricter emissions control regulations.

Conversely, in situations where emissions are reduced or removed, the system generates carbon credits that can be sold, thereby generating revenue. It is worth highlighting practical real scenarios capable of reducing and/or removing emissions in the industry.

The utilization of mitigation equipment, such as carbon capture and storage systems (CCS), enables emission removal. In this case, the investment value (external cost per hour due to capital cost, operation, and maintenance of this equipment, Z) must be taken into account to assess and compare the cost of the investment with the potential revenue generated from carbon credits due to avoided emissions. Another feasible scenario for emission reduction involves substituting a more environmentally friendly fuel (e.g., natural gas with biogas or biomethane).

The H&S Model involves the internalization of the carbon credit value within the environmental device, represented as $(Z_{env} > 0)$ for expenses and $(Z_{env} < 0)$ for revenue. On the other hand, the E Model utilizes the internalization process through a combustion chamber (CC), with $(Z_{carbon\ credit} > 0)$ for expenses and $(Z_{carbon\ credit} < 0)$ for revenue.

To simulate the impact of these hypotheses, scenarios involving various percentages of emission increase (ranging from 10% to 50%) and reduction (ranging from −10% to −50%) concerning the base case were considered. The resulting Table 4 presents the unit costs of final products c_{Q_U} and c_{W_N} for each of these situations, along with the corresponding amount of carbon credits that would be generated and the associated costs (revenue and expense).

Table 4. Monetary unit cost [USD/MWh] and carbon credit for the simulated situations.

Emissions	E Model		H&S Model		Carbon Credit/Day	USD/Day
	(c_{Q_U})	(c_{W_N})	(c_{Q_U})	(c_{W_N})		
+50%	120.33	87.66	102.57	96.32	26.7	−2273
+40%	115.64	84.21	98.41	92.37	21.4	−1818
+30%	110.95	80.76	94.26	88.42	16.0	−1364
+20%	106.26	77.31	90.10	84.47	10.7	−909
+10%	101.57	73.86	85.95	80.52	5.3	−455
Base case	96.88	70.41	81.79	76.57	0	0
−10%	92.19	66.96	77.64	72.62	5.3	455
−20%	87.51	63.51	73.48	68.67	10.7	909
−30%	82.82	60.06	69.33	64.72	16.0	1364
−40%	78.14	56.6	65.17	60.77	21.4	1818
−50%	73.45	53.15	61.02	56.82	26.7	2273

When conducting an analysis of c_{Q_U} and c_{W_N}, it has been observed that within the E Model, the costs of final products exhibit an approximate variation of 5% and 25% when there is a 10% and 50% increase or reduction in emissions, respectively, compared to the base case. Similarly, in the case of the H&S Model, these same scenarios lead to cost variations of approximately 5% and 26% for the final products. The observed cost fluctuations are attributed to the distinct criteria employed by each model, including the internalization of carbon credits in the CC and the environmental device.

Upon examining the carbon credits for the simulated scenarios, it has been noted that a 10% change in emissions results in the generation of 5.3 credits per day, equivalent to an expense or revenue of USD 455 per day. In the most extreme emission variation scenario (50%), the corresponding expenditure or revenue can reach USD 2273 per day. The utilization of the value derived from the purchase or sale of carbon credits can serve as a valuable indicator for companies in making decisions pertaining to the implementation of environmental equipment or the acquisition of carbon credits.

In all results, the ordered pair belongs to the straight solution, specific to the operating conditions of the plant. In addition, the models are consistent from a thermodynamic point of view according to the efficiencies and irreversibilities obtained.

Note that the primary objective of this study is to illustrate the application of the thermoeconomic methodology for integrating carbon pricing into cogeneration systems analysis. It is important to note that the study does not focus on analyzing the system's behavior in terms of its capacity to increase or decrease emissions, nor does it involve defining the specific emission parameters that would govern the carbon market.

Examining the incorporation of environmental costs, such as carbon pricing, reveals an intriguing phenomenon denoted by Equation (12). This equation demonstrates that the Z term, and consequently Z_{env}, influence the B coefficient of the linear equation (Equation (13)). Consequently, this adjustment leads to a parallel displacement of the initial straight line (base case) in comparison to the original condition. Specifically, when revenue is generated ($Z_{env} < 0$) through the sale of carbon credits, the straight line approaches the origin, resulting in a reduction in product costs. Conversely, for expenses ($Z_{env} > 0$) incurred in the purchase of carbon credits, the straight line shifts away from the origin, leading to an increase in production costs. It is essential to emphasize that the straight line transitions to distinct yet parallel positions. However, the slope remains unaltered due to the unchanged coefficient A.

Moreover, as the system conditions are defined and environmental costs are integrated, various thermoeconomic methodologies that account for the environmental aspect establish ordered pairs of power and heat costs lying on the same straight-line solution.

4. Conclusions

This research presents a comprehensive thermoeconomic approach to incorporate monetary environmental costs into multiproduct system assessments, demonstrated through a case study involving a gas turbine cogeneration system. The primary cost aspect addressed in this paper pertains to the valuation of carbon emissions. Nevertheless, the methodology also encompasses the internalization of other environmental expenses, including licensing, permits, and the procurement of environmental treatment and control equipment. Furthermore, the proposed methodology is adaptable for the allocation of both cost and emissions stemming from fuel utilization to final products across various categories of multiproduct systems, exemplified by industries such as sugarcane and alcohol, pulp and paper, and steel production.

In this study, two models, namely Models E and H&S, were employed for analysis. However, the primary focus is on H&S, as it offers a methodology that defines a crucial device responsible for representing the environment in the diagrams. This particular device plays a significant role in both the dissipation of cycle waste and the internalization of environmental costs, thereby enabling a systematic redistribution of costs across the system's remaining components and final products. It is worth noting that any other thermoeconomic methodology, based on exergy and coherently defining this environmental device, could also be adopted by following similar methods. Moreover, models that do not explicitly define the ENV device in the diagram can internalize carbon credits within the equipment generating emissions. However, in the case of costs associated with the acquisition of environmental treatment/control equipment, these models might be limited in their analyses due to their inability to isolate this type of equipment (dissipative) within the productive structure.

The study presents the H&S Model as a viable instrument to achieve the objective by elucidating the inclusion of the carbon market and the internalization of carbon pricing and other environmental expenses in the analysis. In addition to explaining the calculation methodology, the study also examines the variations in monetary costs of cogeneration systems' final products.

In conclusion, this research establishes that the proposed adapted methodology is consistent with the theoretical principles of thermodynamics and thermoeconomics. As a result, it can effectively facilitate the allocation of carbon credits to both the internal processes and final products of multiproduct systems.

By incorporating carbon credits into thermoeconomic modeling, it becomes possible: to evaluate the financial implications of carbon emissions and incentivize the reduction of greenhouse gases; to help in understanding the environmental impact of the thermal system; and to optimize the thermal system's design and operation to maximize the economic benefits while minimizing emissions. It also can help in designing and implementing effective carbon pricing mechanisms and environmental policies to achieve emission reduction targets. Thus, companies and policymakers can gain a better understanding of the true costs and benefits of different thermal system configurations and make informed decisions that prioritize both economic and environmental sustainability.

Author Contributions: Conceptualization, P.R.d.F., I.C.B., R.G.d.S., M.A.B. and J.J.C.S.S.; methodology, P.R.d.F., I.C.B. and R.G.d.S.; software, I.C.B. and P.R.d.F.; validation, R.G.d.S. and J.J.C.S.S.; formal analysis, I.C.B., R.G.d.S. and M.A.B.; investigation, P.R.d.F., I.C.B. and R.G.d.S.; writing—original draft preparation, P.R.d.F. and R.G.d.S.; writing—review and editing, P.R.d.F., I.C.B., R.G.d.S., M.A.B. and J.J.C.S.S.; supervision, J.J.C.S.S.; project administration, J.J.C.S.S.; funding acquisition, J.J.C.S.S. All authors have read and agreed to the published version of the manuscript.

Funding: The APC was funded by "Fundação de Amparo a Pesquisa e Inovação do Espírito Santo—FAPES, T.O: 033/2024–P: 2024-J4KRS".

Data Availability Statement: Data will be made available on request.

Acknowledgments: The authors would like to thank FAPES <Number T.O. 033/2024–P: 2024-J4KRS>, IFES, PPGEM-UFES, CNPq and CAPES.

Conflicts of Interest: The authors declare no conflict of interest.

Abbreviations

AC	Air compressor
c	Monetary unit cost (USD/MWh)
CC	Combustion chamber
CCS	Carbon capture and storage
CEPCI	Chemical Engineering Cost Index
CHP	Combined heat and power
E	Exergy Flow (kW)
ENV	Environmental device
GHG	Greenhouse gas
GT	Gas turbine
IPCC	Intergovernmental Panel on Climate Change
JB	Junction–bifurcation
k^*	Exergetic unit cost (kW/kW)
Q	Heat (exergy) (kW)
RB	Recovery boiler
W	Power (kW)
Y	Generic thermodynamic magnitude (kW)
Z	Hourly equipment cost (USD/h)
Greek symbols	
λ	Specific CO_2 emission (g/MWh)
Subscripts and superscripts	
0	Reference conditions
CH	Chemical exergy (kW)
Env	Environmental
F	Fuel
H	Enthalpic flow (kW)
i; j	Indexes for productive components
in	Inlet
N	Net
out	Outlet
PH	Physical exergy (kW)
S	Entropic flow (kW)
U	Useful heat

References

1. IPCC. Mitigation of Climate Change. 2022. Available online: https://report.ipcc.ch/ar6wg3/pdf/IPCC_AR6_WGIII_FinalDraft_FullReport.pdf (accessed on 9 May 2022).
2. IEA. CO_2 Emissions from Fuel Combustion: Overview-International Energy Agency; Paris. 2020. Available online: https://www.iea.org/reports/co2-emissions-from-fuel-combustion-overview (accessed on 28 April 2023).
3. Instituto E+ Transição Energética. Manual de Termos e Conceitos: Transição Energética; Rio de Janeiro/RJ—Brasil. 2020. Available online: https://emaisenergia.org/publicacao/manual-de-termos-e-conceitos-transicao-energetica/ (accessed on 12 April 2023).
4. SendeCO_2. Available online: https://www.sendeco2.com/es/ (accessed on 18 March 2023).
5. The World Bank. State and Trends of Carbon Pricing 2022; Washington, DC. 2022. Available online: https://openknowledge.worldbank.org/entities/publication/a1abead2-de91-5992-bb7a-73d8aaaf767f (accessed on 20 March 2023).
6. IPCC. AR6 Synthesis Report Climate Change 2023. 2023. Available online: https://www.ipcc.ch/report/ar6/syr/ (accessed on 19 March 2023).
7. Valero, A.; Serra, L.; Uche, J. Fundamentals of Exergy Cost Accounting and Thermoeconomics. Part I: Theory. *J. Energy Resour. Technol.* **2006**, *128*, 1–8. [CrossRef]
8. von Spakovsky, M.R. Application of Engineering Functional Analysis to the Analysis and Optimization of the CGAM Problem. *Energy* **1994**, *19*, 343–364. [CrossRef]
9. Dincer, İ.; Rosen, M.A. *Exergy: Energy, Environment, and Sustainable Development*, 1st ed.; Elsevier Science: Amsterdam, The Netherlands, 2007; ISBN 9780080445298.
10. Palacios-Bereche, M.C.; Palacios-Bereche, R.; Ensinas, A.V.; Gallego, A.G.; Modesto, M.; Nebra, S.A. Brazilian Sugar Cane Industry—A Survey on Future Improvements in the Process Energy Management. *Energy* **2022**, *259*, 124903. [CrossRef]
11. Man, Y.; Yan, Y.; Wang, X.; Ren, J.; Xiong, Q.; He, Z. Overestimated Carbon Emission of the Pulp and Paper Industry in China. *Energy* **2023**, *273*, 127279. [CrossRef]
12. Mishra, R.; Singh, R.K.; Gunasekaran, A. Adoption of Industry 4.0 Technologies for Decarbonisation in the Steel Industry: Self-Assessment Framework with Case Illustration. *Ann. Oper Res.* **2023**. [CrossRef]
13. Faria, P.R.; dos Santos, R.G.; Santos, J.J.; Barone, M.A.; Miotto, B.M. On the Allocation of Residues Cost Using Conventional and Comprehensive Thermoeconomic Diagrams. *Int. J. Thermodyn.* **2021**, *24*, 134–149. [CrossRef]
14. Barone, M.A.; dos Santos, R.G.; Faria, P.R.; Lorenzoni, R.A.; Lourenço, A.B.; Santos, J.J.C.S. On the Arbitrariness and Complexity in Thermoeconomics Due to Waste Cost and Supplementary Firing Treatment. *Eng. Sci.* **2022**, *17*, 328–342. [CrossRef]
15. Faria, P.R.; dos Santos, R.G.; Santos, J.J.C.S.; Barone, M.A.; Miotto, B.M.F. On the Allocation of Residues Cost in Thermoeconomics Using a Comprehensive Diagram. In Proceedings of the ECOS 2020—The 33rd International Conference on Efficiency, Cost, Optimization, Simulation and Environmental Impact of Energy Systems, Osaka, Japan, 29 June–3 July 2020; pp. 2444–2456.
16. Cerqueira, S.A.A.G.; Nebra, S.A. Cost Attribution Methodologies in Cogeneration Systems. *Energy Convers. Manag.* **1999**, *40*, 1587–1597. [CrossRef]
17. Modesto, M.; Nebra, S.A. Analysis of a Repowering Proposal to the Power Generation System of a Steel Mill Plant through the Exergetic Cost Method. *Energy* **2006**, *31*, 3261–3277. [CrossRef]
18. Santos, R.G.; Faria, P.R.; Santos, J.J.C.S.; da Silva, J.A.M. Thermoeconomic Modeling for CO_2 Allocation in Steam and Gas Turbine Cogeneration Systems. In Proceedings of the ECOS 2015—The 28th International Conference on Efficiency, Cost, Optimization, Simulation and Environmental Impact of Energy Systems, Pau, France, 29 June–3 July 2015.
19. Faria, P.R.; Barone, M.A.; Santos, R.G.; Santos, J.J.C.S. The Environment as a Thermoeconomic Diagram Device for the Systematic and Automatic Waste and Environmental Cost Internalization in Thermal Systems. *Renew. Sustain. Energy Rev.* **2023**, *171*, 113011. [CrossRef]
20. Faria, P.R.; Torres, C.; Santos, J.J.; Valero, A. On the Thermoeconomic Modelling of Thermal Waste Cost Allocation and Diagnosis of Malfunctions for Environmental Aspects Assessment. In Proceedings of the ECOS 2022—The 35th International Conference on Efficiency, Cost, Optimization, Simulation and Environmental Impact of Energy Systems, Copenhagen, Denmark, 3–7 July 2022.
21. Fortes, A.F.C.; Carvalho, M.; da Silva, J.A.M. Environmental Impact and Cost Allocations for a Dual Product Heat Pump. *Energy Convers. Manag.* **2018**, *173*, 763–772. [CrossRef]
22. Lozano, M.A.; Valero, A.; Serra, L.M. Theory of Exergetic Cost and Thermoeconomic Optimization. In *Energy Systems and Ecology, Proceedings of the ENSEC' 93: Energy Systems and Ecology: Proceedings of the International Conference: Cracow, Poland, 5–9 July 1993*; Szargut, J., Kolenda, Z., Tsatsaronis, G., Ziebik, A., Eds.; University of Zaragoza: Zaragoza, Spain; Volume 1, pp. 339–350.
23. Frangopoulos, C.A. Application of the Thermoeconomic Functional Approach to the CGAM Problem. *Energy* **1994**, *19*, 323–342. [CrossRef]
24. Torres, C.; Valero, A.; Rangel, V.; Zaleta, A. On the Cost Formation Process of the Residues. *Energy* **2008**, *33*, 144–152. [CrossRef]
25. F-Chart Software Engineering Equation Solver—EES 2017. Available online: http://www.fchart.com (accessed on 17 August 2022).
26. Santos, J.J.C.S. Aplicação Da Neguentropia Na Modelagem Termoeconômica de Sistemas. Ph.D. Dissertation, Federal University of Itajubá, Itajubá, Brazil, 2009. (In Portuguese).
27. Trading Economics. Available online: https://pt.tradingeconomics.com/commodity/natural-gas (accessed on 19 March 2023).
28. Chemical Engineering CEPCI. Available online: https://www.chemengonline.com/category/business-economics/ (accessed on 19 March 2023).

29. Valero, A.; Royo, J.; Lozano, M.A. The Characteristic Equation and Second Law Efficiency of Thermal Energy Systems. In Proceedings of the International Conference Second Law Analysis of Energy Systems: Towards the 21st Century, Rome, Italy, 5–7 July 1995; Sciubba, E., Moran, M.J., Eds.; La Sapienza: Rome, Italy, 1995; pp. 99–112.
30. de Oliveira Junior, S. *Exergy: Production, Cost and Renewability*; Green Energy and Technology; Springer: London, UK, 2013; ISBN 978-1-4471-4164-8.
31. Lazzaretto, A.; Tsatsaronis, G. SPECO: A Systematic and General Methodology for Calculating Efficiencies and Costs in Thermal Systems. *Energy* **2006**, *31*, 1257–1289. [CrossRef]
32. Santos, R.G.; Faria, P.R.; Barone, M.A.; Lorenzoni, R.A.; Orozco, D.J.R.; Santos, J.J. On the Thermoeconomic Diagnosis through the Localized Physical Exergy Disaggregation for Dissipative Component Isolation. In Proceedings of the ECOS 2021—The 34th International Conference on Efficiency, Cost, Optimization, Simulation and Environmental Impact of Energy Systems, Taormina, Italy, 27 June–2 July 2021; Volume 1, pp. 450–461.
33. dos Santos, R.G.; Lourenço, A.B.; Faria, P.R.; Barone, M.A.; Santos, J.J.C.S. A New Exergy Disaggregation Approach for Complexity Reduction and Dissipative Equipment Isolation in Thermoeconomics. *Entropy* **2022**, *24*, 1672. [CrossRef] [PubMed]
34. Abreu, R.P.; Correia, V.H.L.; Lourenço, A.B.; Marques, A.d.S.; Carvalho, M. Thermoeconomic and Thermoenvironmental Analysis of the Chilled Water System in a Shopping Mall. *Int. J. Refrig.* **2022**, *134*, 304–311. [CrossRef]
35. Santos, R.G.; Faria, P.R.; Santos, J.J.C.S.; Da Silva, J.A.M.; Donatelli, J.L.M. The Effect of the Thermodynamic Models on the Thermoeconomic Results for Cost Allocation in a Gas Turbine Cogeneration System. *Rev. Eng. Térmica* **2015**, *14*, 47. [CrossRef]
36. Santos, R.G.; Lourenço, A.; Faria, P.R.; Barone, M.A.; Santos, J.J. A Comparative Study of a New Exergy Disaggregation Approach with Conventional Thermoeconomic Methodologies for Cost Allocation in an Organic Rankine Cycle Powered Vapor Compression Refrigeration System. In Proceedings of the ECOS 2022—The 35th International Conference on Efficiency, Cost, Optimization, Simulation and Environmental Impact of Energy Systems, Copenhagen, Denmark, 3–7 July 2022; pp. 1057–1068.
37. Bhering Trindade, A.; Luiza Grillo Renó, M.; José Rúa Orozco, D.; Martín Martinez Reyes, A.; Aparecido Vitoriano Julio, A.; Carlos Escobar Palacio, J. Comparative Analysis of Different Cost Allocation Methodologies in LCA for Cogeneration Systems. *Energy Convers. Manag.* **2021**, *241*, 114230. [CrossRef]
38. de Araújo, L.R.; Morawski, A.P.; Barone, M.A.; Donatelli, J.L.M.; Santos, J.J.C.S. On the Effects of Thermodynamic Assumptions and Thermoeconomic Approaches for Optimization and Cost Allocation in a Gas Turbine Cogeneration System. *J. Braz. Soc. Mech. Sci. Eng.* **2020**, *42*, 323–341. [CrossRef]

processes

Article

Interdependent Expansion Planning for Resilient Electricity and Natural Gas Networks

Weiqi Pan, Yang Li, Zishan Guo and Yuanshi Zhang *

School of Electrical Engineering, Southeast University, Nanjing 210096, China; wpan@seu.edu.cn (W.P.); li_yang@seu.edu.cn (Y.L.); zishanguo@seu.edu.cn (Z.G.)
* Correspondence: yuanshizhang@seu.edu.cn

Abstract: This study explores enhancing the resilience of electric and natural gas networks against extreme events like windstorms and wildfires by integrating parts of the electric power transmissions into the natural gas pipeline network, which is less vulnerable. We propose a novel integrated energy system planning strategy that can enhance the systems' ability to respond to such events. Our strategy unfolds in two stages. Initially, we devise expansion strategies for the interdependent networks through a detailed tri-level planning model, including transmission, generation, and market dynamics within a deregulated electricity market setting, formulated as a mixed-integer linear programming (MILP) problem. Subsequently, we assess the impact of extreme events through worst-case scenarios, applying previously determined network configurations. Finally, the integrated expansion planning strategies are evaluated using real-world test systems.

Keywords: expansion planning; resilience networks; electric power grid; natural gas network

Citation: Pan, W.; Li, Y.; Guo, Z.; Zhang, Y. Interdependent Expansion Planning for Resilient Electricity and Natural Gas Networks. *Processes* **2024**, *12*, 775. https://doi.org/10.3390/pr12040775

Academic Editor: Takuya Oda

Received: 21 February 2024
Revised: 6 April 2024
Accepted: 9 April 2024
Published: 12 April 2024

1. Introduction

The resilience of energy infrastructure against extreme events like windstorms and wildfires has gained increasing attention recently [1], and this is particularly highlighted by the 2021 Texas blackouts. Enhancing energy infrastructure resilience against extreme events, such as windstorms and wildfires, to mitigate their impacts on interdependent energy systems, including electric and natural gas infrastructures, is a critical yet challenging task [2].

Traditional energy resource planning is reliability oriented, typically employing a probabilistic model focused on high-chance and low-impact events [3]. The introduction of the "defender–attacker–defender" framework, incorporating $N-k$ criteria, marks a shift toward improving system resilience against more severe scenarios [4]. In the expansion planning problem, conventional models with an $N-k$ static set are effective but often fail to account for the interactions among energy systems and the low-chance high-impact events. Meanwhile, extreme events usually come with unlikely high-impact consequences. Thus, there is a growing need for resilience-focused operational and planning methodologies.

Some researchers have delved into the concept of interdependent planning. Reference [5] proposed a co-optimization planning model that targets the long-term interdependency of electricity and natural gas systems with both economic and security constraints. Reference [6] offered an optimal planning model with reliability constraints for an integrated energy hub with multiple types of energy systems. Reference [7] developed a stochastic day-ahead hourly scheduling algorithm to dispatch both supply- and demand-side resources, and it can utilize ramping flexibility to deal with the variability of renewable energy resources over time. Reference [8] presented a flexible stochastic security-constrained unit commitment (SCUC) model to accommodate the high integration of wind energy in the midterm allocations of natural gas and hydrosystems optimally. Reference [9] presented a co-optimization algorithm to schedule the coordinated operation of the two types of energy systems in a robust way. Reference [10] focused on the optimal operation of the integrated electric and gas network with bidirectional energy conversion. However, these studies often overlooked the synergistic potential and complementary nature of combining electric and natural gas systems, especially in mitigating impacts from extreme events.

Moreover, the dynamics of the wholesale energy market are essential for effective interdependent planning [11,12]. This market, overseen by an Independent System Operator (ISO), includes transmission limitations and the process for energy market clearing, significantly affecting decisions in resource planning [13]. For example, a Generation Company (GENCO) might postpone investment if faced with severe transmission congestion [14], which can lead to a decrease in generation capacity. Alternatively, excessively expanding the transmission capacity could negatively impact generation investment since the potential for high profits often directs investments toward achieving favorable settlement prices. Therefore, optimal planning decisions require integrating transmission planning with generation planning in a market-based framework [15,16].

Therefore, this paper proposes a two-stage, robust optimization framework aimed at coordinated expansion planning to enhance resilience against extreme events. The first stage combines transmission and generation planning with market clearing to devise an optimal expansion strategy for electricity and natural gas systems. The second stage applies resilience constraints to these systems in anticipation of extreme events, utilizing robust modeling to address potential impacts. Through case studies, we demonstrate the efficacy of our integrated planning approach in enhancing energy system resilience.

The contributions of this paper are as follows:

(1) A novel framework of centralized transmission planning with decentralized generation planning is proposed. It is formulated as an integrated resource tri-level optimization model in a deregulated market environment.
(2) Detailed profiles, constraints, and market dynamics are considered in the interdependent expansion planning of energy systems to enhance its resilience.
(3) To identify promising expansion plans, a Complementary Problem (CP) reformulation is employed to address the original tri-level programming expansion challenge, while a decomposition method is applied to manage the two-stage resilience problem.

The remainder of the paper is organized as follows. In Section 2, a thorough problem description is presented to illustrate the background and framework. In Section 3, the optimal expansion model considering resilience constraints for each level is proposed, covering transmission planning, generation planning, and the market clearing process. Section 4 presents a case study of an IEEE standard test system. Finally, Section 5 summarizes the main conclusions of the paper.

2. Problem Description

The vulnerability of energy infrastructures to extreme weather events, such as windstorms and wildfires, has led to a re-evaluation of traditional energy system designs. Notably, the 2021 winter storm in Texas highlighted the comparative resilience of underground infrastructures over their overhead counterparts [17,18]. Conventionally, gas transportation networks are constructed underground, offering inherent protection against such events. Similarly, underground cable installations represent a strategic defense against the vulnerabilities faced by overhead energy facilities. While these underground solutions entail higher initial costs, their potential to significantly enhance system resilience, particularly in regions prone to extreme events, is substantial.

Thus, by considering the partial integration of power transmission infrastructure with underground gas transportation systems, this study aimed to create a more robust response mechanism to extreme events that threaten utility services. Figure 1 illustrates a system where natural gas units serve as conduits between electric power and natural gas systems, demonstrating the potential for natural gas pipelines to supplement power transmission capacities. This not only meets increased heat demands but also supports electric power delivery, presenting a comprehensive strategy for decision making in energy system planning.

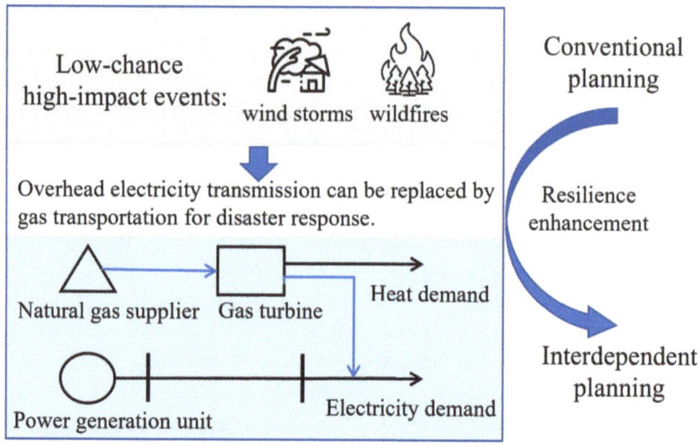

Figure 1. Interconnection of natural gas and electricity systems.

Operating a natural gas system involves complex considerations around safety, reliability, and the physical properties of gas flow. While the compressibility of natural gas offers certain operational flexibilities, such as line packing and storage, the detailed modeling of these aspects falls outside the scope of this study, which is focused on broader planning strategies rather than time-domain transient modeling or stability analysis. Moreover, it is supposed that the non-convex steady-state equations for the gas flow are excluded as well in order to reduce computational burdens in real-world integrated planning [19]. Accordingly, this work set forth specific assumptions to streamline the integrated planning process:

- The initial planning stage conceptualizes a typical hour of operation, setting the groundwork for further detailed scenario analysis in subsequent stages. This approach allows for the possibility of expanding the model to encompass various time frames and stochastic scenarios, including renewable energy outputs and demand variability.
- This work focuses on expanding the capacity of existing lines rather than constructing new ones, assuming constant impedance for any such expansions to simplify the analysis.
- Cost functions for both transmission and generation expansion are assumed as linear to minimize the computational complexity.
- Within the competitive electricity market, Generation Companies (GENCOs) are modeled to make strategic and reasonable decisions. Although each GENCO is initially represented by a single unit for simplicity, the model allows for extension to more complex configurations involving multiple units per GENCO.
- The model presumes a perfectly competitive electricity market, with the capability for all buses to support both load and generation.
- The market equilibrium is simultaneously solved through ISO's market clearing process, and the GENCOs can bid in the market.
- For Case 7 in the analysis, the installed capacities of the candidate transmission lines are set 20% lower than in Case 1.
- The scope of this study is confined to considering expansions in the natural gas transportation system's pipeline infrastructure.

3. Model Formulation

The structured two-stage optimization framework, as illustrated in Figure 2, is formulated for interdependent planning. In Stage-1 , the optimal expansion planning strategies are explored using a tri-level planning model, including a transmission level, generation level, and market level in a deregulated electricity market environment. Subsequently, Stage-2 subjects these identified expansion strategies to rigorous analysis under extreme conditions, employing worst-case scenario assessments to evaluate their robustness and effectiveness.

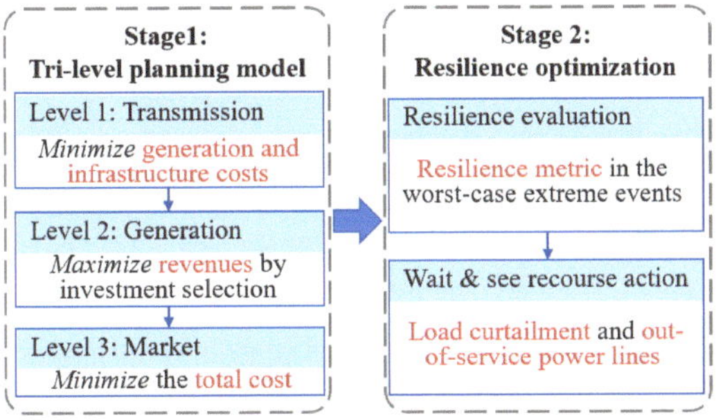

Figure 2. Configuration of two-stage problem.

3.1. Stage-1: Planning Strategy Development

The hierarchical structure of our proposed optimization framework for Stage-1 is depicted in Figure 3. This structure combines the complexities of generation and transmission planning with market clearing processes into a unified optimization model. In this framework, individual GENCOs would simulate the future power prices settled by an energy market and then obtain its investment options and corresponding operational dispatches with the new resource. These strategic decisions are related with transmission planning, as the availability of an adequate transmission capacity is crucial for GENCOs to secure additional revenue by supplying energy from new sources to consumers.

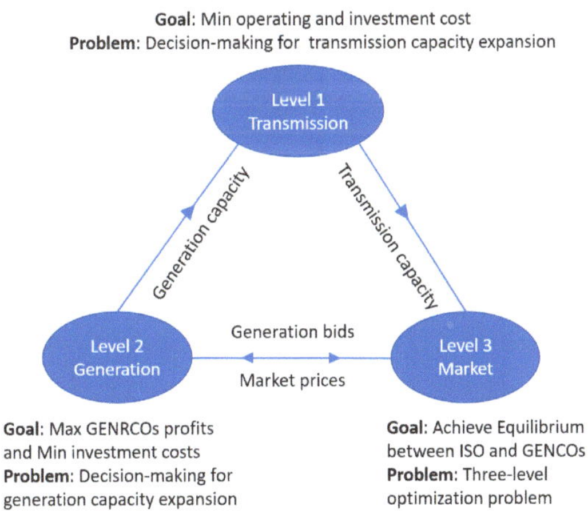

Figure 3. Hierarchical structure of the tri-level model.

Despite being in a deregulated environment, transmission system planning still remains centralized to assure the reliability of the bulk electric power grid since the ownership of transmission facilities lies with Transmission Companies (TRANSCOs). Consequently, the model assigns the responsibility of centralized transmission planning and the reliability management of the electricity market to an Independent System Operator (ISO). The generation and transmission planning issue is structured as a mixed-integer programming (MIP) problem within a tri-level framework, where centralized transmission planning decisions form the first level, followed by GENCOs' expansion strategies at the second level, and an energy market equilibrium problem at the third level.

To streamline this complex tri-level model into a more manageable form, we apply the Karush–Kuhn–Tucker (KKT) conditions to transform it into a single-level convex formulation [20]. Thus, this problem can be solved using commercially available solvers, such as CPLEX [21], directly.

3.1.1. Level-1: Transmission Expansion Planning

At Level-1, we consider enhancing the original system by introducing potential transmission lines and natural gas pipelines. These candidate lines are expected to augment the capacity of existing infrastructure. The objective function of this level, Equation (1), encompasses the operational costs of generation, capital expenditure for the addition of transmission lines and pipelines, and the investment costs from Level-2:

$$\begin{aligned}\min_{f_l^P, f_j^Q} & \sum_i (a_i - b_i \bar{P}_i^{GC})(P_i^{GC} + P_i^{GE}) \\ & + \sum_e a_e P_e^{GE} + \sum_i K_i \bar{P}_i^{GC} \\ & + \sum_l K_l (f_l^P - f_l^{P,0}) + \sum_j K_j (f_j^Q - f_j^{Q,0})\end{aligned} \tag{1}$$

subject to

$$f_l^{P,0} \le f_l^P \le f_l^{P,max} \tag{2}$$

$$f_j^{Q,0} \le f_j^Q \le f_j^{Q,max} \tag{3}$$

$$\text{Level-2 solution} \tag{4}$$

Note that the decision variables $l \in L^{P,inv}$ and $j \in L^{Q,inv}$ are the transmission and pipe line capacity limits after the decisions at Level-1 are obtained. They are constants for Level-2 and Level-3. f_l^P and f_j^Q are defined as continuous ones with limits $[f_l^{P,0}, f_l^{P,max}]$ and $[f_j^{Q,0}, f_j^{Q,max}]$, respectively.

3.1.2. Level-2: Generation Expansion Equilibrium

Here, each GENCO determines its investment strategy to maximize revenue as depicted in (5). The capacity expansion of GENCO G is given by the following profit maximization expression:

$$\begin{aligned}\max_{\bar{P}_i^{GC}} U = & \\ & \sum_{i \in N^{inv}} \pi_i (P_i^{GC} + P_i^{GE}) - [a_i - b_i \bar{P}_i^{GC}](P_i^{GC} + P_i^{GE}) \\ & + \sum_{e \in N^{fix}} \pi_k P_e^{GE} - a_e P_e^{GE} - \sum_{i \in N^{inv}} K_i \bar{P}_i^{GC}\end{aligned} \tag{5}$$

subject to

$$\text{Equilibrium Constraints} \tag{6}$$

The optimization goal is to balance the revenue from energy sales against the costs of capacity expansion, as the first term is the revenues obtained from the market, and the second term is the investment cost for capacity expansion. This formulation is a Mathematical Program with Equilibrium Constraints (MPEC) model, where (6) represents the equilibrium constraints that are obtained from Level-3.

More specifically, each GENCO can decide whether to expand its capacity. However, since these expansion decisions will consider other GENCOs' expansion decisions, we plan to employ the Nash equilibrium to determine the decisions on the equilibrium of the entire power grid in the deregulated wholesale market environment. In this regard, the Nash equilibrium includes all GENCO expansion equilibrium strategies $\bar{P}_i^{GC,EQ}$, where each equilibrium strategy renders more revenues than any other methodology, $\bar{P}_i^{GC}$. It is supposed that the other GENCOs are constants in their equilibrium strategies $\bar{P}_{-i}^{GC,EQ}$. The equilibrium problem at Level-2 is subject to equilibrium constraints obtained through a perfectly competitive equilibrium; i.e., all GENCOs solve their max profit problems at the same time. Constraint (7) presents the Nash equilibrium condition.

$$\begin{aligned} &U_G^E(\bar{P}_i^{GC,E}, \forall i \in N^{inv}) \\ &\quad \geq \max_{\bar{P}_i^{GC}} U_G(\bar{P}_i^{GC}, \bar{P}_{-i}^{GC,EQ}, \forall i \in N_G^{inv}, \forall - i \in N_{-G}^{inv}) \end{aligned} \tag{7}$$

3.1.3. Level-3: Market Clearing Process

Level-3 clears the market with the resources determined previously. The model takes into account transmission constraints with a lossless DC approximation while considering competitive generators, locational marginal prices (LMPs), and inelastic demands.

The market model is divided into two components: (1) candidate generating units eligible for resource capacity expansion and (2) resources that cannot have their capacities increased through investment. In this formulation, dual variables are denoted on the right side of the equations. The problem facing the ISO in the wholesale market aims to minimize total costs, as specified in Equation (8), while adhering to constraints related to resources and the system. Specifically, Equations (9) and (10) detail the output capacities for both candidate and established units. Equation (11) accounts for the total power imported to or exported from each bus, while Equation (12) outlines the constraints on power flow. LMPs are computed based on the dual variables associated with the power balance constraints at each bus, as shown in Equation (13):

$$\min_{P_e^{GE}, P_i^{GC}, P_n^{inj}} \sum_i (a_i - b_i \bar{P}_i^{GC})(P_i^{GC} + P_e^{GE}) + \sum_e a_e P_e^{GE} \tag{8}$$

subject to

$$0 \leq P_i^{GC} \leq \bar{P}_i^{GC} \quad : \xi_i^-, \xi_i^+ \tag{9}$$

$$0 \leq P_e^{GE} \leq \bar{P}_e^{GE} \quad : \gamma_e^-, \gamma_e^+ \tag{10}$$

$$\sum_n P_n^{inj} = 0 \quad : \alpha \tag{11}$$

$$-f_l^P \leq \sum_{n \in N_{n,l}} \varphi_{l,n} P_n^{inj} \leq f_l^P \quad : \lambda_l^-, \lambda_l^+ \tag{12}$$

$$\sum_{i \in N_{i,n}} P_i^{GC} + \sum_{i \in N_{e,n}} P_e^{GE} + P_n^{inj} = P_n^D \quad : \pi_n \tag{13}$$

$$\begin{aligned} &S_m - Q_m^D - c_e P_e^{GE} - c_i P_i^{GC} \\ &\quad = \sum_{n \in N_{m,j}} Q_j^L \quad : \mu_m \end{aligned} \tag{14}$$

$$0 \leq S_m \leq S_m^{max} \quad : \sigma_m^-, \sigma_m^+ \tag{15}$$

$$-f_j^Q \leq Q_j^L \leq f_j^Q \quad : \delta_j^-, \delta_j^+ \tag{16}$$

$$\sum_j Q_j^L = 0 \quad : \beta \tag{17}$$

Constraints (14)–(17) incorporate operational protocols of the natural gas network. Specifically, Constraint (14) addresses the balance of natural gas loads at nodes and the consumption of gas by generation units following capacity expansion. Constraint (15) sets limits on the amount of fuel that can be sourced from individual natural gas nodes. Following expansion, Constraint (16) applies flow restrictions to the natural gas pipelines, assuming that these pipelines operate according to a linear transport model, rather than a nonlinear one, to simplify the planning process without sacrificing significant accuracy. Constraint (17) details the total gas imported to or exported from each node. This setup posits that gas-fired units maintain a consistent ratio of fuel usage to electricity generation, simplifying fuel usage into a piecewise linear function of power generation with minimal impact on computational complexity.

The market clearing model employed for both the market operator and participants is framed linearly, with the Karush–Kuhn–Tucker (KKT) conditions providing a sufficient framework to identify the globally optimal solution. Thus, the original multi-level problem is converted into a single-level model, streamlining the solution process. Further insights into this methodology can be found in [22]. It should be noted that the perfectly competitive environment for maximizing the revenues of generators as a whole is consistent with the economic dispatch model that is used by the market operator, and thus, it is an accurate, equivalent form of the original problems.

Accordingly, Equations (18)–(46) are an equivalent KKT formulation of the original problems set out in Equations (8)–(17). The Fortuny-Amat and McCarl linearization [23] is employed to convert the slackness conditions into linear constraints, which enhances the tractability of the optimization problem and ensures its alignment with the practical demands of energy system planning.

$$a_i - b_i \bar{P}_i^{GC} - \pi_n - \xi_i^- + \xi_i^+ + c_i \mu_m = 0 \tag{18}$$

$$a_e - \pi_n + c_e \mu_m - \gamma_e^- + \gamma_e^+ = 0 \tag{19}$$

$$\alpha - \pi_n + \sum_{l \in L} (\lambda_l^+ - \lambda_l^-) \varphi_{l,i} = 0 \tag{20}$$

$$-\mu_m - \sigma_m^- + \sigma_m^+ = 0 \tag{21}$$

$$\mu_m + \beta - \delta_j^- + \delta_j^+ = 0 \tag{22}$$

$$0 \le \xi_i^- \le M^{\xi_i^-} (\eta_i^{\xi^-}) \tag{23}$$

$$0 \le P_i^{GC} \le M^{\xi_i^-} (1 - \eta_i^{\xi^-}) \tag{24}$$

$$0 \le \xi_i^+ \le M^{\xi_i^+} (\eta_i^{\xi^+}) \tag{25}$$

$$0 \le g_i - P_i^{GC} \le M^{\xi_i^+} (1 - \eta_i^{\xi^+}) \tag{26}$$

$$0 \le \gamma_e^- \le M^{\gamma_e^-} (\eta_e^{\gamma^-}) \tag{27}$$

$$0 \le P_e^{GE} \le M^{\gamma_e^-} (1 - \eta_e^{\gamma^-}) \tag{28}$$

$$0 \le \gamma_e^+ \le M^{\gamma_e^+} (\eta_e^{\gamma^+}) \tag{29}$$

$$0 \le g_e^0 - P_i^{GE} \le M^{\gamma_e^+} (1 - \eta_e^{\gamma^+}) \tag{30}$$

$$0 \le \lambda_l^- \le M^{\lambda_l} (\eta_l^{\lambda^-}) \tag{31}$$

$$0 \le f_l^P + \sum_{n \in N} \varphi_{l,n} P_n^{inj} \le M^{f_l^P} (1 - \eta_l^{\lambda^-}) \tag{32}$$

$$0 \le \lambda_l^+ \le M^{\lambda_l} (\eta_l^{\lambda^+}) \tag{33}$$

$$0 \le f_l^P - \sum_{n \in N} \varphi_{l,n} P_n^{inj} \le M^{f_l^P} (1 - \eta_l^{\lambda^+}) \tag{34}$$

$$0 \le \sigma_m^- \le M^{\sigma_m^-} (\eta_m^{\sigma^-}) \tag{35}$$

$$0 \le S_m \le M^{\sigma_m^-} (1 - \eta_m^{\sigma^-}) \tag{36}$$

$$0 \le \sigma_m^+ \le M^{\sigma_m^+}(\eta_m^{\sigma^+}) \tag{37}$$
$$0 \le S_m^{max} - S_m \le M^{\sigma_m^+}(1 - \eta_m^{\sigma^+}) \tag{38}$$
$$0 \le \delta_j^- \le M^{\delta_j}(\eta_j^{\delta^-}) \tag{39}$$
$$0 \le f_j^Q + Q_j^L \le M^{f_j^Q}(1 - \eta_j^{\delta^-}) \tag{40}$$
$$0 \le \delta_j^+ \le M^{\delta_j}(\eta_j^{\delta^+}) \tag{41}$$
$$0 \le f_j^Q - Q_j^L \le M^{f_j^Q}(1 - \eta_j^{\delta^+}) \tag{42}$$
$$\sum_n P_n^{inj} = 0 \tag{43}$$
$$\sum_{i \in N_{i,n}} P_i^{GC} + \sum_{i \in N_{e,n}} P_e^{GE} + P_n^{inj} = P_n^D \tag{44}$$
$$S_m - Q_m^D - c_e P_e^{GE} - c_i P_i^{GC} = \sum_{n \in N_{m,j}} Q_j^L \tag{45}$$
$$\sum_j Q_j^L = 0 \tag{46}$$

With binary expansion [24] and Fortuny-Amat linearization [20], variable P_i^{GC} can be discretized, and the nonlinear product $\xi_i \bar{P}_i^{GC}$ can be transferred to a linear expression (48) with constraints (49) and (50) as follows:

$$\bar{P}_i^{GC} = \Delta_{g_i} \sum_{k=0}^{\Lambda_i} 2^k y_{ki} \tag{47}$$
$$\xi_i \bar{P}_i^{GC} = \Delta_{g_i} \sum_{k=0}^{\Lambda_i} 2^k \hat{y}_{ki} \tag{48}$$
$$0 < \xi_i - \hat{y}_{ki} \le M^{\xi_i}(1 - y_{ki}) \tag{49}$$
$$0 \le \hat{y}_{ki} \le M^{\xi_i} y_{ki} \tag{50}$$

Hence, the complex tri-level framework is rebuilt into a compact MILP model (A1–A56), which enables the accommodation of complex, large-scale systems with numerous decision variables.

3.2. Stage-2: Resilience Optimisation

In this stage, the planning options derived from Stage-1 are subjected to analysis under scenarios of low-probability but high-impact extreme events. This evaluation employs a deterministic approach, utilizing a resilience metric (RM) that quantifies the minimum load curtailment during the most severe scenarios as follows:

$$RM = \max_{z \in Z} \min_{(P,Q)} \sum_n F_n(P_n^{LC}) \le RM_{max} \tag{51}$$

Load curtailment alters the energy flow across gas pipelines and power lines. To streamline our analysis, we omit the impact on gas pipelines due to their resilience against extreme events compared to overhead power lines. The effects on power lines are encapsulated by the constraints below, which take into account the possibility of line outages during extreme events:

$$-f_l^P * (1 - z_l) \le \sum_{n \in N} \varphi_{l,n} P_n^{inj} \le f_l^P * (1 - z_l) \tag{52}$$
$$\sum_{i \in N_{i,n}} P_i^{GC} + \sum_{i \in N_{e,n}} P_e^{GE} + P_n^{inj} = P_n^D - P_n^{LC} \tag{53}$$

In addition, the number of out-of-service power lines due to extreme events is used to measure the impact of extreme events [6], and those failed lines are modeled within the following uncertainty set.

The significance of extreme events is gauged using the number of power lines that go out of service, modeled within the uncertainty set Z, where the confidence level k represents the duration, destructive path, and other characteristics and indicates the severity of extreme events.

$$Z = \{Z_{LC} | \sum_{LC} z_{LC} \leq k\} \tag{54}$$

3.3. Problem Formulation and Solution

The two-stage optimization model is expressed mathematically as follows, with linear formulations accommodating generation costs and load curtailments. The model anticipates uncertain extreme events in Stage-1, then selects the worst-case scenarios based on the initial network configuration, and proposes responses to mitigate load curtailment Stage-2.

$$\begin{aligned} \min_{P^0,Q^0,f_j,f_l} \quad & F^0(P^0, Q^0) \\ \text{s.t. } & A^0P^0 + B^0Q^0 \leq C^0u + D^0f_l \\ & \max_z \min_{P,Q} F(P) \leq RM_{max} \\ & AP + BQ \leq Cu + Dz \end{aligned} \tag{55}$$

In the formulation, RM_{max} offers a mechanism for adjusting the resilience against critical loads proportionately to the total load, enhancing the model's robustness.

This can be solved with the column-and-constraint generation (CCG) method, which has been widely utilized to deal with robust optimization problems [25].

This two-stage max–min optimal problem (55) is tackled using the column-and-constraint generation (CCG) method, effectively addressing robust optimization [25]. The problem is bifurcated into a master problem and a sub-problem, as depicted in (56) and (57). The master problem provides a relaxed version of the original problem, while the sub-problem, focused on resilience metrics, integrates new worst-case scenarios into the model iteratively:

$$\begin{aligned} \min_{P^0,Q^0,f_j,f_l} \quad & F^0(P^0, Q^0) \\ \text{s.t. } & A^0P^0 + B^0Q^0 \leq C^0f_j + D^0f_l \\ & F(P^s) \leq RM_{max} \\ & AP^s + BQ^s \leq Cu + D\hat{z}^s \end{aligned} \tag{56}$$

$$\begin{aligned} \max_z \min_{P,Q} \quad & F(P) \\ & AP + BQ \leq C\hat{u}^s + Dz \end{aligned} \tag{57}$$

The iterative process, detailed in Algorithm 1, continues until the sub-problem's objective function falls below RM_{max}, ensuring that all resilience constraints are met.

Hence, the sub-problem can be transformed into a bilinear problem through the dualization of the inner minimization problem and then further reformulated as an MILP problem using the Big-M method. The master problem is an MILP problem as well. In this regard, these problems can both be solved with available commercial solvers like CPLEX [21].

Algorithm 1 Two-stage resilient planning.

Initialisation: Set the resilience metric $RM \leftarrow \infty$, iteration index $s = 0$, and extreme event scenario $\hat{z}^0 \leftarrow 0$.
while *not converged* **do**
  Solve the primary problem and determine the optimal expansion decisions.
  Solve the secondary problem and calculate the optimal solution $\hat{z}$ and update the related resilience metric RM.
  Build dispatch variables P^s and Q^s, as well as the associated operation constraints, on the basis of $\hat{z}$, and then substitute those variables and constraints into the primary problem.
  Update index $s \leftarrow s + 1$.
end while
return planning decisions

4. Case Studies

4.1. System Settings

In this section, we undertake a comprehensive analysis of the ISO New England power grid, modeled across eight zones, interconnected with a six-node natural gas system, including Maine (ME), Vermont (VT), New Hampshire (NH), NE Mass & Boston (NEMAB), WC Mass (WCMA), SE Mass (SEMA), Rhode Island (RI), and Connecticut (CT) [26]. The map of the system is shown in Figure 4. All eight zones contain electric demands, and all zones except for NEMAB are eligible with existing or candidate generation capacities.

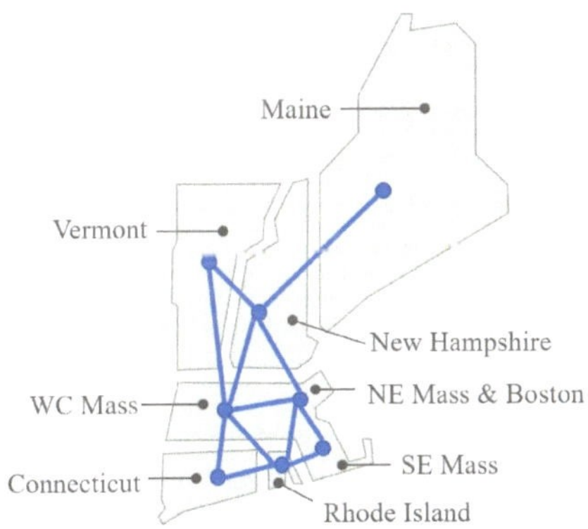

Figure 4. ISO New England system.

The case studies evaluate the interconnected system's response to various scenarios, including cost adjustments, capacity changes, and extreme event simulations, thereby highlighting the robustness and adaptability of our proposed planning approach.

The parameters of existing and candidate generic resources are adopted from [27]. It is assumed that each candidate's resource capacity can be expanded up to 1.5 GW. In addition, we assume that existing and candidate transmission capacities can be expanded up to 1.5 GW. An installation cost of USD 45 millionis imposed for the candidate lines. The network contains two natural gas sources, a large gas source connected to node 1 and a relatively small gas source linked to node 6. The existing capacity of the gas pipelines is adopted from [27]. Moreover, the gas pipeline can be expanded to a maximum capacity of 100,000 MBTU/h at an installation cost of USD 100,000 per MBTU/h. The investment cost is set according to [27]. The electric and non-generation-associated gas heat loads are adopted from [28,29] and are scaled on the basis of the assumed resource and natural gas pipeline capacities. A fixed natural gas price of USD 3 per MBTU is imposed for the whole system.

4.2. General Analysis

To analyze the impact of various costs on the planning results, we evaluate the following cases. Table 1 and Figures 5–8 show the results for all eight cases. For cost analysis, this indicates that lower investment costs are produced by lower-priced power system equipment. The natural gas pipeline investment cost has a significant impact on the total investment. Various equipment investment costs have no impact on the operation costs. For capacity analysis, lower candidate component capacities generally result in higher investment costs. However, in Case 7, with a different generation expansion strategy, a lower investment cost is achieved. The transmission line candidate capacities have no impact on the operating costs, whereas the natural gas pipeline candidate capacities and generation unit candidate capacities both have positive effects on the operating costs. In addition, lower costs are obtained using the proposed method, even with a higher contingency level. The eight cases are as follows:

- Case 1 (Benchmark): the data summarized in Section 4.1 were used to form the benchmark case.
- Case 2: the investment costs of a natural gas pipeline capacity expansion are set 50% lower than in Case 1.
- Case 3: the investment costs of a resource capacity expansion are set 50% higher than in Case 1.
- Case 4: the investment costs of a transmission line capacity expansion are set 50% higher than in Case 1.
- Case 5: the installed capacities of the candidate natural gas pipelines are set 10% lower than in Case 1.
- Case 6: the installed capacities of the candidate generation units are set 20% higher than in Case 1.
- Case 7: the installed capacities of the candidate transmission lines are set 20% lower than in Case 1.
- Case 8: in Stage-1, the simulated loss of transmission lines is reduced by 66%.

Figure 6 lists the generation unit investment decisions. In the cost analysis, there is no difference between Cases 2, 3, and 4. For the capacity analysis, only Cases 5 and 6 have different investment capacities, both of which are higher than in the other cases, i.e., 4000 MW. The higher cost is caused by lower natural gas pipeline candidate capacities or higher generation unit candidate capacities.

Figure 7 summarizes the investment in the transmission system. For the cost analysis, various equipment prices have no impact on the expansion of the transmission network. For the capacity analysis, the various natural gas pipeline candidate capacities and generation unit candidate capacities lead to different expansion strategies.

Figure 8 depicts the investment in natural gas pipeline capacity expansions. A lower candidate capacity results in a lower installed capacity, whereas lower-priced pipelines lead to the installation of more gas pipelines. Moreover, using the proposed method, the load curtailment is reduced from 8688 MW to 7587 MW for the same level of extreme event.

Table 1. Optimal costs for all cases.

	Objective Function Value [USD Billion]	Investment Cost [USD Billion]	Operating Cost [USD Million]
Case 1	30.167	30.166	1.343
Case 2	17.189	17.188	1.343
Case 3	32.217	32.216	1.343
Case 4	30.222	30.22	1.343
Case 5	30.276	30.275	1.283
Case 6	28.352	28.351	1.383
Case 7	29.367	29.366	1.343
Case 8	30.305	30.304	1.303

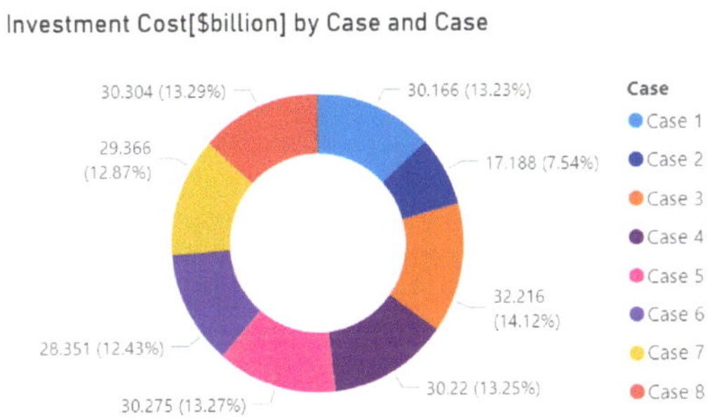

Figure 5. Optimal objective function values for all cases.

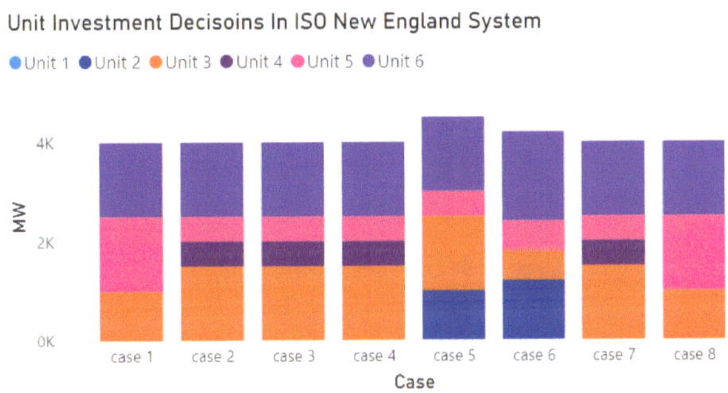

Figure 6. Unit investment decisions in ISO New England system.

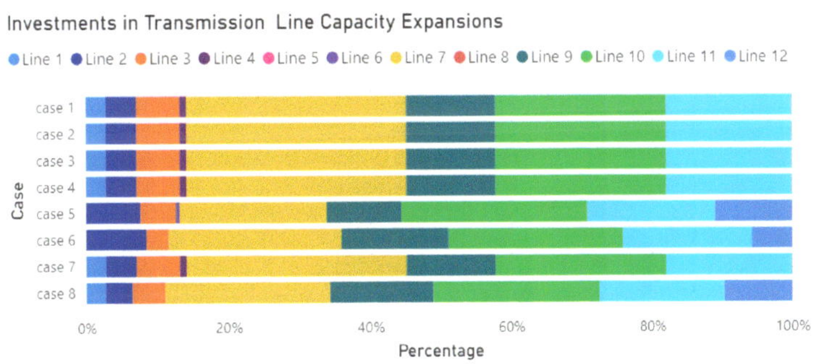

Figure 7. Investments in transmission line capacity expansions.

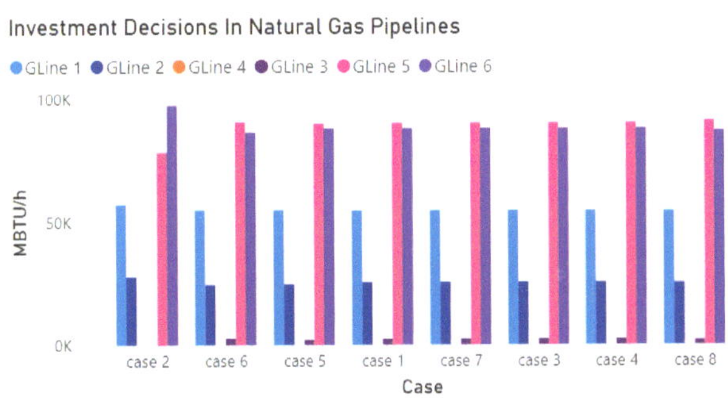

Figure 8. Investment decisions in natural gas pipelines.

4.3. Cost Sensitivity Analysis

Our findings reveal that adjustments in investment costs directly influence the overall planning outcomes. Notably, a reduction in natural gas pipeline expansion costs (Case 2) significantly decreases the total investment required, underscoring the substantial financial impact of pipeline infrastructure on regional energy systems. Conversely, increases in the costs associated with resource and transmission line expansions (Cases 3 and 4) slightly elevate the total project costs.

4.4. Capacity Sensitivity Analysis

Adjusting the capacities of candidate components yields insightful trends; notably, reducing natural gas pipeline capacities (Case 5) leads to marginally higher investment costs, reflecting the critical role of gas infrastructure in ensuring system resilience. On the other hand, enhancing generation unit capacities (Case 6) or reducing transmission line capacities (Case 7) influences both investment and operational costs, highlighting the intricate balance between generation, transmission, and demand in maintaining system efficiency.

4.5. Load Curtailment Sensitivity

The reduction in simulated transmission line losses (Case 8) further validates the resilience of our planning approach. By effectively managing infrastructure vulnerabilities, our model demonstrates a capacity to mitigate the impacts of severe disruptions, ensuring reliable energy delivery even under stringent conditions.

4.6. Operational Implications and Strategic Insights

The diverse investment decisions across generation units, transmission systems, and gas pipeline expansions, as visualized in Figures 6–8, encapsulate the strategic nuances inherent in our planning model. Specifically, the model's adaptability to various pricing and capacity scenarios illustrates its potential to guide strategic investment decisions, ensuring that infrastructure development aligns with regional energy needs and resilience objectives.

Moreover, the resilience optimization stage plays a pivotal role in enhancing system robustness against extreme events. By quantifying the minimum load curtailment achievable under severe conditions, our model not only informs infrastructure investment decisions but also contributes to the strategic planning necessary to withstand unforeseen disruptions.

5. Conclusions

In conclusion, this paper emphasizes the vital importance of integrating pipeline planning within the broader strategy for enhancing utility resilience. Underground pipelines, which are less affected by extreme weather events compared to overhead power lines, provide a sturdy solution for maintaining electricity supply during adverse conditions.

We developed a two-stage model to strengthen the combined resilience of electricity and natural gas systems. This approach, practical through the use of mixed-integer programming, was designed to meet current planning needs while being adaptable for future changes. The insights from our case studies on cost, capacity, and load curtailment illustrate the model's ability to navigate the intricacies of modern energy systems effectively, offering valuable strategies for utilities to ensure economic efficiency alongside resilience to extreme events.

Future work will focus on incorporating renewable energy and adjusting to evolving market dynamics, enhancing our model's readiness for a renewable-powered future.

Author Contributions: Conceptualization, W.P. and Y.L.; methodology, W.P.; software, W.P.; validation, W.P.; formal analysis, W.P., Z.G., and Y.Z.; investigation, W.P. and Y.L.; resources, W.P.; data curation, W.P.; writing—original draft preparation, W.P.; writing—review and editing, Z.G.; visualization, W.P.; supervision, Y.L. and Y.Z. All authors have read and agreed to the published version of the manuscript.

Funding: This research is supported by the Jiangsu Provincial Key Research & Development Plan (BE2020081 2).

Data Availability Statement: The code used in this study is available from the authors upon request.

Conflicts of Interest: The authors declare no conflicts of interest.

Nomenclature

Indices	
e	Index of existing generation units.
i	Index of candidate generation units.
j	Index of natural gas pipelines.
l	Index of transmission lines.
m	Index of natural gas system nodes.
n	Index of power system buses.
G	Index of generation companies.
Sets	
N^{inv}	Set of candidate generation units.
N^{fix}	Set of fixed generation units.
$L^{P,inv}$	Set of candidate transmission lines.
$L^{Q,inv}$	Set of candidate gas pipelines.
Constants	
a_i, b_i	Parameters of the candidate generation operational cost function of unit i.
g_i^0	Generation capacity of unit i available before *Level-2*.
K_i	Annual unit cost of investment in capacity expansion for candidate generation unit i.
K_j	Annual unit cost of investment in capacity expansion for gas pipeline j.
K_l	Annual unit cost of investment in capacity expansion for transmission line l.
$f_l^{P,0}, f_l^{P,max}$	Initial/maximum capacity of transmission line l.
$f_j^{Q,0}, f_j^{Q,max}$	Initial/maximum capacity of gas pipeline j.
M	Large constant.
$\varphi_{l,n}$	Power transfer distribution factor associated with line l with respect to unit injection/withdrawal at bus n.
Δ_{g_i}	Size of the step used to discretize generation capacity i.
P_n^D	Inelastic demand at bus n.
Λ_i	Parameter used to the discretize generation capacity g_i expansion associated with the number of binary variables. The total number of binary variables is $\Lambda_i + 1$.
Variables	
f_l^P	Thermal capacity limit of transmission line after decisions have been made at *Level-1*. This remains constant for *Level-2* and *Level-3*.
f_j^Q	Capacity limit of gas pipeline after decisions have been made at *Level-1*. This remains constant for *Level-2* and *Level-3*.
g_i	Generation capacity available at node after decisions have been made at *Level-2*.
P_e^{GE}	Power generated by existing generation units.
P_i^{GC}	Power generated by candidate generation units.
P_n^{inj}	Import/export power from/to bus i.
S_m	Natural gas extracted from node m.
Q_j^L	Natural gas flow through pipeline j.
$\bar{P}_i^{GC,EQ}$	Expansion equilibrium strategies for all GENCOs.
$\bar{P}_i^{GC}$	Each equilibrium strategy renders more revenue than any other one.
$\bar{P}_{-i}^{GC,EQ}$	Other GENCOs are considered constants in their equilibrium strategies.
Dual Variables	
α	Dual variable related to constraint P_n^{inj}.
β	Dual variable related to the nodal natural gas flow balance constraint.
λ_l^-, λ_l^+	Dual variable of the thermal capacity bounds of transmission line l.
δ_j^-, δ_j^+	Dual variable of the capacity bounds of gas pipeline j.
γ_e^-, γ_e^+	Dual variable of the operating range for existing unit e.

μ_m	Dual variable of the gas balance equation at node m.
π_n	Locational marginal price reflecting the dual variable of the electricity balance equation at bus n.
σ_m^-, σ_m^+	Dual variable of the production capacity for gas source m.
ξ_i^-, ξ_i^+	Dual variable of the production operating range for candidate resource i.
Ancillary Variables	
η	Binary variables from the Fortuny-Amat linearization at *Level-3*.
y_{ki}	Binary variable that is equal to 1 if the kth step of the discretization of g_i is considered and is equal to 0 otherwise.
$\hat{y}_{ki}$	Product of ξ^+ by y_{ki}.
$\tilde{y}_{ki}$	Product of P_i^{GC} by y_{ki}.

Appendix A

We present the complete model of the transmission planning formulated as an MILP problem subject to the EPEC-MILP and market equilibrium constraints:

$$\begin{aligned}\min_{f_l^P, f_j^Q} \sum_{i\in N^{inv}} & \left[a_i(P_i^{GE,E} + P_i^{GC,E}) - b_i(\Delta_{g_i} \sum_{k=0}^{\Lambda_i} 2^k \tilde{y}_{ki}^E)\right] \\ & + \sum_{i\in N^{fix}} a_i P_i^{GE,E} + \sum_{i\in N^{inv}} K_i(\Delta_{g_i} \sum_{k=0}^{\Lambda_i} 2^k y_{ki}^E) \\ & + \sum_{l\in L^{inv}} K_l(f_l^P - f_l^{P,0}) + \sum_{j\in J^{inv}} K_j(f_j^Q - f_j^{Q,0})\end{aligned} \tag{A1}$$

subject to

$$f_l^{P,0} \le f_l^P \le f_l^{P,max} \tag{A2}$$

$$f_j^{Q,0} \le f_j^Q \le f_j^{Q,max} \tag{A3}$$

$$0 \le P_i^{GC,E} - \tilde{y}_{ki}^E \le M^{g_i}(1 - y_{ki}^E) \tag{A4}$$

$$0 \le \tilde{y}_{ki}^E \le M^{g_i} y_{ki}^E \tag{A5}$$

Equilibrium and Profit Definition

$$U_G^E \ge U_G^S \tag{A6}$$

$$\begin{aligned}U_G^E = \sum_e P_e^{GE,max} \gamma_e^{+,E} + \sum_{i\in N_G^{inv}} & \Big\{ (P_i^{GC,max} \xi_i^{+,E} + \\ \Delta_{g_i} \sum_{k=0}^{\Lambda_i} 2^k \hat{y}_{ki}^E) - K_i(\Delta_{g_i} & \sum_{k=0}^{\Lambda_i} 2^k y_{ki}^E) \Big\}\end{aligned} \tag{A7}$$

$$\begin{aligned}U_G^S = \sum_e P_e^{GE,max} \gamma_e^{+,S} + \sum_{i\in N_G^{inv}} & \Big\{ (P_i^{GC,max} \xi_i^{+,S} + \\ \Delta_{g_i} \sum_{k=0}^{\Lambda_i} 2^k \hat{y}_{ki}^S) - K_i(\Delta_{g_i} & \sum_{k=0}^{\Lambda_i} 2^k y_{ki}^S) \Big\}\end{aligned} \tag{A8}$$

Left-Hand-Side Constraints

$$a_e - \pi_n^E - \gamma_e^{-,E} + \gamma_e^{+,E} + c_e \mu_e^E = 0 \tag{A9}$$

$$\begin{aligned}& a_i - b_i(\Delta_{g_i} \sum_{k=0}^{\Lambda_i} 2^k y_{ki}^E) - \pi_n^E \\ & - \xi_i^{-,E} + \xi_i^{+,E} + c_i \mu_i^E = 0\end{aligned} \tag{A10}$$

$$\alpha^{E} - \pi_{n}^{E} + \sum_{l \in L} (\lambda_{l}^{+,E} - \lambda_{l}^{-,E}) \varphi_{l,n} = 0 \tag{A11}$$

$$-\mu_{m}^{E} - \sigma_{m}^{-,E} + \sigma_{m}^{+,E} = 0 \tag{A12}$$

$$\mu_{m}^{E} + \beta^{E} - \delta_{j}^{-,E} + \delta_{j}^{+,E} = 0 \tag{A13}$$

$$0 \le \xi_{i}^{-,E} \le M^{\xi_{i}^{-,E}} (\eta_{i}^{\xi^{-,E}}) \tag{A14}$$

$$0 \le P_{i}^{GC,E} \le M^{\xi_{i}^{-,E}} (1 - \eta_{i}^{\xi^{-,E}}) \tag{A15}$$

$$0 \le \xi_{i}^{+,E} \le M^{\xi_{i}^{+,E}} (\eta_{i}^{\xi^{+,E}}) \tag{A16}$$

$$0 \le \Delta_{g_i} \sum_{k=0}^{\Lambda_i} 2^{k} y_{ki}^{E} - P_{i}^{GC,E} \le M^{\xi_{i}^{+,E}} (1 - \eta_{i}^{\xi^{+,E}}) \tag{A17}$$

$$0 \le \gamma_{e}^{-,E} \le M^{\gamma_{e}^{-,E}} (\eta_{e}^{\gamma^{-,E}}) \tag{A18}$$

$$0 \le P_{e}^{GE,E} \le M^{\gamma_{e}^{-,E}} (1 - \eta_{e}^{\gamma^{-,E}}) \tag{A19}$$

$$0 \le \gamma_{e}^{+,E} \le M^{\gamma_{e}^{+,E}} (\eta_{e}^{\gamma^{+,E}}) \tag{A20}$$

$$0 \le P_{e}^{GE,max} - P_{e}^{GE,E} \le M^{\gamma_{e}^{+,E}} (1 - \eta_{e}^{\gamma^{+,E}}) \tag{A21}$$

$$0 \le \lambda_{l}^{-,E} \le M^{\lambda_{l}} (\eta_{l}^{\lambda^{-,E}}) \tag{A22}$$

$$0 \le f_{l}^{P} + \sum_{n \in N} \varphi_{l,n} P_{n}^{inj,E} \le M^{f_{l}^{P}} (1 - \eta_{l}^{\lambda^{-,E}}) \tag{A23}$$

$$0 \le \lambda_{l}^{+,E} \le M^{\lambda_{l}} (\eta_{l}^{\lambda^{+,E}}) \tag{A24}$$

$$0 \le f_{l}^{P} - \sum_{n \in N} \varphi_{l,n} P_{n}^{inj,E} \le M^{f_{l}^{P}} (1 - \eta_{l}^{\lambda^{+,E}}) \tag{A25}$$

$$0 \le \sigma_{m}^{-,E} \le M^{\sigma_{m}^{-,E}} (\eta_{m}^{\sigma^{-,E}}) \tag{A26}$$

$$0 \le S_{m} \le M^{\sigma_{m}^{-,E}} (1 - \eta_{m}^{\sigma^{-,E}}) \tag{A27}$$

$$0 \le \sigma_{m}^{+,E} \le M^{\sigma_{m}^{+,E}} (\eta_{m}^{\sigma^{+,E}}) \tag{A28}$$

$$0 \le S_{m}^{max} - S_{m} \le M^{\sigma_{m}^{+,E}} (1 - \eta_{m}^{\sigma^{+,E}}) \tag{A29}$$

$$0 \le \delta_{j}^{-,E} \le M^{\delta_{j}} (\eta_{j}^{\delta^{-,E}}) \tag{A30}$$

$$0 \le f_{j}^{Q} + Q_{j}^{L} \le M^{f_{j}^{Q}} (1 - \eta_{j}^{\delta^{-,E}}) \tag{A31}$$

$$0 \le \delta_{j}^{+,E} \le M^{\delta_{j}} (\eta_{j}^{\delta^{+,E}}) \tag{A32}$$

$$0 \le f_{j}^{Q} - Q_{j}^{L} \le M^{f_{j}^{Q}} (1 - \eta_{j}^{\delta^{+,E}}) \tag{A33}$$

$$0 \le \xi_{i}^{+,E} - \hat{y}_{ki}^{E} \le M^{\xi_{i}^{+}} (1 - y_{ki}^{E}) \tag{A34}$$

$$0 \le \hat{y}_{ki}^{E} \le M^{\xi_{i}^{+}} y_{ki}^{E} \tag{A35}$$

$$\sum_{n} P_{n}^{inj,E} = 0 \tag{A36}$$

$$\sum_{i \in N_{i,n}} P_{i}^{GC,E} + \sum_{i \in N_{e,n}} P_{e}^{GE,E} + P_{n}^{inj,E} = P_{n}^{D} \tag{A37}$$

$$S_{m}^{E} - Q_{m}^{D} - c_{e} P_{e}^{GE,E} - c_{i} P_{i}^{GC,E} = \sum_{n \in N_{m,j}} Q_{j}^{L,E} \tag{A38}$$

$$\sum_{j} Q_{j}^{L,E} = 0 \tag{A39}$$

Right -Hand-Side Constraints

$$a_e - \pi_n^S - \gamma_e^{-,S} + \gamma_e^{+,S} + c_e\mu_e^S = 0 \tag{A40}$$

$$a_i - b_i(\bar{P}_i^{GC.S}) - \pi_n^S - \xi_i^{-,S} + \xi_i^{+,S} + c_i\mu_i^S = 0 \tag{A41}$$

$$a_{-i} - b_{-i}(\Delta_{g_{-i}} \sum_{k=0}^{\Lambda_{-i}} 2^k y_{k_{-i}}^E) - \pi_n^S - \xi_{-i}^{-,S} + \xi_{-i}^{+,S} + c_{-i}\mu_{-i}^S = 0 \tag{A42}$$

$$\alpha^S - \pi_n^S + \sum_{l \in L} (\lambda_l^{+,S} - \lambda_l^{-,S})\varphi_{l,n} = 0 \tag{A43}$$

$$\mu_m^S - \sigma_m^{-,S} + \sigma_m^{+,S} = 0 \tag{A44}$$

$$\mu_m^S + \beta^S - \delta_j^{-,S} + \delta_j^{+,S} = 0 \tag{A45}$$

$$0 \le \xi_i^{-,S} \le M^{\xi_i^{-,S}}(\eta_i^{\xi^{-,S}}) \tag{A46}$$

$$0 \le P_i^{GC,S} \le M^{\xi_i^{-,S}}(1 - \eta_i^{\xi^{-,S}}) \tag{A47}$$

$$0 \le \xi_i^{+,S} \le M^{\xi_i^{+,S}}(\eta_i^{\xi^{+,S}}) \tag{A48}$$

$$0 \le \Delta_{g_i} \sum_{k=0}^{\Lambda_i} 2^k y_{ki}^E - P_i^{GC,S} \le M^{\xi_i^{+,S}}(1 - \eta_i^{\xi^{+,S}}) \tag{A49}$$

$$0 \le \gamma_e^{-,S} \le M^{\gamma_e^{-,S}}(\eta_e^{\gamma^{-,S}}) \tag{A50}$$

$$0 \le P_e^{GE,S} \le M^{\gamma_e^{-,S}}(1 - \eta_e^{\gamma^{-,S}}) \tag{A51}$$

$$0 \le \gamma_e^{+,S} \le M^{\gamma_e^{+,S}}(\eta_e^{\gamma^{+,S}}) \tag{A52}$$

$$0 \le P_e^{GE,max} - P_e^{GE,S} \le M^{\gamma_e^{+,S}}(1 - \eta_e^{\gamma^{+,S}}) \tag{A53}$$

$$0 \le \lambda_l^{-,S} \le M^{\lambda_l}(\eta_l^{\lambda^{-,S}}) \tag{A54}$$

$$0 \le f_l^P + \sum_{n \in N} \varphi_{l,n} P_n^{inj,S} \le M^{f_l^P}(1 - \eta_l^{\lambda^{-,S}}) \tag{A55}$$

$$0 \le \lambda_l^{+,EQ} \le M^{\lambda_l}(\eta_l^{\lambda^{+,S}}) \tag{A56}$$

$$0 \le f_l^P - \sum_{n \in N} \varphi_{l,n} P_n^{inj,S} \le M^{f_l^P}(1 - \eta_l^{\lambda^{+,S}}) \tag{A57}$$

$$0 \le \sigma_m^{-,S} \le M^{\sigma_m^{-,S}}(\eta_m^{\sigma^{-,S}}) \tag{A58}$$

$$0 \le S_m \le M^{\sigma_m^{-,S}}(1 - \eta_m^{\sigma^{-,S}}) \tag{A59}$$

$$0 \le \sigma_m^{+,S} \le M^{\sigma_m^{+,S}}(\eta_m^{\sigma^{+,S}}) \tag{A60}$$

$$0 \le S_m^{max} - S_m \le M^{\sigma_m^{+,S}}(1 - \eta_m^{\sigma^{+,S}}) \tag{A61}$$

$$0 \le \delta_j^{-,S} \le M^{\delta_j}(\eta_j^{\delta^{-,S}}) \tag{A62}$$

$$0 \le f_j^{Q,S} + Q_j^{L,S} \le M^{f_j^Q}(1 - \eta_j^{\delta^{-,S}}) \tag{A63}$$

$$0 \le \delta_j^{+,S} \le M^{\delta_j}(\eta_j^{\delta^{+,S}}) \tag{A64}$$

$$0 \le f_j^{Q,S} - Q_j^{L,S} \le M^{f_j^Q}(1 - \eta_j^{\delta^{+,S}}) \tag{A65}$$

$$0 \le \xi_{-i}^{+,S} - \hat{y}_{k-i}^S \le M^{\xi_{-i}^+}(1 - y_{k-i}^E) \tag{A66}$$

$$0 \le \hat{y}_{k-i}^S \le M^{\xi_{-i}^+} y_{k-i}^E \tag{A67}$$

$$\sum_n P_n^{inj,S} = 0 \tag{A68}$$

$$\sum_{i\in N_{i,n}} P_i^{GC,S} + \sum_{i\in N_{e,n}} P_e^{GE,S} + P_n^{inj,S} = P_n^D \tag{A69}$$

$$S_m^S - Q_m^D - c_e P_e^{GE,S} - c_i P_i^{GC,S} = \sum_{n\in N_{m,j}} Q_j^{L,S} \tag{A70}$$

$$\sum_j Q_j^{L,S} = 0 \tag{A71}$$

References

1. Wang, C.; Ju, P.; Wu, F.; Pan, X.; Wang, Z. A systematic review on power system resilience from the perspective of generation, network, and load. *Renew. Sustain. Energy Rev.* **2022**, *167*, 112567. [CrossRef]
2. Xu, Y.; Huang, C.; Li, X.; Li, F. A Novel Automatic Generation Control for Thermal and Gas Power Plants. In Proceedings of the 2018 IEEE Power Energy Society General Meeting (PESGM), Portland, OR, USA, 5–10 August 2018; pp. 1–5.
3. Wang, X.; McDonald, J.R. *Modern Power System Planning*; McGraw-Hill: NewYork, NY, USA, 1994.
4. Yao, Y.; Edmunds, T.; Papageorgiou, D.; Alvarez, R. Trilevel Optimization in Power Network Defense. *IEEE Trans. Syst. Man Cybern. Part C Appl. Rev.* **2007**, *37*, 712–718. [CrossRef]
5. Zhang, X.; Shahidehpour, M.; Alabdulwahab, A.; Abusorrah, A. Security-constrained co-optimization planning of electricity and natural gas transportation infrastructures. *IEEE Trans. Power Syst.* **2015**, *30*, 2984–2993. [CrossRef]
6. Zhang, X.; Che, L.; Shahidehpour, M.; Alabdulwahab, A.; Abusorrah, A. Reliability-Based Optimal Planning of Electricity and Natural Gas Interconnections for Multiple Energy Hubs. *IEEE Trans. Smart Grid* **2017**, *8*, 1658–1667. [CrossRef]
7. Zhang, X.; Che, L.; Shahidehpour, M.; Alabdulwahab, A.; Abusorrah, A. Electricity-Natural Gas Operation Planning with Hourly Demand Response for Deployment of Flexible Ramp. *IEEE Trans. Sustain. Energy* **2016**, *7*, 996–1004. [CrossRef]
8. Kamalinia, S.; Wu, L.; Shahidehpour, M. Stochastic Midterm Coordination of Hydro and Natural Gas Flexibilities for Wind Energy Integration. *IEEE Trans. Sustain. Energy* **2014**, *5*, 1070–1079. [CrossRef]
9. He, C.; Wu, L.; Liu, T.; Shahidehpour, M. Robust Co-Optimization Scheduling of Electricity and Natural Gas Systems via ADMM. *IEEE Trans. Sustain. Energy* **2017**, *8*, 658–670. [CrossRef]
10. Fang, J.; Zeng, Q.; Ai, X.; Chen, Z.; Wen, J. Dynamic Optimal Energy Flow in the Integrated Natural Gas and Electrical Power Systems. *IEEE Trans. Sustain. Energy* **2018**, *9*, 188–198. [CrossRef]
11. Nourollahi, R.; Gholizadeh-Roshanagh, R.; Feizi-Aghakandi, H.; Zare, K.; Mohammadi-Ivatloo, B. Power distribution expansion planning in the presence of wholesale multimarkets. *IEEE Syst. J.* **2022**, *17*, 1684–1693. [CrossRef]
12. Hakimi, S.M.; Hasankhani, A.; Shafie-khah, M.; Catalão, J.P. Stochastic planning of a multi-microgrid considering integration of renewable energy resources and real-time electricity market. *Appl. Energy* **2021**, *298*, 117215. [CrossRef]
13. Jin, S.; Ryan, S.M. A Tri-Level Model of Centralized Transmission and Decentralized Generation Expansion Planning for an Electricity Market—Part I. *IEEE Trans. Power Syst.* **2014**, *29*, 132–141. [CrossRef]
14. Gómez, S.; Olmos, L. Coordination of generation and transmission expansion planning in a liberalized electricity context—coordination schemes, risk management, and modelling strategies: A review. *Sustain. Energy Technol. Assessments* **2024**, *64*, 103731. [CrossRef]
15. Asgharian, V.; Abdelaziz, M. A low-carbon market-based multi-area power system expansion planning model. *Electr. Power Syst. Res.* **2020**, *187*, 106500. [CrossRef]
16. Dini, A.; Azarhooshang, A.; Pirouzi, S.; Norouzi, M.; Lehtonen, M. Security-Constrained generation and transmission expansion planning based on optimal bidding in the energy and reserve markets. *Electr. Power Syst. Res.* **2021**, *193*, 107017. [CrossRef]
17. Maliszewski, P.J.; Perrings, C. Factors in the resilience of electrical power distribution infrastructures. *Appl. Geogr.* **2012**, *32*, 668–679. [CrossRef]
18. Daeli, A.; Mohagheghi, S. Power grid infrastructural resilience against extreme events. *Energies* **2022**, *16*, 64. [CrossRef]
19. He, C.; Zhang, X.; Liu, T.; Wu, L. Distributionally Robust Scheduling of Integrated Gas-Electricity Systems With Demand Response. *IEEE Trans. Power Syst.* **2019**, *34*, 3791–3803. [CrossRef]
20. Fortuny-Amat, J.; McCarl, B. A representation and economic interpretation of a two-level programming problem. *J. Oper. Res. Soc.* **1981**, *32*, 783–792. [CrossRef]
21. IBM. *Getting Started with CPLEX for MATLAB Version 12.4*; IBM Corp.: Armonk, NY, USA , 2014.
22. Chen, Z.; Wu, M.; Zhao, Z. Evaluations of Aggregators and DERs in Distribution System Operations with Uncertainties. In Proceedings of the 2018 IEEE International Conference on Probabilistic Methods Applied to Power Systems (PMAPS), Boise, ID, USA, 24–28 June 2018; pp. 1–6. [CrossRef]
23. Kazempour, S.J.; Conejo, A.J.; Ruiz, C. Strategic Bidding for a Large Consumer. *IEEE Trans. Power Syst.* **2015**, *30*, 848–856. [CrossRef]
24. Pereira, M.V.; Granville, S.; Fampa, M.H.; Dix, R.; Barroso, L.A. Strategic bidding under uncertainty. *IEEE Trans. Power Syst.* **2005**, *20*, 180–188. [CrossRef]
25. Zeng, B.; Zhao, L. Solving two-stage robust optimization problems using a column-and-constraint generation method. *Oper. Res. Lett.* **2013**, *41*, 457–461. [CrossRef]

26. Eight Zone Test Bed. Available online: https://bitbucket.org/kdheepak/eightbustestbedrepo/src/master/ (accessed on 11 April 2024).
27. Zhao, B.; Conejo, A.J.; Sioshansi, R. Coordinated Expansion Planning of Natural Gas and Electric Power Systems. *IEEE Trans. Power Syst.* **2018**, *33*, 3064–3075. [CrossRef]
28. Energy, Load, and Demand Reports. Available online: https://www.iso-ne.com/isoexpress/web/reports/load-and-demand/-/tree/zone-info (accessed on 11 April 2024).
29. Natural Gas Consumption by End Use. Available online: https://www.eia.gov/dnav/ng/ng_cons_sum_a_EPG0_vgt_mmcf_m.htm (accessed on 11 April 2024).

processes MDPI

Article

Regional Operation of Electricity-Hythane Integrated Energy System Considering Coupled Energy and Carbon Trading

Dong Yang, Shufan Wang *, Wendi Wang, Weiya Zhang, Pengfei Yu and Wei Kong

Nanjing Suyi Industrial Co., Ltd., Nanjing 210009, China
* Correspondence: suyi_wangsf@163.com

Abstract: The deepening implementation of the energy and carbon market imposes trading requirements on multiple regional integrated energy system participants, including power generation plants, industrial users, and carbon capture, utilization, and storage (CCUS) facilities. Their diverse roles in different markets strengthen the interconnections among these subsystems. On the other hand, the operation of CCUS, containing carbon capture (CS), power-to-hydrogen (P2H), and power-to-gas (P2G), results in the coupling of regional carbon reduction costs with the operation of electricity and hythane networks. In this paper, we propose a regional economic dispatching model of an integrated energy system. The markets are organized in a centralized form, and their clearing conditions are considered. CCUS is designed to inject hydrogen or natural gas into hythane networks, operating more flexibly. A generalized Nash game is applied to analyze the multiple trading equilibria of different entities. Simulations are carried out to derive a different market equilibrium regarding network scales, seasonal load shifts, and the ownership of CCUS. Simulation results in a 39-bus/20-node coupled network indicate that the regional average carbon prices fluctuate from ¥1078.82 to ¥1519.03, and the organization of independent CCUS is more preferred under the proposed market structure.

Keywords: regional integrated energy system; economic dispatching; generalized Nash equilibrium; energy and carbon trading; carbon capture, utilization, and storage (CCUS)

Citation: Yang, D.; Wang, S.; Wang, W.; Zhang, W.; Yu, P.; Kong, W. Regional Operation of Electricity-Hythane Integrated Energy System Considering Coupled Energy and Carbon Trading. *Processes* **2024**, *12*, 2245. https://doi.org/10.3390/pr12102245

Academic Editors: Adam Smoliński and Dominic C. Y. Foo

Received: 17 August 2024
Revised: 14 September 2024
Accepted: 9 October 2024
Published: 14 October 2024

1. Introduction

Intensifying global warming has aroused widespread concern regarding greenhouse gas emissions [1], particularly carbon dioxide emissions from productive activities [2]. To price the negative externality of carbon emissions, many countries or regions opt to initiate carbon emission markets, aiming to fully harness the market mechanism in emission control and low-carbon transition [3–5]. From the perspective of current carbon market practices, both fossil fuel-based electricity generation and certain carbon-intensive industrial sectors have been incorporated as primary trading entities within carbon market mechanisms. For example, in Europe, participation in the Europe Emissions Trading System (EU ETS) is mandatory for electricity generation and industrial sectors including steel, paper, glass, and so on [6]. China is also preparing to incorporate industries, such as petrochemicals, chemicals, building materials, etc., into carbon control mechanisms [7].

Meanwhile, these carbon market participants are also the major trading entities in the regional integrated energy system [8–11]. Their operations are extensively coupled with trading behaviors. For instance, natural gas power plants not only need to purchase fuels from the gas networks for their production needs but also sell electricity to users or as reserves in the electricity system [12], and emissions from such production drive the need for additional carbon allowance. The industrial sectors also face similar circumstances. This creates interconnections not only within the operation level but also within the trading level. Furthermore, with the participation of carbon capture, utilization and storage (CCUS), the coupling relationship is expected to deepen and complicate. CCUS generally captures carbon dioxide from emissions of fossil fuel combustion. The produced carbon dioxide will either be sequestrated or utilized as the raw material for synthesizing natural gas, and this process requires additional hydrogen. In its operation, it can choose to sell either natural gas or hydrogen according to current gas prices, especially in an electricity-hythane integrated energy system.

In general, three types of carbon markets are considered in the existing research associated with coupled energy and carbon operation and trading. The first type assumes the emission cost is levied in the form of static carbon taxes. The emission changes brought by generator states (start-up or shutdown) and flexible loads and incorporates carbon intensives into real-time electricity balancing market are considered in [13]. Similar modeling has also been adopted in [14], which also takes the clearing of the gas market into consideration. The second type adopts ladder carbon prices. Its basic idea is that the marginal carbon emission price will increase as the total carbon emissions increase, thereby encouraging participants in the carbon market to control the total emissions at a lower level from the integrated energy system [15]. In [16], a demand side carbon trade scheme is proposed based on the ladder type carbon price to enhance the low-carbon demand response for integrated electricity–gas systems, enabling the carbon trades among different participants. In [17], a carbon emission and energy management framework is proposed for trades among industrial clusters. The third type considers the cap-and-trade mechanism and the carbon price signal is often generated by the duality variables of the market clearing condition. In [18], a two-layer policy optimization and evaluation model is developed for emission cap-and-trade in the electricity network. Ref. [19] proposes a complementary clearing condition for the emission market based on the Nash equilibrium. When all emissions of market participants reach the setting cap, the price of emission allowance will be positive; otherwise, the price will be zero. This clearing condition is also applied in [20]. It establishes a conjectural-variation equilibrium model to analyze the interconnections among electricity, gas, and carbon emission markets. It assumes only fossil-fuel units are required to trade carbon allowance. Constrained by the modeling concept, only one carbon price will be formed within the carbon commitment period. In [21], dynamic carbon emission intensity is utilized to analyze GENCO's behaviors in the joint electricity and carbon market. In [22], a novel role of a regional carbon market operator responsible for the centralized trading of carbon allowance is proposed in a regional carbon market to find the operational equilibrium of the electricity, gas, and carbon market. Ref. [23] estimates the Nash Equilibrium in the joint peer-to-peer electricity and carbon market considering the load and price uncertainties.

Some gaps exist between the carbon market assumptions in current research and real practice. On the consumption side of carbon allowance, in the electrical grids, users are drivers of carbon emissions [16], but not direct sources. From the perspective of the energy system operator, these emissions have already been accounted for through the carbon allowance allocated to the generators, thus it is not advisable to subject users to further commitment. On the other hand, direct emissions from fossil fuels consumption by end-users may compete with generators in the carbon market, which is not fully considered in the current research. On the supply side of carbon allowance, the direct source of carbon allowance is managed by the relevant administration overseeing carbon emissions. It formulates the annual allowance based on the government's long-term emission control plan and socio-economic development needs. The allocation of carbon allowance, either through free allocation or auctioning, is determined according to market liquidity and demand. Furthermore, in the near future, entities with carbon capture capabilities, such as carbon capture, utilization, and storage (CCUS) facilities, are also capable of generating carbon allowance, which isolates the emissions from the atmosphere. Therefore, the generated price signals in the carbon market not only serve to facilitate the allocation of carbon emission rights among diverse emitters but also need to reflect the marginal cost of the system's decarbonization. Considering the aforementioned gaps, in this paper, we propose a regional dispatching model of the electricity-hythane integrated energy system considering coupled energy and carbon trading. The contributions can be summarized as:

1. A centralized regional integrated market structure of electricity, hythane, and carbon allowance is established. The carbon market operator is designed to be responsible for assigning carbon allowance following the annual overall control principle of carbon emissions, constituting the supply side of the regional carbon market together with CCUS.
2. The benefits of multiple system participants are considered. Their trading behaviors are formulated by the generalized Nash game model and the complementary form is induced based on the Karush–Kuhn–Tucker (KKT) optimal condition.
3. Comprehensive simulations are conducted to test the proposed market equilibrium. The impact of seasonal load shift and CCUS ownership is analyzed.

The remainder of this paper is organized as follows: Section 2 provides the details of the structure, participants, and assumptions of the integrated market. Section 3 gives the optimization model formulation of the market participants and operators. Section 4 establishes the generalized Nash equilibrium of the integrated market based on the complementary model. Section 5 presents the simulation results. Section 6 concludes the paper.

2. System Structures, Participants, and Assumptions

Figure 1 illustrates the structure of the regionally integrated market. It is formulated by three local markets: the electricity market, the gas market, and the carbon allowance market.

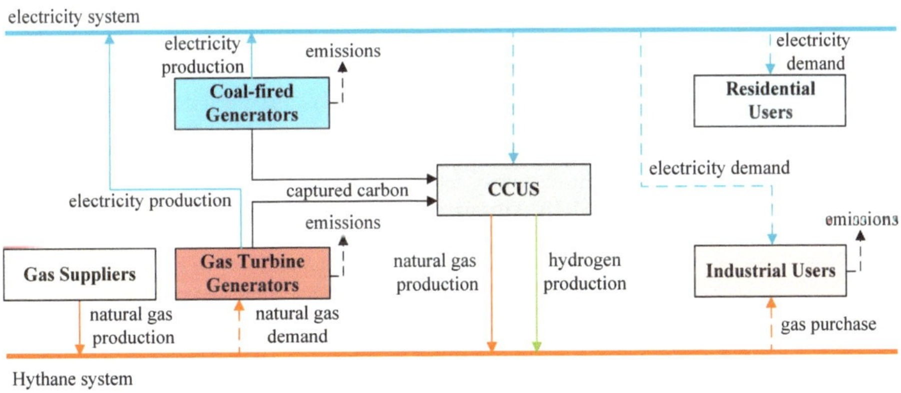

Figure 1. The structure of the regional integrated market of electricity, gas, and carbon allowance.

2.1. Participants Assumptions

Six types of system participants are considered. The roles of different system participants are summarized in Table 1.

Table 1. The Roles of Participants in Different Subsystems.

System Participant	Electricity	Hythane	Carbon
residential user	consumer	—	— *
industrial user	consumer	consumer	consumer
gas turbine generator	producer	consumer	consumer
coal-fired generator	producer	—	consumer
gas supplier	—	producer	—
CCUS	consumer	producer	producer

* Not involved.

The users are divided into two types, residential users and industrial users. We assume that the residential users only consume electricity. None of them produce direct carbon emissions from the combustion of fossil fuels, and they are not regulated by carbon allowance. Industrial users, however, require both electricity and hythane in production. They also need to purchase carbon allowance equal to the emissions from the combustion of hythane gas. It should be noted that although the emissions from the industrial production process should also be regulated in current carbon market practices, they are not considered which is beyond the scope of this study. Therefore, we simplify the carbon emissions of industrial users by only taking the emissions generated from consuming gas into account.

The generators also include two types, gas turbine generators and coal-fired generators. The main difference between them is the used fuels. The gas turbine generators purchase hythane in the gas market to produce electrical power for sale in the electricity market. The coal-fired generators, however, utilize coal to produce electrical power, so they do not participate in the gas market. They are both regulated by carbon allowance.

Gas suppliers here can be viewed as the upstream gas station. They are the main sources to provide natural gas in the gas market.

CCUS consists of three parts: carbon capture system (CC), power-to-hydrogen(P2H), and power-to-gas system (P2G). Generally, there are three types of CCs: pre-combustion capture, post-combustion capture, and oxy-fuel capture. We assume CC is post-combustion capture in this study, so it captures the carbon emissions from the combustion of generators [24]. The collected carbon emissions are stored for sequestration when P2H is operating only. This will happen when electricity prices are higher for further producing natural gas instead of merely storing carbon dioxide. Or they will be utilized as the raw materials to produce natural gas [25].

2.2. Market Assumptions

Three independent market operators are considered in the regional integrated energy system. They are the electrical market operator (EMO), the gas market operator (GMO), and the carbon market operator (CMO). We adopt locational marginal prices (LMPs) for the settlement in the electricity and gas market. These prices are obtained by solving the DC power flow model and the natural gas flow model [26–28]. Thus, the losses that happened in energy distribution are ignored.

As for the carbon price, we require that all carbon emitters purchase the amount of allowance equal to their emissions at each hour to avoid punishment for excessive emissions during the end of commitment periods. Furthermore, we note that the total amount of the available carbon allowance distributed by the corresponding administration is typically fixed during an annual carbon accounting period in current practices. This is mainly because carbon emissions are generally regulated by overall quantity control [29]. Thus, we assume that the CMO provides a certain amount of carbon allowance which is limited by the emission control goals and gains revenue from the settlement of the carbon market. In this way, CMO and CCUS together provide carbon allowance and affect carbon prices in the market. To focus on the coupling of the three markets, we do not consider the allowance distributed freely to emitters, which is a common practice in the cap-and-trade mechanism, and assume that all allowance needs to be purchased at the settlement prices.

3. Model Formulation

This section introduces the behavior model of the aforementioned system participants and market operators.

3.1. The Model of Market Participants

3.1.1. Residential Users

The behavior model of residential user k, $k \in \Omega_{\text{res}}$, is formulated as

$$\max \quad \sum_{t \in T} \left(a_{k,t}^{\text{res,e}} \left(q_{i(k),t}^{\text{res,e}} \right)^2 + b_{k,t}^{\text{res,e}} q_{i(k),t}^{\text{res,e}} \right)$$

$$-\sum_{t\in\mathcal{T}} \pi^{\rm e}_{i(k),t} q^{\rm res,e}_{i(k),t} \tag{1a}$$

$$\text{s.t.} \quad \underline{q}^{\rm res,e}_{k,t} \le q^{\rm res,e}_{k,t} \le \overline{q}^{\rm res,e}_{k,t}, \ \left[\underline{\mu}^{\rm res,1}_{k,t}, \overline{\mu}^{\rm res,1}_{k,t}\right]. \tag{1b}$$

The objective (1a) is the quadratic monetary utility of consuming electricity minus the cost of purchasing electricity [30]. (1b) imposes the minimum and maximum constraints on the electrical demands. The associated dual variables of each constraint are denoted in square brackets.

3.1.2. Industrial Users

The behavior model of industrial user k, $k \in \Omega_{\rm ind}$, is formulated as

$$\begin{aligned}
\max \quad & \sum_{t\in\mathcal{T}} \left(a^{\rm ind,e}_{k,t} \left(q^{\rm ind,e}_{i(k),t}\right)^2 + b^{\rm ind,e}_{k,t} q^{\rm ind,e}_{i(k),t} \right) \\
& + \sum_{t\in\mathcal{T}} \left(a^{\rm ind,g}_{k,t} \left(q^{\rm ind,g}_{j(k),t}\right)^2 + b^{\rm ind,g}_{k,t} q^{\rm ind,g}_{j(k),t} \right) \\
& - \sum_{t\in\mathcal{T}} \left(\pi^{\rm e}_{i(k),t} q^{\rm ind,e}_{i(k),t} + \pi^{\rm g}_{j(k),t} q^{\rm ind,g}_{j(k),t} + \pi^{\rm c}_{t} q^{\rm ind,c}_{k,t} \right) \qquad (2a) \\
\text{s.t.} \quad & \underline{q}^{\rm ind,e}_{k,t} \le q^{\rm ind,e}_{k,t} \le \overline{q}^{\rm ind,e}_{k,t}, \ \left[\underline{\mu}^{\rm ind,1}_{k,t}, \overline{\mu}^{\rm ind,1}_{k,t}\right], t \in \mathcal{T}, \qquad (2b) \\
& \underline{q}^{\rm ind,g}_{k,t} \le q^{\rm ind,g}_{k,t} \le \overline{q}^{\rm ind,g}_{k,t}, \ \left[\underline{\mu}^{\rm ind,2}_{k,t}, \overline{\mu}^{\rm ind,2}_{k,t}\right], t \in \mathcal{T}, \qquad (2c) \\
& q^{\rm ind,c}_{k,t} = \delta^{\rm gas} q^{\rm ind,g}_{k,t}, \ \left[\lambda^{\rm ind,3}_{k,t}\right], t \in \mathcal{T}. \qquad (2d)
\end{aligned}$$

The first and the second term in the objective (2a) is the utility of consuming electricity and natural gas, respectively. The third term is the cost of purchasing electricity, natural gas, and carbon allowance. Industrial users emit carbon dioxide when combusting hythane gas, so they need to purchase a carbon allowance equal to the amount of the emissions. (2b) and (2c) limits the minimum and maximum values of electrical and gas demands, respectively. (2d) converts the quantity of natural gas to carbon dioxide emissions. The conversion coefficient $\delta^{\rm gas}$ is calculated by

$$\delta^{\rm gas} = (1 - \Gamma_{\rm H_2}) \frac{m_{\rm CO_2}}{m_{\rm C}} {\rm C}^{\rm gas}_{\rm ar} {\rm OF}^{\rm gas}. \tag{3}$$

$m_{\rm CO_2}$ and $m_{\rm C}$ is the relative molecular mass, which takes 44 and 12, respectively. ${\rm C}^{\rm gas}_{\rm ar}$ is the carbon content as the received basis of natural gas. ${\rm OF}^{\rm gas}$ is the carbon oxidation rate. $\Gamma_{\rm H_2}$ is the proportion of enriched hydrogen.

3.1.3. Gas Turbine Generators

The behavior model of a gas turbine generator (GT) k, $k \in \Omega_{\rm gt}$, is formulated as

$$\begin{aligned}
\max \quad & \sum_{t\in\mathcal{T}} \left(\pi^{\rm e}_{i(k),t} q^{\rm gt,e}_{i(k),t} - \pi^{\rm g}_{j(k),t} q^{\rm gt,g}_{j(k),t} - \pi^{\rm c}_{t} q^{\rm gt,c}_{k,t} \right) \qquad (4a) \\
\text{s.t.} \quad & q^{\rm gt,g}_{j(k),t} = b^{\rm gt}_{k} q^{\rm gt,e}_{i(k),t} + c^{\rm gt}_{k}, \ \left[\lambda^{\rm gt,1}_{k,t}\right], t \in \mathcal{T}, \qquad (4b) \\
& \underline{q}^{\rm gt,e}_{k} \le q^{\rm gt,e}_{i(k),t} \le \overline{q}^{\rm gt,e}_{k}, \ \left[\underline{\mu}^{\rm gt,2}_{k,t}, \overline{\mu}^{\rm gt,2}_{k,t}\right], t \in \mathcal{T}, \qquad (4c) \\
& q^{\rm gt,c}_{k,t} = \delta^{\rm gas} q^{\rm gt,g}_{j(k),t}, \ \left[\lambda^{\rm gt,3}_{k,t}\right], t \in \mathcal{T}, \qquad (4d) \\
& -\overline{q}^{\rm gt,r}_{k} \le q^{\rm gt,e}_{i(k),t} - q^{\rm gt,e}_{i(k),t-1} \le \overline{q}^{\rm gt,r}_{k}, \\
& \left[\underline{\mu}^{\rm gt,4}_{k,t}, \overline{\mu}^{\rm gt,4}_{k,t}\right], 2 \le t \le T. \qquad (4e)
\end{aligned}$$

The objective (4a) maximizes the profit of k-th GT. The three terms represent the revenue of selling electrical power, the cost of purchasing natural gas, and the cost of purchasing carbon allowance, respectively. Constraint (4b) describes the fuel consumption of GT. (4c) limits the minimum and maximum output of GT. (4d) has the same meaning as (2d). (4e) limits the ramp power of the generator.

3.1.4. Coal-Fired Generators

The behavior model of coal-fired generators (GC) is similar to the model of GTs. The only difference is that coal-fired generators purchase coal as their fuel rather than natural gas. The model of coal-fired generator k, $k \in \Omega_{gc}$, is formulated as

$$\max \sum_{t\in\mathcal{T}} \left(\pi^{e}_{i(k),t} q^{gc,e}_{i(k),t} - \kappa^{coal} q^{gc,f}_{k,t} - \pi^{c}_{t} q^{gc,c}_{k,t} \right) \tag{5a}$$

$$\text{s.t.} \quad q^{gc,f}_{j(k),t} = b^{gc}_{k} q^{gc,e}_{i(k),t} + c^{gc}_{k}, \ \left[\lambda^{gc,1}_{k,t}\right], t \in \mathcal{T} \tag{5b}$$

$$\underline{q}^{gc,e}_{k} \le q^{gc,e}_{i(k),t} \le \overline{q}^{gc,e}_{k}, \ \left[\underline{\mu}^{gc,2}_{k,t}, \overline{\mu}^{gc,2}_{k,t}\right], t \in \mathcal{T} \tag{5c}$$

$$q^{gc,c}_{k,t} = \delta^{coal} q^{gc,f}_{j(k),t}, \ \left[\lambda^{ind,3}_{k,t}\right], t \in \mathcal{T} \tag{5d}$$

$$-\overline{q}^{gc,r}_{k} \le q^{gc,e}_{i(k),t} - q^{gc,e}_{i(k),t-1} \le \overline{q}^{gc,r}_{k}, \ \left[\underline{\mu}^{gc,4}_{k,t}, \overline{\mu}^{gc,4}_{k,t}\right], 2 \le t \le T. \tag{5e}$$

3.1.5. Gas Suppliers

The behavior model of the gas supplier k, $k \in \Omega_{gs}$ is formulated as

$$\max \sum_{t\in\mathcal{T}} \left(\pi^{g}_{j(k),t} q^{gs,g}_{j(k),t} - \kappa^{gas}_{k} q^{gs,g}_{k,t} \right) \tag{6a}$$

$$\text{s.t.} \quad \underline{q}^{gs,g}_{k} \le q^{gs,g}_{k,t} \le \overline{q}^{gs,g}_{k}, \ \left[\underline{\mu}^{gs,g}_{k,t}, \overline{\mu}^{gs,g}_{k,t}\right], t \in \mathcal{T}. \tag{6b}$$

The objective (6a) is the revenue of selling natural gas minus the cost of supplying natural gas. (6b) limits the minimum and maximum output of the gas supplier.

3.1.6. CCUS

As mentioned before, the CCUS contains three parts, the carbon capture system (CC), power-to-hydrogen (P2H) and the power-to-gas (P2G) system. The carbon emissions from combustion are captured by CC, and then utilized to produce methane (CH_4) by $CO_2 + 4H_2 \rightarrow CH_4 + 2H_2O$ or sequestrated as carbon storage (CS). Thus, the CCUS purchases electrical power and sells produced hydrogen or natural gas and carbon allowance during the carbon capture process. The behavior model of CCUS k, $k \in \Omega_{ccus}$, is formulated as

$$\max \sum_{t\in\mathcal{T}} \left(-\pi^{e}_{i(k),t} q^{ccus,e}_{i(k),t} + \pi^{g}_{j(k),t} q^{ccus,g}_{j(k),t} \right) + \sum_{t\in\mathcal{T}} \left(\pi^{c}_{t} q^{ccus,c}_{k,t} - \kappa^{CO_2} q^{ccus,CO_2}_{k,t} - \kappa^{cs} q^{cs,c}_{k,t} \right) \tag{7a}$$

$$\text{s.t.} \quad q^{ccus,e}_{k,t} = q^{cc,e}_{k,t} + q^{p2h,e}_{k,t} + q^{p2g,e}_{k,t}, \ \left[\lambda^{ccus,1}_{k,t}\right], t \in \mathcal{T}, \tag{7b}$$

$$q^{ccus,g}_{j(k),t} = q^{p2g,g}_{k,t} + \gamma q^{p2h,g2}_{k,t}, \ \left[\lambda^{ccus,2}_{k,t}\right], t \in \mathcal{T}, \tag{7c}$$

$$\underline{q}^{cc,e}_{k} \le q^{cc,e}_{k,t} \le \overline{q}^{cc,e}_{k}, \ \left[\underline{\mu}^{ccus,3}_{k,t}, \overline{\mu}^{ccus,3}_{k,t}\right], t \in \mathcal{T}, \tag{7d}$$

$$\underline{q}^{p2g,e}_{k} \le q^{p2g,e}_{k,t} \le \overline{q}^{p2g,e}_{k}, \ \left[\underline{\mu}^{ccus,4}_{k,t}, \overline{\mu}^{ccus,4}_{k,t}\right], t \in \mathcal{T}, \tag{7e}$$

$$\underline{q}_k^{\text{p2h,e}} \leq q_{k,t}^{\text{p2h,e}} \leq \overline{q}_k^{\text{p2h,e}}, \left[\underline{\mu}_{k,t}^{\text{ccus,5}}, \overline{\mu}_{k,t}^{\text{ccus,5}}\right], t \in \mathcal{T}, \tag{7f}$$

$$q_{k,t}^{\text{cc,e}} = \eta_k^{\text{cc,e}} q_{k,t}^{\text{cc,c}}, \left[\lambda_{k,t}^{\text{ccus,6}}\right], t \in \mathcal{T}, \tag{7g}$$

$$q_{k,t}^{\text{p2h,g}} = \eta_k^{\text{p2h,g}} q_{k,t}^{\text{p2h,e}}, \left[\lambda_{k,t}^{\text{ccus,7}}\right], t \in \mathcal{T}, \tag{7h}$$

$$q_{k,t}^{\text{p2g,g}} = \eta_k^{\text{p2g,g}} q_{k,t}^{\text{p2g,e}}, \left[\lambda_{k,t}^{\text{ccus,8}}\right], t \in \mathcal{T}, \tag{7i}$$

$$q_{k,t}^{\text{p2h,g1}} = \frac{4m_{\text{H}_2}}{m_{\text{CO}_2}} q_{k,t}^{\text{p2g,c}}, \left[\lambda_{k,t}^{\text{ccus,9}}\right], t \in \mathcal{T}, \tag{7j}$$

$$q_{k,t}^{\text{p2h,g}} = q_{k,t}^{\text{p2h,g1}} + q_{k,t}^{\text{p2h,g2}}, \left[\lambda_{k,t}^{\text{ccus,10}}\right], t \in \mathcal{T}, \tag{7k}$$

$$q_{k,t}^{\text{p2g,g}} = \frac{m_{\text{CH}_4}}{m_{\text{CO}_2}} q_{k,t}^{\text{p2g,c}}, \left[\lambda_{k,t}^{\text{ccus,11}}\right], t \in \mathcal{T}, \tag{7l}$$

$$q_{k,t}^{\text{p2g,c}} \geq 0, \left[\mu_{k,t}^{\text{ccus,12}}\right], t \in \mathcal{T}, \tag{7m}$$

$$q_{k,t}^{\text{p2g,c}} = q_{k,t}^{\text{cu,c}} + q_{k,t}^{\text{ccus,CO}_2}, \left[\lambda_{k,t}^{\text{ccus,13}}\right], t \in \mathcal{T}, \tag{7n}$$

$$q_{k,t}^{\text{ccus,c}} \geq 0, \left[\mu_{k,t}^{\text{ccus,14}}\right], t \in \mathcal{T}, \tag{7o}$$

$$q_{k,t}^{\text{cc,c}} = q_{k,t}^{\text{cu,c}} + q_{k,t}^{\text{cs,c}}, \left[\lambda_{k,t}^{\text{ccus,15}}\right], t \in \mathcal{T}, \tag{7p}$$

$$q_{k,t}^{\text{cu,c}} \geq 0, \left[\mu_{k,t}^{\text{ccus,16}}\right], t \in \mathcal{T}, \tag{7q}$$

$$q_{k,t}^{\text{cs,c}} \geq 0, \left[\mu_{k,t}^{\text{ccus,17}}\right], t \in \mathcal{T}, \tag{7r}$$

$$q_{k,t}^{\text{ccus,c}} = q_{k,t}^{\text{cc,c}}, \left[\lambda_{k,t}^{\text{ccus,18}}\right], t \in \mathcal{T}, \tag{7s}$$

$$q_{k,t}^{\text{ccus,CO}_2} \geq 0, \left[\mu_{k,t}^{\text{ccus,19}}\right], t \in \mathcal{T}, \tag{7t}$$

$$\sum_{k \in \Omega_{\text{ccus}}} q_{k,t}^{\text{cc,c}} \leq \sum_{k \in \Omega_{\text{gc}}} q_{k,t}^{\text{gc,c}} + \sum_{k \in \Omega_{\text{gt}}} q_{k,t}^{\text{gt,c}},$$

$$\left[\mu_t^{\text{ccus,20}}\right], t \in \mathcal{T}. \tag{7u}$$

$$q_{k,t}^{\text{p2h,g1}} \geq 0, \left[\mu_{k,t}^{\text{ccus,21}}\right], t \in \mathcal{T}, \tag{7v}$$

$$q_{k,t}^{\text{p2h,g2}} \geq 0, \left[\mu_{k,t}^{\text{ccus,22}}\right], t \in \mathcal{T}, \tag{7w}$$

(7x)

The objective (7a) maximizes the profit of CCUS in the three markets. The five terms are the cost of purchasing electrical power, the revenue of selling natural gas, the revenue of selling carbon allowance, the cost of purchasing additional CO_2 for producing natural gas, and the cost of carbon storage, respectively. κ^{CO_2} is the marginal cost for external CO_2. κ^{cs} is the marginal cost for carbon storage. (7b), (7c) and (7s) states the connection between the trade amounts of CCUS and its internal operating variables. Especially, (7c) specifies the trading gas of CCUS contains the produced natural gas and hydrogen with equal calorific value. (7d)–(7f) constrains the upper and lower electrical power of CC, P2H, and P2G. (7g)–(7i) describes the energy conversion of CC, P2H, and P2G. According to the conservation of mass, (7j), (7l) states the mass conversion between natural gas production and CO_2 demand. (7k) indicates the produced hydrogen from P2H either be used as the raw material for P2G or be sold. (7n) states that the CO_2 source of P2G comprises the external procurement and the captured emissions for utilization. (7p) divides the total amount of the captured emissions into two parts: utilization and storage. (7m), (7o), (7q), (7r), (7t), (7v), and (7w) are the non-negative constraints. Under the assumption of post-combustion capture, (7q) limits the maximum amount of captured emissions no larger than the total amount of emitted CO_2 from all generators.

3.2. The Model of Market Operators

3.2.1. Electrical Market Operator

The behavior of EMO is formulated as

$$\max \quad \sum_{t\in\mathcal{T}}\sum_{m\in\mathcal{V}^{\mathrm{e}}} \pi^{\mathrm{e}}_{m,t} q^{\mathrm{e}}_{m,t} \tag{8a}$$

$$\text{s.t.} \quad \sum_{(mn)\in\mathcal{E}^{\mathrm{e}}} B_{mn}(\theta_{m,t}-\theta_{n,t}) + q^{\mathrm{e}}_{m,t} = 0, \quad \left[\lambda^{\mathrm{e},1}_{m,t}\right], \forall m \in \mathcal{V}^{\mathrm{e}}, t \in \mathcal{T}, \tag{8b}$$

$$\sum_{i(k)=m} \left(q^{\mathrm{res,e}}_{i(k),t} + q^{\mathrm{ind,e}}_{i(k),t} + q^{\mathrm{ccus,e}}_{i(k),t}\right) - \sum_{i(k)=m} \left(q^{\mathrm{gc,e}}_{i(k),t} + q^{\mathrm{gt,e}}_{i(k),t}\right) - q^{\mathrm{e}}_{m,t} = 0, \quad \left[\lambda^{\mathrm{e},2}_{m,t}\right], \forall m \in \mathcal{V}^{\mathrm{e}}, t \in \mathcal{T}, \tag{8c}$$

$$-\overline{P}^{\mathrm{e}}_{mn} \le B_{mn}(\theta_{m,t}-\theta_{n,t}) \le \overline{P}^{\mathrm{e}}_{mn}, \quad \left[\underline{\mu}^{\mathrm{e},3}_{mn,t}, \overline{\mu}^{\mathrm{e},3}_{mn,t}\right], (mn) \in \mathcal{E}^{\mathrm{e}}, t \in \mathcal{T}. \tag{8d}$$

The objective (8a) maximizes the social welfare in the electrical power market. (8b) keeps the power balance at node m. (8c) defines the net electrical demand at node m. (8d) limits the lower/upper limits of line transmission capacity.

3.2.2. Gas Market Operator

The behavior of GMOs is formulated as

$$\max \quad \sum_{t\in\mathcal{T}}\sum_{m\in\mathcal{V}^{\mathrm{g}}} \pi^{\mathrm{g}}_{m,t} q^{\mathrm{g}}_{m,t} \tag{9a}$$

$$\text{s.t.} \quad \sum_{n\in\mathcal{V}^{\mathrm{g}}_{\mathrm{to}}(m)} f_{mn,t} - \sum_{n\in\mathcal{V}^{\mathrm{g}}_{\mathrm{from}}(m)} f_{mn,t} - \sum_{m\in\mathcal{V}^{\mathrm{com}}_{\mathrm{from}}(c)} f^{\mathrm{com}}_{c,t} + \sum_{m\in\mathcal{V}^{\mathrm{com}}_{\mathrm{to}}(c)} (1+\delta^{\mathrm{com}}_{c}) f^{\mathrm{com}}_{c,t} + q^{\mathrm{g}}_{m,t} = 0, \quad \left[\lambda^{\mathrm{g},1}_{m,t}\right], \forall m \in \mathcal{V}^{\mathrm{g}}, t \in \mathcal{T}, \tag{9b}$$

$$\sum_{j(k)=m} \left(q^{\mathrm{ind,g}}_{j(k),t} + q^{\mathrm{gc,g}}_{j(k),t}\right) - \sum_{j(k)=m} \left(q^{\mathrm{gas,g}}_{j(k),t} + q^{\mathrm{ccus,g}}_{j(k),t}\right) - q^{\mathrm{g}}_{m,t} = 0 \quad \left[\lambda^{\mathrm{g},2}_{m,t}\right], \forall m \in \mathcal{V}^{\mathrm{g}}, t \in \mathcal{T}, \tag{9c}$$

$$\left\| \begin{matrix} 2f_{mn,t}/S_{mn} \\ \Pi_{m,t}-\Pi_{n,t}-1 \end{matrix} \right\|_2 \le \Pi_{m,t} - \Pi_{n,t} + 1, \quad \left[\Lambda^{\mathrm{g},1}_{mn,t}, \Lambda^{\mathrm{g},2}_{mn,t}, \Lambda^{\mathrm{g},3}_{mn,t}\right], \forall (mn) \in \mathcal{E}^{\mathrm{g}}, t \in \mathcal{T}, \tag{9d}$$

$$f_{mn,t} \ge 0, \left[\mu^{\mathrm{g},4}_{mn,t}\right], \forall (mn) \in \mathcal{E}^{\mathrm{g}}, t \in \mathcal{T}, \tag{9e}$$

$$\underline{\Pi}_{m} \le \Pi_{m,t} \le \overline{\Pi}_{m}, \quad \left[\underline{\mu}^{\mathrm{g},5}_{m,t}, \overline{\mu}^{\mathrm{g},5}_{m,t}\right], \forall m \in \mathcal{V}^{\mathrm{g}}, t \in \mathcal{T}, \tag{9f}$$

$$\underline{\rho}^{\mathrm{com}}_{c} \Pi^{\mathrm{com,to}}_{c,t} \le \Pi^{\mathrm{com,from}}_{c,t} \le \overline{\rho}^{\mathrm{com}}_{c} \Pi^{\mathrm{com,to}}_{c,t}, \quad \left[\underline{\mu}^{\mathrm{g},6}_{c,t}, \overline{\mu}^{\mathrm{g},6}_{c,t}\right], \forall c \in \Omega_{\mathrm{com}}, t \in \mathcal{T}, \tag{9g}$$

$$0 \leq f_{c,t}^{\text{com}} \leq \overline{f}_c^{\text{com}},$$
$$\left[\underline{\mu}_{c,t}^{g,7}, \overline{\mu}_{c,t}^{g,7}\right], \forall c \in \Omega_{\text{com}}, t \in \mathcal{T}. \quad (9h)$$

The objective (9a) maximizes the social welfare in the regional gas market. (9b) keeps the gas balance at node m. (9c) calculates the net demand at node m. (9c) is the second-order cone relaxation of the Weymouth equation $(f_{mn,t}/S_{mn})^2 = \Pi_{m,t} - \Pi_{n,t}$, which defines the relationship between pipeline flows and nodal pressure. At the same time, we assume the gas flow directions are known to avoid the introduction of binary variables. (9d) is the non-negative constraint of gas flows. We assume that GMO knows the direction of gas flows in advance to avoid introducing the binary variables. (9e) limits the nodal pressure. (9f) poses constraints on the input/output pressure of the compressors. (9h) limits the gas flow through the compressor.

3.2.3. Carbon Market Operator

The behavior of the CMO is formulated as

$$\max \quad \sum_{t\in\mathcal{T}} \pi_t^{c} q_t^{\text{cmo,c}} \quad (10a)$$
$$\text{s.t.} \quad \underline{q}^{\text{cmo,c}} \leq \sum_{t\in\mathcal{T}} q_t^{\text{cmo,c}} \leq \overline{q}^{\text{cmo,c}},$$
$$\left[\underline{\mu}_t^{c,1}, \overline{\mu}_t^{c,1}\right], t \in \mathcal{T}, \quad (10b)$$
$$q_t^{\text{cmo,c}} \geq 0, \left[\mu_t^{c,2}\right], t \in \mathcal{T}, \quad (10c)$$
$$\sum_{k\in\Omega_{\text{ind}}} q_{k,t}^{\text{ind,c}} + \sum_{k\in\Omega_{\text{gc}}} q_{k,t}^{\text{gc,c}} + \sum_{k\in\Omega_{\text{gt}}} q_{k,t}^{\text{gt,c}}$$
$$= \sum_{k\in\Omega_{\text{ccus}}} q_{k,t}^{\text{ccus,c}} + q_{k,t}^{\text{cmo,c}}, \left[\lambda_t^{c,2}\right], t \in \mathcal{T}, \quad (10d)$$

The objective (10a) maximizes the revenue of allocating carbon allowance to the market. (10b) limits the total amount of available carbon allowance from the carbon market administration, the upper limit of which is also the periodic emission control target. (10c) is the non-negative constraint. We assume that there is no buy-back of carbon allowance. (10d) clears the carbon market. The carbon emitters are required to purchase a carbon allowance equal to their emissions to avoid excessiveness or insufficiency.

Additionally, a set of price inequality constraints is added to avoid unreasonable energy and carbon prices.

$$\pi_{m,t}^{e} \geq \underline{\pi}^{e}, t \in \mathcal{T}, m \in \mathcal{V}^{e}, \quad (11a)$$
$$\pi_{m,t}^{g} \geq \underline{\pi}^{g}, t \in \mathcal{T}, m \in \mathcal{V}^{g}, \quad (11b)$$
$$\pi_t^{c} \geq \underline{\pi}^{c}, t \in \mathcal{T}. \quad (11c)$$

4. Generalized Nash Equilibrium and Solution Methodology

The multiple mutually dependent constraints can be observed in the previous participant and market models, such as (7q), (8c), (9c), and (10d). These constraints involve variables from different market participants. Thus, the available strategy sets of market participants not only depend on their own operational characteristics but also the strategies the other take [31]. Based on that, we formulate the market equilibrium problem as a generalized Nash equilibrium problem (GNEP). To find out the market equilibrium, the convex optimization problem of behavior models (1)–(10) are converted into complementary models based on Karush–Kuhn–Tucker (KTT) conditions [32].

For simplicity, the condition $t \in \mathcal{T}$ is omitted in the following analysis.

The KKT conditions of (1) are as follows:

$$\pi^{\text{e}}_{i(k),t} - 2a^{\text{res,e}}_{k,t}q^{\text{res,e}}_{i(k),t} - b^{\text{res,e}}_{k,t} - \underline{\mu}^{\text{res,1}}_{k,t} + \overline{\mu}^{\text{res,1}}_{k,t} = 0, \tag{12a}$$

$$0 \le \underline{\mu}^{\text{res,1}}_{k,t} \perp \underline{q}^{\text{res,e}}_{k,t} - q^{\text{res,e}}_{k,t} \le 0, \tag{12b}$$

$$0 \le \overline{\mu}^{\text{res,1}}_{k,t} \perp q^{\text{res,e}}_{k,t} - \overline{q}^{\text{res,e}}_{k,t} \le 0. \tag{12c}$$

The KKT conditions of (2) are as follows:

$$\pi^{\text{e}}_{i(k),t} - 2a^{\text{ind,e}}_{k,t}q^{\text{ind,e}}_{i(k),t} - b^{\text{ind,e}}_{k,t} - \underline{\mu}^{\text{ind,1}}_{k,t} + \overline{\mu}^{\text{ind,1}}_{k,t} = 0, \tag{13a}$$

$$\begin{aligned}&\pi^{\text{g}}_{j(k),t} - 2a^{\text{ind,g}}_{k,t}q^{\text{ind,g}}_{j(k),t} - b^{\text{ind,g}}_{k,t}\\ &- \underline{\mu}^{\text{ind,2}}_{k,t} + \overline{\mu}^{\text{ind,2}}_{k,t} + \delta^{\text{gas}}\lambda^{\text{ind,3}}_{k,t} = 0,\end{aligned} \tag{13b}$$

$$\pi^{\text{c}}_{t} - \lambda^{\text{ind,3}}_{k,t} = 0, \tag{13c}$$

$$0 \le \underline{\mu}^{\text{ind,1}}_{k,t} \perp \underline{q}^{\text{ind,e}}_{k,t} - q^{\text{ind,e}}_{k,t} \le 0, \tag{13d}$$

$$0 \le \overline{\mu}^{\text{ind,1}}_{k,t} \perp q^{\text{ind,e}}_{k,t} - \overline{q}^{\text{ind,g}}_{k,t} \le 0, \tag{13e}$$

$$0 \le \underline{\mu}^{\text{ind,2}}_{k,t} \perp \underline{q}^{\text{ind,g}}_{k,t} - q^{\text{ind,e}}_{k,t} \le 0, \tag{13f}$$

$$0 \le \overline{\mu}^{\text{ind,2}}_{k,t} \perp q^{\text{ind,g}}_{k,t} - \overline{q}^{\text{ind,g}}_{k,t} \le 0. \tag{13g}$$

The KKT conditions of (4) are as follows:

$$\begin{aligned}&-\pi^{\text{e}}_{i(k),t} - b^{\text{gt,e}}_{k,t}\lambda^{\text{gt,1}}_{k,t} - \underline{\mu}^{\text{gt,2}}_{k,t}\\ &+ \overline{\mu}^{\text{gt,2}}_{k,t} - \underline{\mu}^{\text{gt,4}}_{k,t} + \overline{\mu}^{\text{gt,4}}_{k,t} = 0,\ t = T,\end{aligned} \tag{14a}$$

$$\begin{aligned}&-\pi^{\text{e}}_{i(k),t} - b^{\text{gt,e}}_{k,t}\lambda^{\text{gt,1}}_{k,t} - \underline{\mu}^{\text{gt,2}}_{k,t} + \overline{\mu}^{\text{gt,2}}_{k,t}\\ &- \underline{\mu}^{\text{gt,4}}_{k,t} + \overline{\mu}^{\text{gt,4}}_{k,t} + \underline{\mu}^{\text{gt,4}}_{k,t+1} - \overline{\mu}^{\text{gt,4}}_{k,t+1} = 0,\ t \le T-1,\end{aligned} \tag{14b}$$

$$\pi^{\text{g}}_{j(k),t} + \lambda^{\text{gt,1}}_{k,t} + \delta^{\text{gas}}\lambda^{\text{gt,3}}_{k,t} = 0, \tag{14c}$$

$$\pi^{\text{c}}_{t} - \lambda^{\text{gt,3}}_{k,t} = 0, \tag{14d}$$

$$0 \le \underline{\mu}^{\text{gt,2}}_{k,t} \perp \underline{q}^{\text{gt,e}}_{k} - q^{\text{gt,e}}_{i(k),t} \le 0, \tag{14e}$$

$$0 \le \overline{\mu}^{\text{gt,2}}_{k,t} \perp q^{\text{gt,e}}_{i(k),t} - \overline{q}^{\text{gt,e}}_{k} \le 0, \tag{14f}$$

$$0 \le \underline{\mu}^{\text{gt,4}}_{k,t} \perp -\overline{q}^{\text{gt,r}}_{k} - q^{\text{gt,e}}_{i(k),t} + q^{\text{gt,e}}_{i(k),t-1} \le 0, \tag{14g}$$

$$0 \le \overline{\mu}^{\text{gt,4}}_{k,t} \perp q^{\text{gt,e}}_{i(k),t} - q^{\text{gt,e}}_{i(k),t-1} - \overline{q}^{\text{gt,r}}_{k} \le 0, \tag{14h}$$

The KKT conditions of (5) are as follows:

$$\begin{aligned}&-\pi^{\text{e}}_{i(k),t} - b^{\text{gc,e}}_{k,t}\lambda^{\text{gc,1}}_{k,t} - \underline{\mu}^{\text{gc,2}}_{k,t}\\ &+ \overline{\mu}^{\text{gc,2}}_{k,t} - \underline{\mu}^{\text{gc,4}}_{k,t} + \overline{\mu}^{\text{gc,4}}_{k,t} = 0,\ t = T,\end{aligned} \tag{15a}$$

$$\begin{aligned}&-\pi^{\text{e}}_{i(k),t} - b^{\text{gc,e}}_{k,t}\lambda^{\text{gc,1}}_{k,t} - \underline{\mu}^{\text{gc,2}}_{k,t} + \overline{\mu}^{\text{gc,2}}_{k,t}\\ &- \underline{\mu}^{\text{gc,4}}_{k,t} + \overline{\mu}^{\text{gc,4}}_{k,t} + \underline{\mu}^{\text{gc,4}}_{k,t+1} - \overline{\mu}^{\text{gc,4}}_{k,t+1} = 0,\ t \le T-1,\end{aligned} \tag{15b}$$

$$\kappa^{\text{coal}} + \lambda^{\text{gc,1}}_{k,t} + \delta^{\text{coal}}\lambda^{\text{gc,3}}_{k,t} = 0, \tag{15c}$$

$$\pi^{\text{c}}_{t} - \lambda^{\text{gc,3}}_{k,t} = 0, \tag{15d}$$

$$0 \le \underline{\mu}^{\text{gt,2}}_{k,t} \perp \underline{q}^{\text{gc,e}}_{k} - q^{\text{gc,e}}_{i(k),t} \le 0, \tag{15e}$$

$$0 \le \overline{\mu}^{\text{gt,2}}_{k,t} \perp q^{\text{gc,e}}_{i(k),t} - \overline{q}^{\text{gt,e}}_{k} \le 0, \tag{15f}$$

$$0 \le \underline{\mu}_{k,t}^{\text{gc},4} \perp -\overline{q}_{k}^{\text{gc,r}} - q_{i(k),t}^{\text{gc,e}} + q_{i(k),t-1}^{\text{gc,e}} \le 0, \tag{15g}$$

$$0 \le \overline{\mu}_{k,t}^{\text{gc},4} \perp q_{i(k),t}^{\text{gc,e}} - q_{i(k),t-1}^{\text{gc,e}} - \overline{q}_{k}^{\text{gc,r}} \le 0, \tag{15h}$$

The KKT conditions of (6) are as follows:

$$-\pi_{j(k),t}^{\text{g}} + \kappa_{k}^{\text{g}} - \underline{\mu}_{k,t}^{\text{gas},1} + \overline{\mu}_{k,t}^{\text{gas},1} = 0, \tag{16a}$$

$$0 \le \underline{\mu}_{k,t}^{\text{gas},1} \perp \underline{q}_{k}^{\text{gas,g}} - q_{j(k),t}^{\text{gas,g}} \le 0, \tag{16b}$$

$$0 \le \overline{\mu}_{k,t}^{\text{gas},1} \perp q_{j(k),t}^{\text{gas,g}} - \overline{q}_{k}^{\text{gas,g}} \le 0, \tag{16c}$$

$$0 \le \overline{\mu}_{k,t}^{\text{gas},1} \perp q_{j(k),t}^{\text{gas,g}} - \overline{q}_{k}^{\text{gas,g}} \le 0, \tag{16d}$$

The KKT conditions of (7) are as follows:

$$-\pi_{i(k),t}^{\text{e}} + \lambda_{k,t}^{\text{ccus},1} = 0, \tag{17a}$$

$$\pi_{j(k),t}^{\text{g}} + \lambda_{k,t}^{\text{ccus},2} = 0, \tag{17b}$$

$$\pi_{k,t}^{\text{c}} - \mu_{k,t}^{\text{ccus},14} + \lambda_{k,t}^{\text{ccus},18} = 0, \tag{17c}$$

$$-\kappa^{\text{CO}_2} - \lambda_{k,t}^{\text{ccus},13} - \mu_{k,t}^{\text{ccus},19} = 0, \tag{17d}$$

$$-\kappa^{\text{cs}} - \lambda_{k,t}^{\text{ccus},15} + \mu_{k,t}^{\text{ccus},17} = 0, \tag{17e}$$

$$-\lambda_{k,t}^{\text{ccus},1} - \underline{\mu}_{k,t}^{\text{ccus},3} + \overline{\mu}_{k,t}^{\text{ccus},3} + \lambda_{k,t}^{\text{ccus},6} = 0, \tag{17f}$$

$$-\lambda_{k,t}^{\text{ccus},1} - \underline{\mu}_{k,t}^{\text{ccus},5} + \overline{\mu}_{k,t}^{\text{ccus},5} - \eta_{k}^{\text{p2h,g}} \lambda_{k,t}^{\text{ccus},7} = 0, \tag{17g}$$

$$-\lambda_{k,t}^{\text{ccus},1} - \underline{\mu}_{k,t}^{\text{ccus},4} + \overline{\mu}_{k,t}^{\text{ccus},4} - \eta_{k}^{\text{p2h,g}} \lambda_{k,t}^{\text{ccus},8} = 0, \tag{17h}$$

$$-\lambda_{k,t}^{\text{ccus},2} + \lambda_{k,t}^{\text{ccus},8} + \lambda_{k,t}^{\text{ccus},11} = 0, \tag{17i}$$

$$-\gamma \lambda_{k,t}^{\text{ccus},2} - \lambda_{k,t}^{\text{ccus},10} - \mu_{k,t}^{\text{ccus},22} - 0, \tag{17j}$$

$$\begin{aligned} &0 \le \underline{\mu}_{k,t}^{\text{ccus},3} \perp \underline{q}_{k}^{\text{cc,e}} - q_{k,t}^{\text{cc,e}} \le 0, \\ &0 \le \overline{\mu}_{k,t}^{\text{ccus},3} \perp q_{k,t}^{\text{cc,e}} - \overline{q}_{k}^{\text{cc,e}} \le 0, \end{aligned} \tag{17k}$$

$$\begin{aligned} &0 \le \underline{\mu}_{k,t}^{\text{ccus},4} \perp \underline{q}_{k}^{\text{p2g,e}} - q_{k,t}^{\text{p2g,e}} \le 0, \\ &0 \le \overline{\mu}_{k,t}^{\text{ccus},4} \perp q_{k,t}^{\text{p2g,e}} - \overline{q}_{k}^{\text{p2g,e}} \le 0, \end{aligned} \tag{17l}$$

$$\begin{aligned} &0 \le \underline{\mu}_{k,t}^{\text{ccus},5} \perp \underline{q}_{k}^{\text{p2h,e}} - q_{k,t}^{\text{p2h,e}} \le 0, \\ &0 \le \overline{\mu}_{k,t}^{\text{ccus},5} \perp q_{k,t}^{\text{p2h,e}} - \overline{q}_{k}^{\text{p2h,e}} \le 0, \end{aligned} \tag{17m}$$

$$-\eta^{\text{cc,e}_k} \lambda_{k,t}^{\text{ccus},6} + \lambda_{k,t}^{\text{ccus},15} - \lambda_{k,t}^{\text{ccus},18} + \mu_{k,t}^{\text{ccus},20} = 0, \tag{17n}$$

$$\lambda_{k,t}^{\text{ccus},7} + \lambda_{k,t}^{\text{ccus},10} = 0, \tag{17o}$$

$$\lambda_{k,t}^{\text{ccus},9} - \lambda_{k,t}^{\text{ccus},10} - \mu_{k,t}^{\text{ccus},21} = 0, \tag{17p}$$

$$-\frac{4m_{\text{H}_2}}{\text{CO}_2} \lambda_{k,t}^{\text{ccus},9} - \frac{m_{\text{CH}_4}}{\text{CO}_2} \lambda_{k,t}^{\text{ccus},11} - \mu_{k,t}^{\text{ccus},12} + \lambda_{k,t}^{\text{ccus},13} = 0, \tag{17q}$$

$$-\lambda_{k,t}^{\text{ccus},13} - \lambda_{k,t}^{\text{ccus},15} - \mu_{k,t}^{\text{ccus},16} = 0, \tag{17r}$$

$$-\lambda_{k,t}^{\text{ccus},15} - \mu_{k,t}^{\text{ccus},17} = 0, \tag{17s}$$

$$0 \le \mu_{k,t}^{\text{ccus},16} \perp -q_{k,t}^{\text{cu,c}} \le 0, \tag{17t}$$

$$0 \le \mu_{k,t}^{\text{ccus},17} \perp -q_{k,t}^{\text{cs,c}} \le 0, \tag{17u}$$

$$0 \le \mu_{k,t}^{\text{ccus},19} \perp -q_{k,t}^{\text{ccus,CO}_2} \le 0, \tag{17v}$$

$$0 \le \mu_{k,t}^{\text{ccus},20} \perp \sum_{k\in\Omega_{\text{ccus}}} q_{k,t}^{\text{cc,c}} - \sum_{k\in\Omega_{\text{gc}}} q_{k,t}^{\text{gc,c}} - \sum_{k\in\Omega_{\text{gt}}} q_{k,t}^{\text{gt,c}} \le 0, \tag{17w}$$

$$0 \le \mu_{k,t}^{\text{ccus},21} \perp -q_{k,t}^{\text{p2h,g1}} \le 0, \tag{17x}$$

$$0 \le \mu_{k,t}^{\text{ccus},22} \perp -q_{k,t}^{\text{p2h,g2}} \le 0, \tag{17y}$$

$\forall m \in \mathcal{V}^{\text{e}}$, the KKT conditions of (8) are as follows:

$$-\pi_{m,t}^{\text{e}} + \lambda_{m,t}^{\text{e},1} - \lambda_{m,t}^{\text{e},2} = 0, \tag{18a}$$

$$\sum_{(mn)\in\mathcal{E}^{\text{e}}} B_{mn}\left(\lambda_{m,t}^{\text{e},1} - \lambda_{n,t}^{\text{e},1} - \underline{\mu}_{mn,t}^{\text{e},3} + \overline{\mu}_{mn,t}^{\text{e},3}\right) = 0, \tag{18b}$$

$$0 \le \underline{\mu}_{mn,t}^{\text{e},3} \perp -\overline{P}_{mn}^{\text{e}} - B_{mn}(\theta_{m,t} - \theta_{n,t}) \le 0, \tag{18c}$$

$$0 \le \overline{\mu}_{mn,t}^{\text{e},3} \perp B_{mn}(\theta_{m,t} - \theta_{n,t}) - \overline{P}_{mn}^{\text{e}} \le 0. \tag{18d}$$

$\forall m \in \mathcal{V}^{\text{g}}$, the KKT conditions of (9) are as follows:

$$-\pi_{m,t}^{\text{g}} + \lambda_{m,t}^{\text{g},1} - \lambda_{m,t}^{\text{g},2} = 0, \tag{19a}$$

$$\lambda_{m,t}^{\text{g},2} - \lambda_{n,t}^{\text{g},2} - 2\Lambda_{mn,t}^{\text{g},1}/S_{mn} - \underline{\mu}_{m,t}^{\text{g},4} = 0, \tag{19b}$$

$$\begin{aligned} &-\lambda_{m,t}^{\text{g},2}\Big|_{m\in\mathcal{V}_{\text{from}}^{\text{com}}(c)} + (1+\delta_c^{\text{com}})\,\lambda_{m,t}^{\text{g},2}\Big|_{m\in\mathcal{V}_{\text{to}}^{\text{com}}(c)} \\ &-\underline{\mu}_{c,t}^{\text{g},7} + \overline{\mu}_{c,t}^{\text{g},7} = 0, \forall c \in \Omega_{\text{com}}, \end{aligned} \tag{19c}$$

$$\begin{aligned} &-\sum_{(mn)\in\mathcal{E}^{\text{g}}}\left(\Lambda_{mn,t}^{\text{g},2} - \Lambda_{nm,t}^{\text{g},2} + \Lambda_{mn,t}^{\text{g},3} - \Lambda_{nm,t}^{\text{g},3}\right) \\ &-\underline{\mu}^{\text{g},5} + \overline{\mu}^{\text{g},5} + \sum_{m\in\mathcal{V}_{\text{from}}^{\text{com}}(c)}\left(\overline{\mu}_{c,t}^{\text{g},6} - \underline{\mu}_{c,t}^{\text{g},6}\right) \\ &+\sum_{m\in\mathcal{V}_{\text{to}}^{\text{com}}(c)}\left(\underline{\rho}_c^{\text{com}}\underline{\mu}_{c,t}^{\text{g},6} - \overline{\rho}_c^{\text{com}}\overline{\mu}_{c,t}^{\text{g},6}\right) = 0, \end{aligned} \tag{19d}$$

$$2f_{mn,t}\lambda_{mn,t}^{\text{g},1}/S_{mn} = 0, \forall (mn) \in \mathcal{E}^{\text{g}}, \tag{19e}$$

$$(\Pi_{m,t} - \Pi_{n,t} - 1)\Lambda_{mn,t}^{g,2} = 0, \forall (mn) \in \mathcal{E}^{\text{g}}, \tag{19f}$$

$$(\Pi_{m,t} - \Pi_{n,t} + 1)\Lambda_{mn,t}^{g,3} = 0, \forall (mn) \in \mathcal{E}^{\text{g}}, \tag{19g}$$

$$\left\|\begin{matrix}\Lambda_{mn,t}^{\text{g},1} \\ \Lambda_{mn,t}^{\text{g},2}\end{matrix}\right\|_2 \le \Lambda_{mn,t}^{g,3}, \forall (mn) \in \mathcal{E}^{\text{g}}, \tag{19h}$$

$$0 \le \mu_{mn,t}^{\text{g},4} \perp -f_{mn,t} \le 0, \forall (mn) \in \mathcal{E}^{\text{g}}, \tag{19i}$$

$$0 \le \underline{\mu}_{m,t}^{\text{g},5} \perp \underline{\Pi}_m - \Pi_{m,t} \le 0, \tag{19j}$$

$$0 \le \overline{\mu}_{m,t}^{\text{g},5} \perp \Pi_{m,t} - \overline{\Pi}_m \le 0, \tag{19k}$$

$$0 \le \underline{\mu}_{m,t}^{\text{g},6} \perp \underline{\rho}_c^{\text{com}}\Pi_{c,t}^{\text{com,to}} - \Pi_{c,t}^{\text{com,from}} \le 0, \forall c \in \Omega_{\text{com}}, \tag{19l}$$

$$0 \le \overline{\mu}_{m,t}^{\text{g},6} \perp \Pi_{c,t}^{\text{com,from}} - \overline{\rho}_c^{\text{com}}\Pi_{c,t}^{\text{com,to}} \le 0, \forall c \in \Omega_{\text{com}}, \tag{19m}$$

$$0 \le \underline{\mu}_{m,t}^{\text{g},7} \perp -f_{c,t}^{\text{com}} \le 0, \forall c \in \Omega_{\text{com}}, \tag{19n}$$

$$0 \le \overline{\mu}_{m,t}^{\text{g},7} \perp f_{c,t}^{\text{com}} - \overline{f}_c^{\text{com}} \le 0, \forall c \in \Omega_{\text{com}}. \tag{19o}$$

The KKT conditions of (10) are as follows:

$$-\pi_t^{\text{c}} - \underline{\mu}_1^{\text{c},1} + \overline{\mu}_1^{\text{c},1} - \mu_t^{\text{c},2} - \lambda_t^{\text{c},3} = 0, \tag{20a}$$

$$0 \le \underline{\mu}_1^{\text{c},1} \perp \underline{q}^{\text{cmo,c}} - \sum_{t\in\mathcal{T}} q_t^{\text{cmo,c}} \le 0, \tag{20b}$$

$$0 \leq \overline{\mu}_1^{c,1} \perp \sum_{t \in T} q_t^{cmo,c} - \overline{q}^{cmo,c} \leq 0, \tag{20c}$$

$$0 \leq \mu_1^{c,2} \perp -q_t^{cmo,c} \leq 0, \tag{20d}$$

Together, the market equilibrium problem can be formulated as

$$\begin{aligned} \min \quad & \mathbf{1} \\ \text{s.t.} \quad & (2d), (4b), (4d), (5b), (5d), (7b)\text{–}(7c), \\ & (7f)\text{–}(7h), (7j), (7l), (7o), (8b)\text{–}(8c), (9b)\text{–}(9d), \\ & (9b)\text{–}(9d), (10d), (11), (12)\text{–}(20). \end{aligned} \tag{21}$$

5. Simulation Results

This section demonstrates the simulation results of the proposed market equilibrium model on two test systems. First, a small-scale test system consisting of a three-bus electrical system and a three-node gas system is an illustrative example. Second, on the IEEE 39-bus system and the modified 20-node gas system. As for the common parameters, we set the value of δ^{gas} to 2.16 tCO_2/t and the value of δ^{coal} to 3.31 tCO_2/t according the default parameters in [33]. κ^{CO_2} is set to 240 ¥/tCO_2, κ^{cs} 350 ¥/tCO_2, and κ^{coal} 600 ¥/t. $\underline{\pi}^{e}$, $\underline{\pi}^{g}$, and $\underline{\pi}^{c}$ is set to 100, 100, and 0, respectively. Γ is set to 0.05 and γ is 0.3. The other related parameters can be found in [12].

5.1. Test Case 1: On a Three-Bus Electrical System and a Three-Node Gas System

The structure of the three-bus electrical system and the three-node gas system is illustrated in Figure 2. It contains one residential user and two industrial users. The electricity is supplied by a GT and a GC. The gas is supplied by a GS and a CCUS together. The six test settings are considered:

(1) No upper limit for available carbon allowance from CMO;
(2) The upper limit for available carbon allowance is set to **90%** of the allowance quantity in (1);
(3) The upper limit for available carbon allowance is set to **80%** of the allowance quantity in (1);
(4) The upper limit for available carbon allowance is set to **70%** of the allowance quantity in (1);
(5) The upper limit for available carbon allowance is set to **60%** of the allowance quantity in (1);
(6) The upper limit for available carbon allowance is set to **50%** of the allowance quantity in (1).

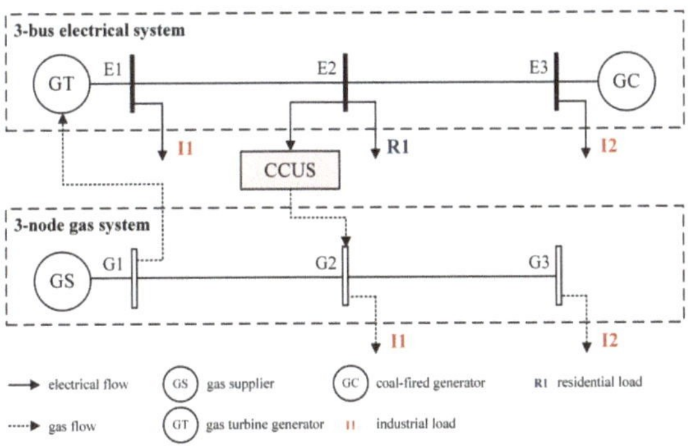

Figure 2. The three-bus electrical system and the three-node gas system.

The summation of available carbon allowance in (1) is considered as the maximum carbon allowance that should be distributed to emitters by CMO, which is 319.78 tCO_2 in this case. In order to maintain consistency in modeling, the upper limit of available carbon allowance in (1) is set to a very large number, such as 1e6.

Figure 3 illustrates the price changes for three commodities. In general, the reduction in available carbon allowance leads to an increase in electricity prices and carbon allowance prices. The price of natural gas remains unaffected due to the presence of a sole natural gas supplier in the system. These changes cause two main shifts: (1) The increasing carbon prices are reducing the competitiveness of the coal-fired generator, while the proportion of electricity generated by the gas turbine generator is steadily increasing, as depicted in Figure 4. When only considering fuel costs, coal-fired generators exhibit lower marginal generation costs. However, their unit carbon emissions are also comparatively higher than gas turbine generators. (2) The share of allowance provided by CCUS increases, in general, as depicted in Figure 5. It can be observed that when there is no restriction on the available carbon allowance, the allowance in the carbon market is predominantly derived from the CMO. With the upper limit decreasing, CCUS begins capturing emissions from GT and GC to provide an additional allowance in the carbon market. The provided allowance first drops when the limit is 80% max and climbs again with the constraint tightens. Two reasons contribute to this phenomenon. First, the elevation of carbon prices results in a reduction in the overall emissions within the system. Second, the increase in electricity prices leads to a rise in carbon capture costs, while the revenue generated from the elevated carbon prices is insufficient to offset the increased electricity costs. Compared with previous work [20], the calculated carbon prices here are on average higher than them. This is mainly because the prices, fluctuating around $20 in previous work, only reflect the scarcity of carbon allowance. The allowance reproduction role played by CCUS has not been introduced and its cost is not reflected in the carbon prices. Thus, the price level in previous work can hardly cover the carbon allowance production cost of CCUS. However, the carbon prices calculated in our work can help CCUS cover their basic marginal costs, illustrated in Table 2, and encourage the further deployment of such key decarbonization equipment.

Table 2 summarizes the benefits gained by market participants under different carbon market settings. It can be observed that the surplus of users, both residential and industrial, decreases with the reduction in the available allowance from the CMO such as in the aspect of social welfare in two energy markets. For the electricity market, this is primarily due to the rise in electricity prices. Meanwhile, for the natural gas market, although gas prices remain stable, the increase in carbon prices indirectly leads to a rise in the cost of gas consumption, which also hampers the demand for natural gas from users. Another point worth noting is that, in the test case, the profitability of the generators has consistently remained negative, whether it is GT or GC. We annotate the profits earned by GT and GC in the energy market only in parentheses, which exclude the carbon allowance costs. It can be observed that with no carbon allowance costs, GC keeps its profit positive for lower unit production costs. GT also turns a positive profit with the gradual increase in electricity prices. This difference in net profits is caused by the absence of a free carbon allowance. In practice, CMO can adjust the profitability or loss levels of fossil fuel generators by determining the quantity of free carbon allowances allocated to the generators based on this value and the carbon price level. As for CCUS, it can hardly make a profit in the carbon market when there is still ample available allowance. However, when the available carbon allowances in the market decrease to a certain level, the profitability of CCUS experiences a sharp increase. The marginal costs associated with carbon capture and storage are not only comparatively high but also linked to the level of electricity prices. Electricity prices, in turn, rise due to the increase in carbon prices. Thus, it is only when the scarcity of carbon allowances reaches a relatively high degree that the rise in carbon prices can drive further profitability for CCUS.

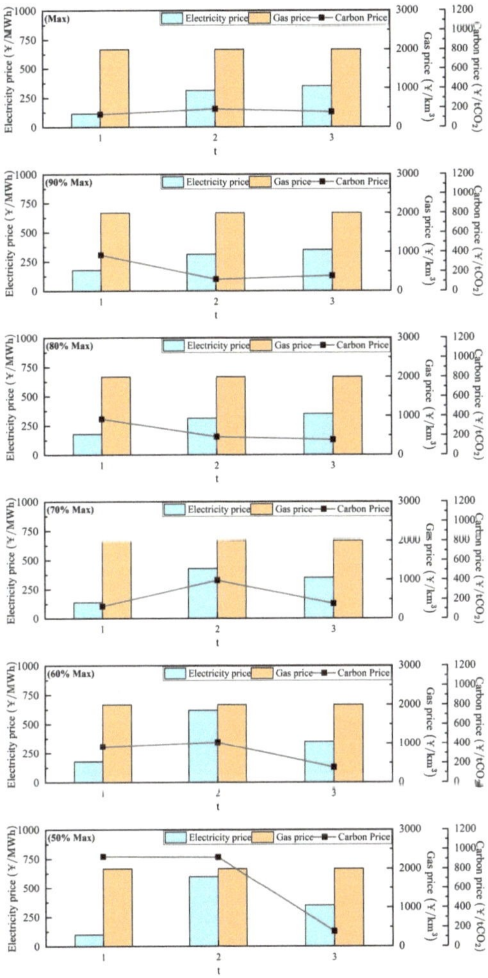

Figure 3. The price result of test case 1.

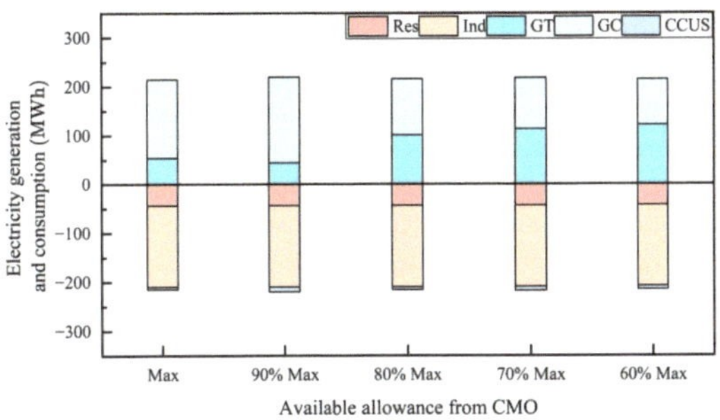

Figure 4. The electricity balance of test case 1 (negative values for electricity consumption).

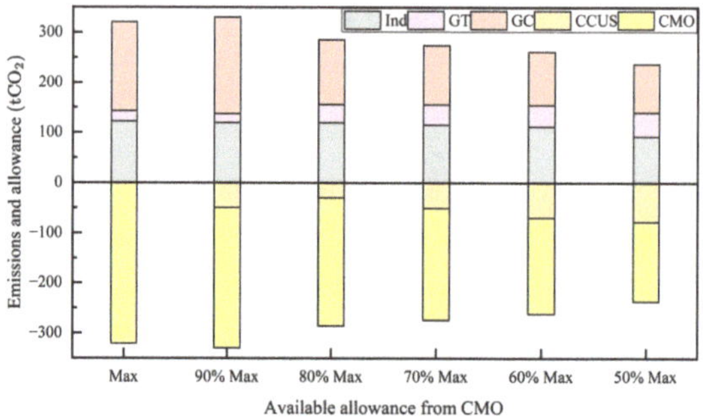

Figure 5. The carbon balance of test case 1 (negative values for carbon allowance).

Table 2. The surplus/profits of different market participant. (Unit: Thousand Chinese Yuan).

Available Allowance from CMO	Surplus		Profit				Social Welfare		Revenue
	Res	Ind	GT	GC	GS	CCUS	Electricity	Gas	CMO
Max	129.80	2102.65	−4.20 * (−0.83) **	−15.19 (12.96)	0.00	0.12	2216.18	80.54	51.40
90% Max	129.43	2102.28	−3.95 (−1.13)	−21.08 (16.69)	0.00	0.00	2216.17	82.78	41.34
80% Max	129.43	2098.12	−8.14 (−1.19)	−17.57 (10.56)	0.00	0.00	2216.17	65.71	46.39
70% Max	128.04	2086.61	−6.85 (2.42)	−13.89 (12.98)	0.00	0.00	2215.92	60.57	43.64
60% Max	125.19	2071.75	−0.29 (11.42)	−12.89 (17.41)	0.00	0.60	2215.31	57.21	44.46
50% Max	125.94	2054.85	−15.96 (9.99)	−42.99 (12.64)	0.00	41.44	2215.41	40.45	48.17

* Profit including carbon allowance costs. ** Profit excluding carbon allowance costs.

5.2. Test Case 2: On a 39-Bus Electrical System and a 20-Node Gas System

Test case 2 consists of the IEEE 39-bus electrical system and the modified Belgian 20-node gas system, as illustrated in Figure 6. It has 19 residential users and 20 industrial users. The electrical demand is supplied by five GCs and five GTs together. Four GSs provide natural gas for GTs and industrial users, playing the role of upstream gas stations. Two CCUSs are responsible for capturing carbon emissions and producing natural gas. The test settings are similar to test case 1, while the ratios are set to **80%**, **60%**, **50%**, **40%**, and **30%** for the broader feasible range of available allowance from CMO. For simplicity, we select four different typical days to simulate the load characteristics in spring, summer, autumn, and winter, respectively. It should be mentioned that the ramp constraints are only effective intraday due to the discontinuity between typical days.

Figure 7 demonstrates the shift in demand curves and carbon prices under different settings. When no upper limit is posed on an available allowance from the CMO, the summation of available carbon allowance from the CMO is 10,000 and carbon prices keep 0. It can be observed the electrical and gas demand are both suppressed by the climbing carbon prices. Table 3 shows the average carbon price of different typical days. Generally speaking, carbon prices tend to be higher during the summer and autumn typical days. Table 4 further summarizes the benefits gained by market participants on different typical days when posed with an 80% allowance limitation. For GCs and GTs, while they did not incur negative profits as observed in test case 1, they still maintained a relatively lower level of profitability. The average carbon price during the winter typical day is relatively low, yet the revenues of CCUS and CMO are higher. This is because the energy consumption levels for both electricity and gas of the winter typical day remain relatively high in all four days, leading to a significant demand for allowances in the carbon market. However, during the summer and autumn typical day, the overall demand for gas load is lower, making it challenging to achieve higher revenues even with higher carbon price levels. This also hints at a seasonal demand shift of carbon allowance.

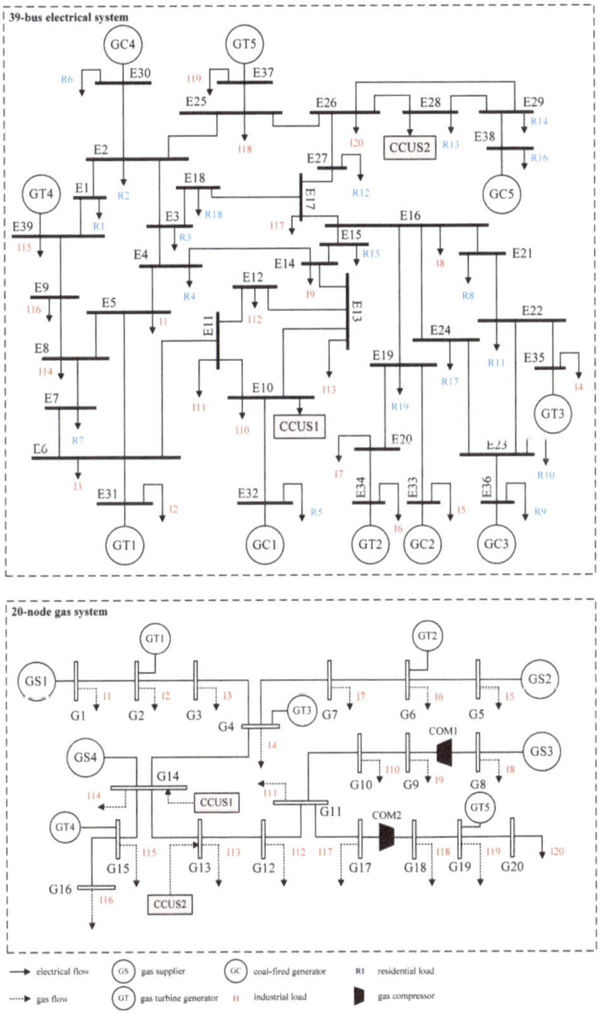

Figure 6. The structure of test case 2.

Table 3. Average carbon price of different days in case 2. (Unit: Chinese Yuan).

Typical Day	80% Max	60% Max	50% Max	40% Max	30% Max
Spring	1126.42	871.70	774.54	1651.34	1340.48
Summer	1315.68	1201.63	1229.40	1499.96	1528.42
Autumn	1126.42	1192.84	1469.95	1143.43	1737.92
Winter	1065.15	1049.10	1038.46	1384.76	1469.31
Total	1158.42	1078.82	1128.09	1419.87	1519.03

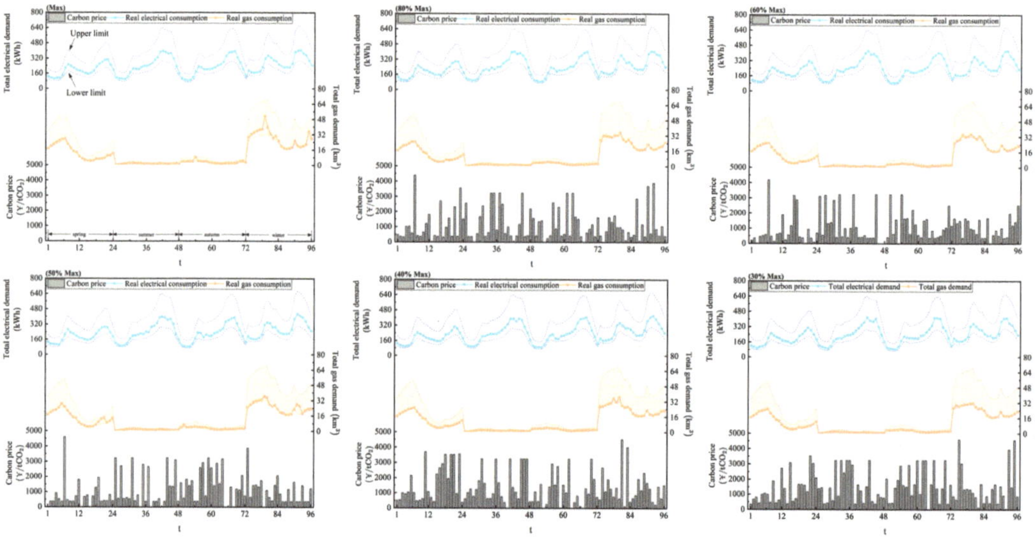

Figure 7. The total demand curves and carbon price results of test case 2.

Table 4. The surplus/profits of different market participants in case 2. (80% Max, Unit: Million Chinese Yuan).

Typical Day	Surplus		Profit				Social Welfare		Revenue
	Res	Ind	GT	GC	GS	CCUS	Electricity	Gas	CMO
Spring	144.99	197.67	1.33	0.74	11.91	17.74	348.55	3.02	1.37
Summer	497.87	359.75	2.50	0.62	2.311	6.32	871.32	2.90	0.56
Autumn	421.92	208.97	2.66	1.03	6.45	8.53	652.98	0.31	1.05
Winter	1648.79	282.60	1.28	0.60	24.50	31.89	1936.61	16.31	2.13
Total	2713.35	1048.99	7.77	3.00	45.18	64.49	3799.45	22.59	5.09

5.3. Comparison: Independent CCUS or Combination with Fossil Fuel Units

In this part, we analyze the impact of CCUS ownership on the market equilibrium. In the analysis before, CCUS is operating as an independent market participant. Nevertheless, occasions may happen that CCUS is built as an emission-reduction facility attached to fossil fuel power plants. Based on test case 2, two CCUSs are allocated separately to GTs and GCs. Modifications are made to the relevant models to accommodate the ownership structure, described as follows. $q_{k,t}^{\mathrm{gt,red}}$ is used to describe the quantity of emission reduction CCUS provides for each generator. Since CCUS is integrated into the units, the emission reduction can be directly allocated to the generators without the need for carbon allowance trading. Thus, (4d) is modified as

$$q_{k,t}^{\mathrm{gt,c}} = \delta^{\mathrm{gas}} q_{j(k),t}^{\mathrm{gt,g}} - q_{k,t}^{\mathrm{gt,red}}, \left[\lambda_{k,t}^{\mathrm{gt,3}}\right], t \in \mathcal{T}, \tag{22a}$$

$$q_{k,t}^{\mathrm{gt,red}} \geq 0, \left[\mu_{k,t}^{\mathrm{gt,5}}\right], t \in \mathcal{T}. \tag{22b}$$

The relative KKT condition is added:

$$-\lambda_{k,t}^{\mathrm{gt,3}} - \mu_{k,t}^{\mathrm{gt,5}} = 0, \tag{23a}$$

$$0 \leq \mu_{k,t}^{\mathrm{gt,5}} \perp q_{k,t}^{\mathrm{gt,red}} \geq 0. \tag{23b}$$

The same modification has also been made to the GC model. As for CCUS, we assume that CCUS 1 is combined with GTs, and CCUS 2 is combined with GCs. Thus, the summation of emission reduction for GTs cannot exceed the captured carbon emissions of CCUS 1. Also, the captured carbon emissions of CCUS 1 should not be larger than the emissions from GTs. Relative constraints, (7o) and (7q), are modified as follows:

$$q_{k,t}^{\text{ccus,c}} + \sum_{k \in \Omega_{\text{gt}}} q_{k,t}^{\text{gt,red}} = q_{k,t}^{\text{cc,c}}, \left[\lambda_{k,t}^{\text{ccus,14}}\right], t \in \mathcal{T}, \tag{24a}$$

$$q_{k,t}^{\text{cc,c}} \leq \sum_{k \in \Omega_{\text{gt}}} q_{k,t}^{\text{gt,c}}, \left[\mu_{k,t}^{\text{ccus,16}}\right], k = 1, t \in \mathcal{T}. \tag{24b}$$

KKT condition (17r) is changed to the following:

$$0 \leq \mu_{k,t}^{\text{ccus,16}} \perp q_{k,t}^{\text{cc,c}} - \sum_{k \in \Omega_{\text{gt}}} q_{k,t}^{\text{gt,c}} \leq 0, k = 1. \tag{25}$$

Similar modifications have also been applied to GCs and CCUS 2.

Table 5 compares the profits of fossil fuel units and CCUSs under different ownership settings. It can be observed that the combination of fossil fuel units and CCUS makes it hard to gain profitability under the current market mechanism. This is because when combined with fossil fuel units, CMO playing as the primary supplier of carbon allowance in the market, the effective formation of carbon price signals which can reflect marginal carbon capturing costs becomes hard. The combination tends to prioritize allocating emissions allowances generated from emission reduction to fossil fuel units, rather than clearing them in the market for additional profits. After all, their profitability from allowance in the carbon market implies the combination will incur additional carbon allowance costs. Consequently, the operational costs of CCUSs would further diminish the profitability of the combination. Additionally, such reconfiguration also reduces the flexibility of operations, increasing the regional demand for carbon allowance. Therefore, from the perspective of CCUS, it is advantageous to participate in the carbon market as an independent entity, as this facilitates the formation of accurate carbon price signals.

Table 5. The Comparison of profits of fossil fuel units and CCUSs under different ownership. (Unit: Million Chinese Yuan).

Available Allowance from CMO	Independent CCUS			Combination	
	GTs	GCs	CCUSs	GTs with CCUS 1	GCs with CCUS 2
80%	7.77	3.00	64.49	−3.07	−0.08
60%	3.86	0.49	70.89	−2.20	−0.76
50%	6.57	0.68	90.75	−6.18	−0.98
40%	5.84	3.31	88.68	— *	—

* Infeasible problem.

6. Conclusions

In this paper, we propose an economic dispatching model of the regionally integrated energy system of electricity and hythane based on the generalized Nash game. To be more aligned with energy and carbon market practice, multiple participants, including industrial users, CCUS, and the carbon market operator, are considered in the regionally integrated energy system. The complementary model of the generalized Nash market equilibrium is induced via the KKT optimal condition. The generated carbon allowance price signals can reflect not only the scarcity of available carbon allowance caused by the carbon emission control target but also the marginal costs of regional carbon reduction, which are highly coupled with the fluctuations in the LMP of electricity and gas. A comprehensive case study is conducted to demonstrate the market equilibrium of different system scales, seasonal loads, and CCUS ownership.

Simulation results indicate that the carbon price shows a trend of rising first and then falling with the tightening of maximum available allowances decreasing. A shift in seasonal changes can also be observed. In some extreme cases, the seasonal highest average price can be about 1.89 times higher than the lowest price. As for the CCUS ownership, the results suggest the organization of independent CCUSs is more beneficial compared with the combination of fossil fuel generators, in the aspect of CCUS profits. The corresponding result can provide insight into carbon allowance allocation, future regional energy and carbon market organization, and coupled evolution of prices.

However, some points are not considered in this paper. For example, the effect on carbon prices caused by transmission losses is not given consideration. Additionally, the role of renewable energy plays in the system is also ignored. These will be discussed in our future research.

Author Contributions: Conceptualization, D.Y. and S.W.; methodology, S.W., W.W. and W.Z.; software, P.Y.; validation, P.Y.; formal analysis, D.Y.; investigation, P.Y.; resources, D.Y.; data curation, W.K.; writing—original draft preparation, D.Y.; writing—review and editing, S.W.; visualization, W.K.; supervision, S.W.; project administration, S.W.; funding acquisition, S.W. All authors have read and agreed to the published version of the manuscript.

Funding: This work was supported by the Science and Technology Project of Industrial Companies of State Grid Jiangsu Electrical Power Co., Ltd. (No. JC2024066).

Data Availability Statement: The data presented in this study are available on request from the corresponding author. The data are not publicly available due to user privacy.

Conflicts of Interest: Authors Dong Yang, Shufan Wang, Wendi Wang, Weiya Zhang, Pengfei Yu and Wei Kong was employed by the company Nanjing Suyi Industrial Co., Ltd. The authors declare that the research was conducted in the absence of any commercial or financial relationships that could be construed as a potential conflict of interest.

Nomenclature

Abbreviations	
ccus	Carbon capture, utlization, and storage.
cc	Carbon capture.
cmo	Carbon market operator.
com	Compressor.
cs	Carbon storage.
cu	Carbon utilization.
c	For carbon.
emo	Electricity market operator.
e	For electricity.
gc	Coal-fired generators.
gmo	Gas market operator.
gs	Gas suppliers.
gt	Gas turbine generators.
g	For gas.
ind	Industrial users.
p2g	Power to gas.
res	Residential users.
r	Ramp power.
Indices	
c	The index of compressor in the gas network.
$i(k)$	The corresponding bus in the electrical network of participant/equipment k.

$j(k)$	The corresponding node in the gas network of participant/equipment k.
k	The index of market participants/equipment.
m, n	The bus/node indices in the electrical/gas network.
t	The index of time.
Parameters	
δ^{com}	The compression coefficient of the compressor in the gas network.
$\delta^{\text{gas}}, \delta^{\text{coal}}$	The carbon emission factor of the fossil fuel, tCO_2/km^3 for gas, tCO_2/t for coal.
γ	The calorific conversion coefficient from hydrogen to natural gas.
$\overline{f}^{\text{com}}$	The gas flow capacity of the compressor, km^3.
$\overline{P}$	The capacity of the transmission lines in the electrical network, MW.
$\underline{\Pi}, \overline{\Pi}$	The lower/upper limit of the nodal pressures in the gas network.
$\underline{\pi}$	The lower limit of the prices.
$\underline{q}, \overline{q}$	The lower/upper limit of the trading quantity.
a	The coefficient of the quadratic term in the utility function, ¥/(MWh)^2 for electricity, ¥/(km^3)^2 for gas.
B	The susceptance of transmission lines in the electrical network, p.u.
$b^{\text{gt}}, b^{\text{gc}}$	The coefficient of the linear term in the fuel cost function, t/MWh for gc, km^3/MWh for gt.
$b^{\text{res}}, b^{\text{ind}}$	The coefficient of the linear term in the utility function, ¥/MWh for electricity, ¥/km^3 for gas.
c	The coefficient of the constant term in the fuel cost function, t for gc, km^3 for gt.
S	The conveyance coefficient in the Weymouth equation, km^3/bar.
Sets	
$\mathcal{E}$	The set of edges in the network.
$\mathcal{T}$	The set of time.
Ω	The set of market participants/equipment.
Variables	
Λ	The dual variable related to conic constraints.
λ	The dual variable related to equality constraints.
$\mu, \underline{\mu}, \overline{\mu}$	The dual variables related to inequality constraints.
Π	The nodal pressure in the gas network, bar.
π	The locational marginal price, ¥/MWh for electricity, ¥/km^3 for gas, ¥/tCO_2 for carbon allowance.
$\Pi^{\text{com,from}}$	The pressure of the node from which gas flow to the compressor, bar.
$\Pi^{\text{com,to}}$	The pressure of the node to which gas flow from the compressor, bar.
θ	Phase angle of the bus in the electrical network, rad.
f	The flow of pipeline/compressor in the gas network, km^3.
q	The trading quantity, kWh for electricity, km^3 for gas, tCO_2 for carbon allowance.

References

1. Rockström, J.; Gaffney, O.; Rogelj, J.; Meinshausen, M.; Nakicenovic, N.; Schellnhuber, H.J. A roadmap for rapid decarbonization. *Science* **2017**, *355*, 1269–1271. [CrossRef] [PubMed]
2. Sinha, R.K.; Chaturvedi, N.D. A review on carbon emission reduction in industries and planning emission limits. *Renew. Sustain. Energy Rev.* **2019**, *114*, 109304. [CrossRef]
3. Redmond, L.; Convery, F. The global carbon market-mechanism landscape: Pre and post 2020 perspectives. *Clim. Policy* **2015**, *15*, 647–669. [CrossRef]
4. Wei, C.; Wang, Y.; Shen, Z.; Xiao, D.; Bai, X.; Chen, H. AUQ–ADMM algorithm-based peer-to-peer trading strategy in large-scale interconnected microgrid systems considering carbon trading. *IEEE Syst. J.* **2023**, *17*, 6248–6259. [CrossRef]
5. Qin, M.; Yang, Y.; Zhao, X.; Xu, Q.; Yuan, L. Low-carbon economic multi-objective dispatch of integrated energy system considering the price fluctuation of natural gas and carbon emission accounting. *Prot. Control Mod. Power Syst.* **2023**, *8*, 1–18. [CrossRef]
6. European Commission. Scope of the EU Emissions Trading System. Available online: https://climate.ec.europa.eu/eu-action/eu-emissions-trading-system-eu-ets/scope-eu-emissions-trading-system_en (accessed on 31 January 2024).
7. Ministry of Ecology and Environment of the People's Republic of China. Notice on the Preparation of Greenhouse Gas Emission Reports and Verification Work for Selected Key Industries' Enterprises for the Period 2023–2025. Available online: https://climate.ec.europa.eu/system/files/2023-11/com_2023_653_glossy_en_0.pdf (accessed on 31 January 2024).

8. Jenkins, S.; Annaswamy, A.; Hansen, J.; Knudsen, J. A dynamic model of the combined electricity and natural gas markets. In Proceedings of the 2015 IEEE Power & Energy Society Innovative Smart Grid Technologies Conference (ISGT), Washington, DC, USA, 18–20 February 2015; pp. 1–5.
9. Oskouei, M.Z.; Mirzaei, M.A.; Mohammadi-Ivatloo, B.; Shafiee, M.; Marzband, M.; Anvari-Moghaddam, A. A hybrid robust-stochastic approach to evaluate the profit of a multi-energy retailer in tri-layer energy markets. *Energy* **2021**, *214*, 118948. [CrossRef]
10. Wu, C.; Gu, W.; Yi, Z. Competitive Equilibrium Analysis for Renewables Integration in Dynamic Combined Heat and Power Trading Market. *IEEE Trans. Power Syst.* **2023**, *39*, 53–65. [CrossRef]
11. Zhang, Y.; Li, J.; Ji, X.; Ye, P.; Yu, D.; Zhang, B. Optimal dispatching of electric-heat-hydrogen integrated energy system based on Stackelberg game. *Energy Convers. Econ.* **2023**, *4*, 267–275. [CrossRef]
12. Chen, S.; Conejo, A.J.; Sioshansi, R.; Wei, Z. Equilibria in electricity and natural gas markets with strategic offers and bids. *IEEE Trans. Power Syst.* **2019**, *35*, 1956–1966. [CrossRef]
13. Cheng, S.; Scholes, S.C.; Kong, W.; Gu, C.; Li, F. Carbon-Oriented Electricity Balancing Market for Dispatchable Generators and Flexible Loads. *IEEE Trans. Power Syst.* **2023**, *38*, 5648–5659. [CrossRef]
14. Xiong, Z.; Luo, S.; Wang, L.; Jiang, C.; Zhou, S.; Gong, K. Bi-level optimal low-carbon economic operation of regional integrated energy system in electricity and natural gas markets. *Front. Energy Res.* **2022**, *10*, 959201. [CrossRef]
15. Wang, R.; Wen, X.; Wang, X.; Fu, Y.; Zhang, Y. Low carbon optimal operation of integrated energy system based on carbon capture technology, LCA carbon emissions and ladder-type carbon trading. *Appl. Energy* **2022**, *311*, 118664. [CrossRef]
16. Yan, Z.; Li, C.; Yao, Y.; Lai, W.; Tang, J.; Shao, C.; Zhang, Q. Bi-Level Carbon Trading Model on Demand Side for Integrated Electricity–gas System. *IEEE Trans. Smart Grid* **2023**, *14*, 2681–2696. [CrossRef]
17. Zhu, D.; Yang, B.; Wu, Y.; Deng, H.; Dong, Z.; Ma, K.; Guan, X. Joint trading and scheduling among coupled carbon-electricity-heat-gas industrial clusters. *IEEE Trans. Smart Grid* **2023**, *Early Access*.
18. Feijoo, F.; Das, T.K. Design of Pareto optimal CO2 cap-and-trade policies for deregulated electricity networks. *Appl. Energy* **2014**, *119*, 371–383. [CrossRef]
19. Ruiz, C.; Conejo, A.J.; Fuller, J.D.; Gabriel, S.A.; Hobbs, B.F. A tutorial review of complementarity models for decision-making in energy markets. *EURO J. Decis. Process.* **2014**, *2*, 91–120. [CrossRef]
20. Chen, S.; Conejo, A.J.; Wei, Z. Conjectural-variations equilibria in electricity, natural-gas, and carbon-emission markets. *IEEE Trans. Power Syst.* **2021**, *36*, 4161–4171. [CrossRef]
21. Jiang, K.; Liu, N.; Yan, X.; Xue, Y.; Huang, J. Modeling Strategic Behaviors for GenCo With Joint Consideration on Electricity and Carbon Markets. *IEEE Trans. Power Syst.* **2023**, *38*, 4724–4738. [CrossRef]
22. Yang, Y.; Shi, J.; Wang, D.; Wu, C.; Han, Z. Identifying Operation Equilibrium in Integrated Electricity, Natural Gas, and Carbon-Emission Markets. *arXiv* **2022**, arXiv:2210.09813.
23. Zhu, Z.; Chan, K.W.; Bu, S.; Zhou, B.; Xia, S. Nash Equilibrium Estimation and Analysis in Joint Peer-to-Peer Electricity and Carbon Emission Auction Market with Microgrid Prosumers. *IEEE Trans. Power Syst.* **2022**, *38*, 5768–5780. [CrossRef]
24. Chao, C.; Deng, Y.; Dewil, R.; Baeyens, J.; Fan, X. Post-combustion carbon capture. *Renew. Sustain. Energy Rev.* **2021**, *138*, 110490. [CrossRef]
25. Chen, S.; Liu, J.; Zhang, Q.; Teng, F.; McLellan, B.C. A critical review on deployment planning and risk analysis of carbon capture, utilization, and storage (CCUS) toward carbon neutrality. *Renew. Sustain. Energy Rev.* **2022**, *167*, 112537. [CrossRef]
26. Liu, H.; Tesfatsion, L.; Chowdhury, A. Locational marginal pricing basics for restructured wholesale power markets. In Proceedings of the 2009 IEEE Power & Energy Society General Meeting, Calgary, AB, Canada, 26–30 July 2009; pp. 1–8.
27. Yang, Z.; Bose, A.; Zhong, H.; Zhang, N.; Lin, J.; Xia, Q.; Kang, C. LMP Revisited: A Linear Model for the Loss-Embedded LMP. *IEEE Trans. Power Syst.* **2017**, *32*, 4080–4090. [CrossRef]
28. Lv, J.; Jiang, Y. Research on generalized locational marginal price of integrated electricity–heat–gas market considering integrated demand response. *J. Renew. Sustain. Energy* **2022**, *14*, 045902. [CrossRef]
29. Ibikunle, G.; Gregoriou, A. *Carbon Markets: Microstructure, Pricing and Policy*; Springer: Berlin/Heidelberg, Germany, 2018.
30. Chai, B.; Chen, J.; Yang, Z.; Zhang, Y. Demand response management with multiple utility companies: A two-level game approach. *IEEE Trans. Smart Grid* **2014**, *5*, 722–731. [CrossRef]
31. Wu, C.; Gu, W.; Bo, R.; MehdipourPicha, H.; Jiang, P.; Wu, Z.; Lu, S.; Yao, S. Energy trading and generalized Nash equilibrium in combined heat and power market. *IEEE Trans. Power Syst.* **2020**, *35*, 3378–3387. [CrossRef]
32. Gabriel, S.A.; Conejo, A.J.; Fuller, J.D.; Hobbs, B.F.; Ruiz, C. *Complementarity Modeling in Energy Markets*; Springer Science & Business Media: Berlin/Heidelberg, Germany, 2012.
33. Ministry of Ecology and Environment of the People's Republic of China. Guidelines for Enterprise Greenhouse Gas Emissions Accounting and Reporting: Power Generation Facilities. Available online: https://www.mee.gov.cn/xxgk2018/xxgk/xxgk06/202212/W020221221671986519778.pdf (accessed on 8 October 2024).

processes

MDPI

Article

Analysis of Heat Transfer of the Gas Head Cover of Diaphragm Compressors for Hydrogen Refueling Stations

Shengdong Ren [1], Xiaohan Jia [1,*], Jiatong Zhang [1], Dianbo Xin [2] and Xueyuan Peng [1,3]

1 School of Energy and Power Engineering, Xi'an Jiaotong University, No. 28 Xianning West Road, Xi'an 710049, China; renshengdong@163.com (S.R.); jiadong0526@stu.xjtu.edu.cn (J.Z.); xypeng@mail.xjtu.edu.cn (X.P.)
2 Qingdao Hisense Hitachi Air-Conditioning Systems Co., Ltd., No. 218, Qianwangang Road, Qingdao 266001, China
3 State Key Laboratory of Multiphase Flow in Power Engineering, Xi'an Jiaotong University, No. 28 Xianning West Road, Xi'an 710049, China
* Correspondence: jiaxiaohan@xjtu.edu.cn

Abstract: The inadequate ability to dissipate heat of the gas head cover of the diaphragm compressor will result in its excessive temperature, which will put the operation of the hydrogen filling station at risk for safety issues and raise operating costs. This paper analyzed the structure and the heat transfer characteristics of the gas head cover, along with the relevant heat transfer boundaries, based on which a finite element simulation model of the temperature distribution was established. A test rig for the temperature test of a 22 MPa diaphragm compressor was built to validate this simulation model. The results indicated that the simulated temperatures agree well with the measured values, and the deviation is within 9.1%. Further, this paper proposed two head cover structures for enhancing the heat transfer according to the temperature field distribution characteristics, and the simulation and experimental verification were carried out, respectively. The findings demonstrate that the method of enhancing heat transfer around the centre area is more effective, reducing the highest temperature by 14.1 °C, because it greatly lowers thermal conduction resistance, which is the principal impediment to the heat dissipation of the gas head cover.

Keywords: diaphragm compressor; temperature distribution; simulation model; heat transfer enhancement

Citation: Ren, S.; Jia, X.; Zhang, J.; Xin, D.; Peng, X. Analysis of Heat Transfer of the Gas Head Cover of Diaphragm Compressors for Hydrogen Refueling Stations. *Processes* **2023**, *11*, 2274. https://doi.org/10.3390/pr11082274

Academic Editors: Chenyu Wu, Zhongkai Yi and Chenhui Lin

Received: 27 June 2023
Revised: 22 July 2023
Accepted: 25 July 2023
Published: 28 July 2023

1. Introduction

Fossil energy consumption, including coal, oil, and natural gas, is the basis for supporting the progress of modern civilization, which leads to excessive global carbon dioxide emissions [1,2]. Hydrogen can be converted into electricity via hydrogen fuel cells, with only water as a by-product, making it an attractive green alternative to fossil energy, especially in transportation field [3–7]. Hydrogen is a never-ending renewable source of energy and thus can be the ideal solution to environmental and energy issues. However, hydrogen's extremely low molecular weight makes it the lowest volumetric energy density of any commonly used fuel [8], which prevents the use of hydrogen as an efficient energy source. The most commonly used method is to pressurize hydrogen to 35 MPa or 70 MPa and store it in a storage tank to increase the energy density and enable the car to have a higher cruising range [9,10]. This makes the hydrogen refueling station an important infrastructure to promote the development of the hydrogen energy industry [11]. The hydrogen compressor is the core equipment in the hydrogen refueling station and dominates the cost [12,13]. The technological development of the hydrogen compressor directly determines the construction process of the refueling infrastructure.

The hydrogen-pressurized equipment mainly includes mechanical compression and non-mechanical compression [14]. The mechanical compressor is the most commonly used type, including the diaphragm compressor, piston compressor, liquid piston compressor, and ionic liquid compressor [14–16]. Diaphragm compressors are the most widely applied in refueling stations due to their unique advantages of no pollution and no leakage [17]. Even so, the technology of diaphragm compressors now is still unlikely to satisfy the targets of large-scale hydrogen refueling station construction.

The low efficiency and short life of the diaphragm are the most concerning issues of diaphragm compressors. The low flow and efficiency of diaphragm compressors result in high energy consumption of hydrogen refueling stations. Altukhov et al. [18] studied the thermodynamic characteristics of diaphragm compressors and researched the effect of the compressibility of the hydraulic fluid and dead volume in the gas cavity on the efficiency. Lei et al. [19] discussed the factors affecting the flow rate of diaphragm compressors and proposed solutions to these issues. Hyun et al. [20] analyzed the influence of oil density, deflection of diaphragm, gas, and oil pressure on the flow rate during a certain period of the compression process. Jia et al. [21] researched the effect of clearance volume on the flow rate of diaphragm compressors through experiments and proposed a method to increase the flow rate by reducing the dead volume of the cavity. Author [22] focused on the influence of hydraulic oil compressibility on volumetric efficiency and proposed a mathematical formula for the calculation of volumetric efficiency.

The short life of the diaphragm compressor, especially the easy fracture of the diaphragm, is a major problem in the application of the diaphragm compressor. To solve this problem, many efforts have been made to improve the stability and life of diaphragm compressors. Li et al. [23] studied the acoustic emission signal of diaphragm compressors in the working process and proposed a test method that can monitor the operating conditions and evaluate the reliability of the compressor. Altukhov et al. [24] found that the size of the exhaust orifice and the oil-gas pressure relationship are the main factors affecting the diaphragm life, and proposed a method to evaluate the design reliability of diaphragm compressors. Lu et al. [25] analyzed the factors affecting the diaphragm life of diaphragm compressors and found that the main factor is the cavity profile. Jia et al. [26] proposed an analysis method of diaphragm stress which is combined with the small deflection and thin-plate large deflection theories and investigated some influence factors of the diaphragm fracture. Hu et al. [27] proposed a new generatrix for cavity profile, which can eliminate the dead volume in the cavity and reduce the maximal and the centric redial stress of the diaphragm by 8.2% and 13.9%, respectively. Li et al. [28] studied the characteristics of diaphragm stress distribution during the working process of diaphragm compressors. Furthermore, Li et al. [29–32] presented two new generatrices of the cavity profile, which can change the stress distribution characteristics of the diaphragm, and verified the design theory through experimental and numerical methods. The new cavity profiles can somewhat improve the stress of the diaphragm, but they do not quite address the problem.

The phenomenon of low flow and high failure of diaphragm compressors has received the majority of research attention, while its fundamental performance, particularly thermal management, has received less attention. The gas head cover of the hydrogen compressor needs to use a hydrogen embrittlement-resistant material, which has poor thermal conductivity, not conducive to the export of compression heat, causing the temperature of the head cover to rise abnormally. A significant failure rate is also brought on by the thermal stress brought on by the gas head cover's abnormally high temperature. The high temperature of the gas head cover can also exacerbate suction gas heating and reduces the volumetric efficiency. Nevertheless, little research has been done on this issue. Only Wang et al. [33,34] discussed that the thermal deformation of the gas head cover had a great influence on the diaphragm fatigue life through the thermal-structure coupled analysis, and proposed a method to improve the structure of the gas headcover. However, they calculated a temperature field result by given parameters and then studied the influence of the temperature field on the structural strength, but did not propose a universal simulation method for the temperature field of the head cover. Analyzing the heat transfer process and researching ways to improve it are both crucial [35–37].

The above research mainly focused on cavity profiles and diaphragm stress. However, the issue of poor heat dissipation of the gas head cover, which has a great impact on the life and efficiency of the compressor, has not been well studied. The purpose of this paper is to analyze the heat transfer process related to the gas head cover, establish a simulation model of the temperature field, and then explore ways to enhance heat dissipation. Analysis and experimental methods were applied to investigate and verify the method to calculate the temperature field of the gas headcover.

2. Analysis of the Heat Transfer Related to the Gas Head Cover

The structure of the diaphragm compressor is shown in Figure 1. Three diaphragms are fixed between the oil head support and the gas head cover. The volume enclosed by the surface of the diaphragm and the gas head cover is the gas cavity, and the other side of the diaphragm and the oil head support enclose the oil cavity. The oil piston is driven by the crank connecting rod mechanism to reciprocate in the oil head support. Then, the oil piston forces diaphragm deformation through hydraulic oil. Further, the volume of the gas cavity is changed as diaphragms move, and the gas can be compressed and discharged. The mechanical energy is converted into the internal energy of the gas during the compression process, and the temperature of the gas will increase. The compression process in the gas cavity of a diaphragm compressor is generally considered to be a nearly isothermal process in the past because the gas head cover of traditional diaphragm compressors is approximately a thin plate and is made of alloy steel with good thermal conductivity. However, for diaphragm compressors used in hydrogen refueling stations, the material of the gas head cover is generally made of stainless steel to prevent hydrogen embrittlement, which has poor thermal conductivity. The pressure of the compressor used in the hydrogen refueling station is relatively high, resulting in a thick gas head cover, which is further not conducive to heat dissipation. The aforementioned factors cause the thermal resistance of the gas head cover to be relatively high and the compression heat released to the environment through the gas head cover to be comparatively low, leading to a higher discharge temperature. The compression process in the gas cavity of hydrogen diaphragm compressors is nearly an adiabatic process. These problems are subjected to the high temperature of the discharge gas, as well as the gas head cover. High gas head cover temperature can lead to large thermal deformation, high thermal stress, and terrible intake heating, which affect the efficiency and life of the compressor. Therefore, it is of great significance to study the heat transfer process and temperature distribution related to the gas head cover to improve the performance of the diaphragm compressor.

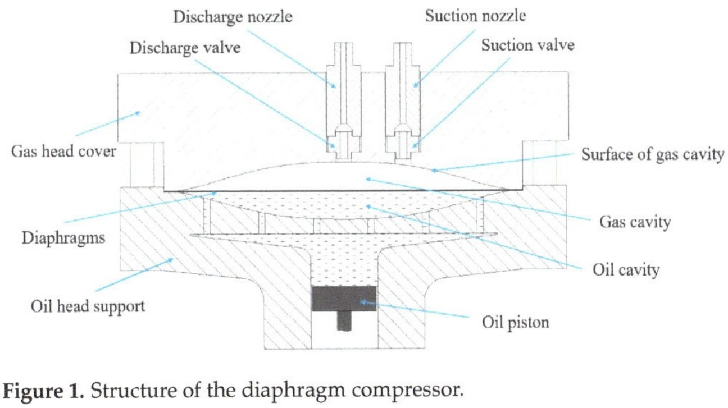

Figure 1. Structure of the diaphragm compressor.

This study focused on the temperature distribution characteristics of the gas head cover. The heat transfer processes related to the gas head cover are shown in Figure 2. The main heat transfer processes and characteristics are as follows:

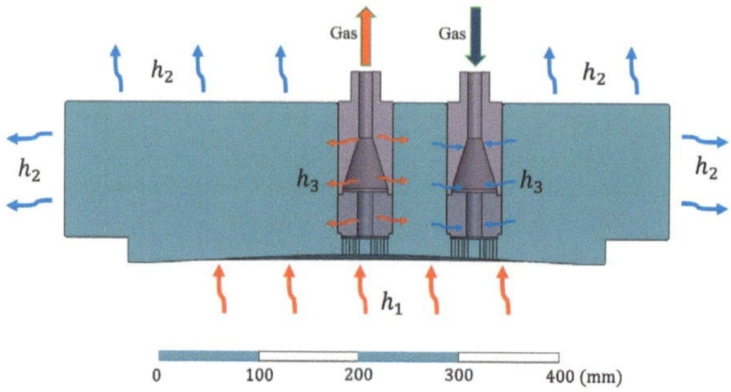

Figure 2. Heat transfer processes of the gas head cover.

2.1. Heat Transfer from Compressed Gas to the Surface of the Gas Cavity

The temperature of the gas in the gas cavity increases as compressed, and the high-temperature gas undergoes forced convection heat transfer with the surface of the gas cavity. It has been analyzed that the compression process in the gas cavity of the hydrogen diaphragm compressor used in refueling stations is nearly an adiabatic process. The temperature of the compressed gas can be calculated as:

$$T_d = T_s\left(\frac{p_d}{p_s}\right)^{\frac{K_T-1}{K_T}} \tag{1}$$

where, T_d and T_s denote the temperature of compressed gas and suction gas, respectively; K; p_d and p_s represent the discharge pressure and suction pressure, respectively, MPa; K_T is the temperature adiabatic index, which is 1.41 for hydrogen.

The cycle process of the compressor includes the expansion process, the suction process, the compression process, and the discharge process. The temperature of the gas in the gas cavity is not always maintained at the highest temperature. The fresh gas with a lower temperature enters the gas cavity during the suction process, and the heat will be transferred from the surface of the gas cavity to the fresh gas. The actual average temperature of the gas in the gas cavity is affected by many factors, such as compressor power, operating conditions, and structural parameters. The average temperature of the gas in the gas chamber of a reciprocating compressor is usually calculated by an empirical formula. The authors verified that the empirical formula is also applicable in diaphragm compressors through experiments. The average temperature can be expressed as:

$$T_a = T_s\left\{1+\frac{2}{3}\left[\left(\frac{p_d}{p_s}\right)^{\frac{K_T-1}{K_T}}-1\right]\right\} \tag{2}$$

where, T_a is the average temperature in the gas cavity.

The convection heat transfer between the gas and the cavity surface is a complex dynamic process. In a cycle process, the convective heat transfer coefficient is constantly changing, but it is not necessary to calculate the heat transfer coefficient at each moment. The equivalent convective heat transfer coefficient of the entire working process can describe the heat transfer between the gas and the cavity surface more intuitively. Because the gas cavity is flat and the discharge holes are in the centre, the closer to the centre, the more heat exchange between the gas and the cavity surface. The equivalent convective heat transfer coefficient can be expressed as:

$$h_1(r) = h_{1max}(1-k\frac{r}{R}) \tag{3}$$

where $h_1(r)$ denotes the convective heat transfer coefficient at radius r, h_{1max} is the convective heat transfer coefficient at the centre, R is the radius of the gas cavity, and k represents the rate of decrease of heat transfer coefficient with radius which is the distance from the gas cavity's centre. The uncertainty around the deformation pattern of the diaphragm results in the shape change pattern of the gas cavity being unclear, which makes the theoretical calculation of h_{1max} and k impossible now. In this study, the simulation model was corrected by the experimental results, and h_{1max} and k are empirical values summarized through experimental measurements [33].

2.2. Heat Transfer from the Outer Surface of the Gas Head Cover to the Environment

The temperature of the outer surface of the gas head cover without cooling can generally reach 80 °C–100 °C when the discharge temperature is greater than 150 °C. Therefore, not only the convective heat transfer but also the heat radiation should be considered when calculating the heat transfer between the outer surface of the gas head cover and the environment. The gas head cover is generally not air-cooled in the design of diaphragm compressors. The form of the convective heat transfer between the outer surface and the environment is air-natural convective heat transfer. The heat transfer coefficient of air natural convective is 5–25 W·m^{-2}·K^{-1}. The heat transfer coefficient of heat radiation can be calculated as: [38]

$$h_r = \varepsilon\sigma\left(T_w{}^3 + T_w{}^2 T_e + T_w T_e{}^2 + T_e{}^3\right) \tag{4}$$

where ε is the blackness of the surface of the gas head cover, σ is the Stefan-Boltzmann constant, T_w and T_e represent the temperature of the outer surface of the gas head cover and the environment, respectively, K. Thus, the total heat transfer coefficient between the surface of the gas head cover and the environment is the sum of the convective heat transfer coefficient and the radiation heat transfer coefficient, and is given by:

$$h_2 = h_c + h_r \tag{5}$$

where h_2 denotes the total heat transfer coefficient, and h_c and h_r are the convective heat transfer coefficient and the radiation heat transfer coefficient, respectively.

2.3. Heat Transfer between the Gas and the Surface of Suction and Discharge Holes

There are some small holes under the suction and discharge valve for gas to enter and exit the gas cavity. As shown in Figure 3, there is forced convective heat transfer between the suction and discharge gas flow and the surfaces of the holes. The convective heat transfer coefficient is associated with the flow velocity and working conditions of the gases.

The flow velocities of the gases flowing through the suction and discharge holes are calculated as follows:

$$u_s = \frac{Q_o p_o T_s Z_s}{p_s T_0 A_s} \tag{6}$$

$$u_d = \frac{Q_o p_o T_d Z_d}{p_d T_0 A_d} \tag{7}$$

where u_s and u_d express the flow velocities through the suction and discharge holes; Q_o, p_o, and T_0 represent the volume flow rate, pressure, and temperature in standard conditions, respectively; p_s, T_s, p_d, and T_d are the pressures and temperatures of suction and discharge gas, respectively; Z_s and Z_d are the compressibility factors in suction and discharge conditions, respectively; and A_s and A_d represent the total sectional areas of suction holes and discharge holes, respectively. Thus, the Reynolds numbers can be calculated as follows:

$$Re = \frac{ud}{v} \tag{8}$$

where u and v are the flow velocity and kinematic viscosity of suction or discharge gas, respectively, and d is the diameter of the small holes. Further, the Nusselt number can be calculated by the Dittus-Boelter equation as follows [38]:

$$Nu = 0.023Re^{0.8}Pr^{n} \tag{9}$$

Therefore, the heat transfer coefficient can be expressed as:

$$h_3 = \frac{Nu\lambda}{d} = \frac{0.023u^{0.8}Pr^{n}\lambda}{v^{0.8}d^{0.2}} \tag{10}$$

where h_3 represents the heat transfer coefficient between the suction gas with the suction holes or between the discharge gas with the discharge holes, u and v are the flow velocity and kinematic viscosity of suction or discharge gas, respectively, d is the diameter of the small holes, Pr, and λ are the Prandtl number and thermal conductivity in suction or discharge conditions, respectively, which can be gained with physical property software, and n is exponent of the Prandtl number, which is 0.3 for discharge gas and 0.4 for suction gas.

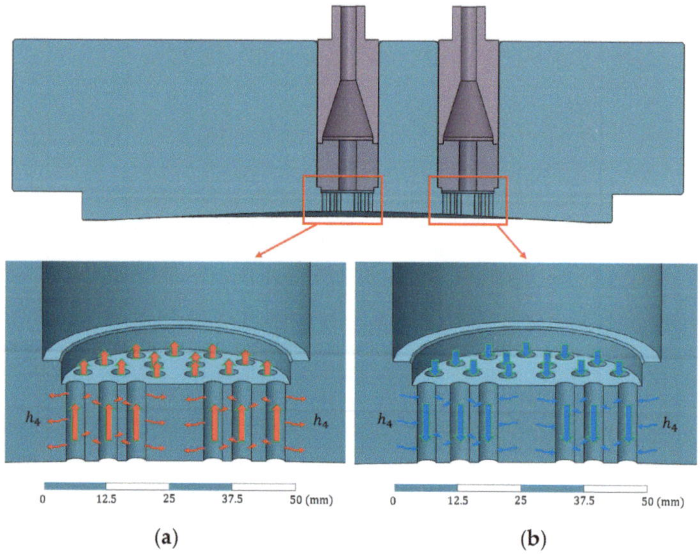

Figure 3. Heat transfers between the gas and the surfaces of the holes. (**a**) Discharge holes. (**b**) Suction holes.

2.4. Heat Transfer between the Suction/Discharge Nozzles and the Gas Head Cover

The temperature of the discharge nozzle is relatively high due to the continuous discharge airflow through it. Similarly, the intake air temperature is relatively low, and thus the temperature of the intake manifold will also be relatively low. The calculation method of heat transfer between the gas and the suction and discharge nozzles is the same as that of the gas and the suction and discharge holes. There is generally a gap of about 1 mm between the suction and discharge holes and the gas head cover. However, as shown in Figure 4, the discharge nozzle still transfers heat to the cylinder head, but the thermal resistance will be larger. Certainly, the gas head cover also transfers heat to the suction nozzle. There are two forms of heat transfer here: thermal radiation and conduction through a thin layer of air. The heat transfer capacity through these two methods can be calculated by the following equations, respectively:

$$\Phi_r = \varepsilon_2 C_0 A_1 \left[\left(\frac{T_1}{100} \right)^4 - \left(\frac{T_2}{100} \right)^4 \right] \tag{11}$$

$$\Phi_c = \frac{2\pi l \lambda_{air}(T_1 - T_2)}{\ln(r_2/r_1)} \tag{12}$$

where Φ_r and Φ_c represent the heat transfer capacity of thermal radiation and heat conduction, respectively, ε_2 is the blackness of the surface of the hole in the gas head cover, C_0 is the black body radiation constant, A_1 is the area of the outer surface of the nozzle, T_1 and T_2 represent the temperature of the nozzle and the surface of the hole in the gas head cover, respectively, r_1 and r_2 represent the diameter of the nozzle and the surface of the hole in the gas head cover, respectively, l represents the length of the nozzle, and λ_{air} expresses the thermal conductivity of air. Then, the total heat transfer capacity can be expressed as follows:

$$\Phi_t = \Phi_r + \Phi_c = h_4(T_1 - T_2)A_1 \tag{13}$$

where h_4 indicates the equivalent heat transfer coefficient between the nozzle and the gas head cover, and can be calculated as follows:

$$h_4 = \frac{\varepsilon_2 C_0 A_1 \left[\left(\frac{T_1}{100} \right)^4 - \left(\frac{T_2}{100} \right)^4 \right] + \frac{2\pi l \lambda_{air}(T_1 - T_2)}{\ln(r_2/r_1)}}{(T_1 - T_2)A_1} \tag{14}$$

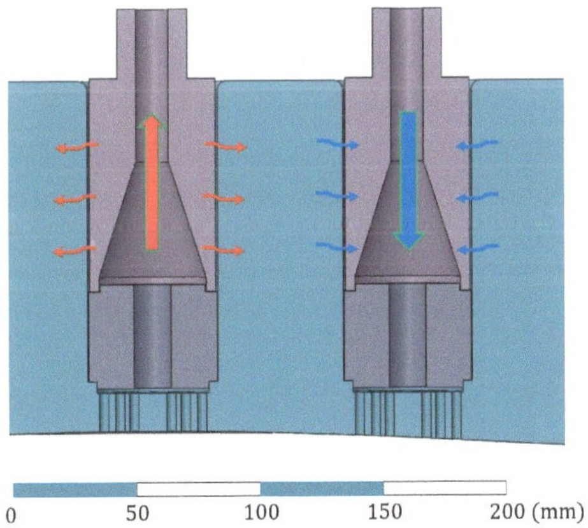

Figure 4. Heat transfers between the surfaces of nozzles and the gas head cover.

The simulation analysis of the temperature field of the gas head cover can be carried out after calculating the heat transfer coefficient of each heat transfer path.

3. Finite Element Analysis of the Temperature Field

3.1. Geometric Model of the Gas Head Cover

The research objects in this study are the gas head cover, the nozzles, and the valves of a diaphragm compressor used in a 22 MPa mother hydrogen refueling station. The gas head cover assembly model was built by 3D modeling software, as shown in Figure 5. There is a stepped hole in the centre of the cylinder for installing the discharge valve. The discharge nozzle presses the discharge valve against the step of the hole, and under the valve, there are some small holes for discharge. The suction step hole is set next to the discharge step hole, and the structure of suction is similar to that of discharge. The structural parameters of this model are shown in Table 1.

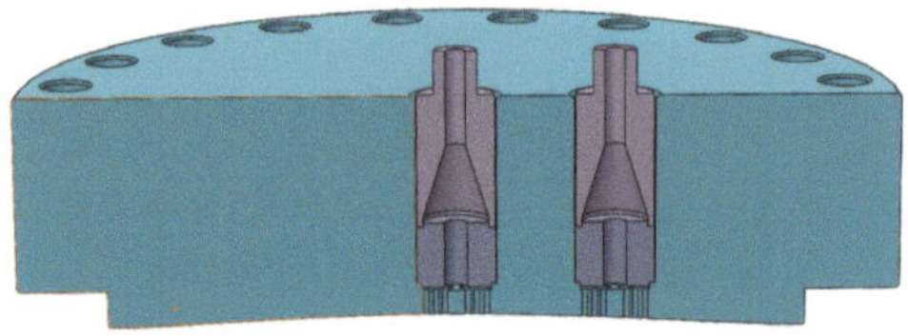

Figure 5. Model of the gas head cover assembly.

Table 1. Main parameters of the gas head cover.

Parameters	Values/mm
Diameter of the gas head cover	615
Thickness of the gas head cover	160
Diameter of the nozzles	54
Diameter of the hole in the gas head cover	56
Diameter of the suction and discharge small holes	4

The effect of hydrogen embrittlement must be considered when selecting the material for the gas head cover assembly [39]. The material of the parts in contact with hydrogen should be made of hydrogen embrittlement-resistant materials, generally stainless steel [40]. The materials of the gas head cover and the nozzles of this model were 1.4418 duplex stainless steel, and the valves were made of 17-4PH martensite stainless steel. Both 1.4418 and 17-4PH have excellent resistance to hydrogen embrittlement, but their thermal conductivities are extremely low, 15 $W{\cdot}(m{\cdot}K)^{-1}$ and 16 $W{\cdot}(m{\cdot}K)^{-1}$, respectively. This is also one of the reasons for the poor heat dissipation of the hydrogen diaphragm compressor.

3.2. Mesh Generation

The dimensions of the different structural features of the gas head cover differ by two orders of magnitude. The diameter of the gas head cover is 615 mm, while the diameter of the suction and discharge small holes is only 4 mm, and it can be predicted that the area around the discharge holes is the hottest part. The temperature of the area around the discharge holes requires special attention. Therefore, the region of the small holes was divided from the gas head cover for a finer mesh as shown in Figure 6. The grid length of the mesh near the small holes was not more than 1 mm, and the other area of the gas head cover adopted a tetrahedral mesh with a size of 2–5 mm. The suction and discharge nozzles and valves adopted tetrahedral meshes with grid lengths of 1–2 mm.

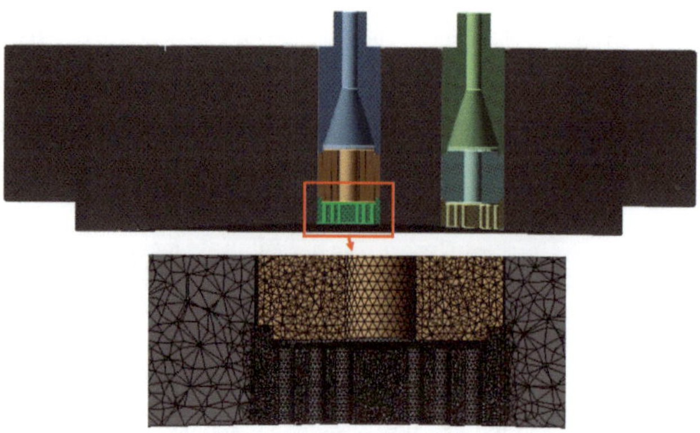

Figure 6. Mesh around the discharge holes.

3.3. Thermal Loads

A steady-state thermal analysis of the gas head cover requires thermal loads to be applied to each boundary. The simulation calculation was carried out in the Steady-State thermal module of ANSYS 2021R2 software. The simulation example in this study is the temperature field of the gas head cover of the hydrogen diaphragm compressor used in the 22 MPa hydrogen filling station. The simulated working conditions are that the suction pressure is 5 Mpa, the suction temperature is 20 °C, the exhaust pressure is 22 Mpa, and the ambient temperature is 20 °C. According to the analysis of the heat transfer of the gas head cover in the previous chapter, the heat transfer coefficient of each heat transfer boundary of the gas head cover can be calculated.

3.4. Mesh Independence Verification

Three meshes were generated with different numbers of elements by adjusting the mesh grid length of the gas head cover, as well as the suction and discharge nozzles and valves, and the number of elements is 2,200,498, 3,421,824, and 4,483,582, respectively. The three models with different mesh numbers were calculated separately. The highest temperature and the temperature of three feature points were taken as the comparison values, and the calculation results under different mesh densities were compared. The positions of the three feature points are shown in Figure 7. Point 1 was located in the middle of the suction and discharge stepped holes, 140 mm deep from the top surface of the gas head cover. Point 2 was located on the other side of the exhaust hole, which is symmetrical with point 1 about the discharge stepped hole. Point 3 was located on the inner wall surface of the discharge stepped hole, on the side away from the suction hole, and is 115 mm deep from the top surface of the gas head cover. The calculation results are shown in Figure 8. The highest temperature and the temperature of the three feature points calculated with three meshes of different densities had no difference, which proved that the mesh density did not affect the calculation results. To save computing resources, this study adopted the mesh density corresponding to the simulation number of 3.42 million.

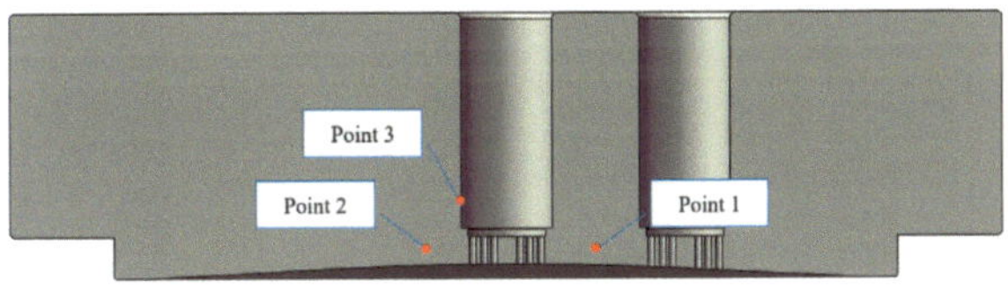

Figure 7. Positions of feature points.

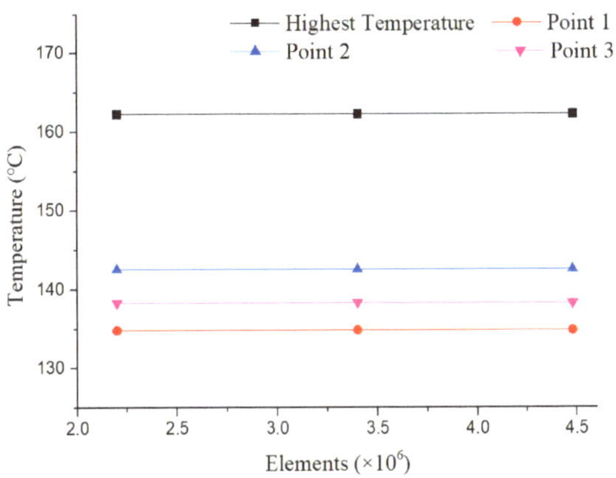

Figure 8. Results of mesh independence verification.

4. Experimental Validation

To verify the correctness of the simulation analysis method of the temperature field, a test rig for the temperature test of the gas head cover of the diaphragm compressor was built, as shown in Figure 9. The experimental equipment is a two-stage horizontal hydrogen diaphragm compressor with a rotational speed of 420 rpm used in a 22 MPa mother hydrogen refueling station. Taking the secondary gas head cover as the research object, the structure size of the gas head cover is the same as that of the simulation analysis model. The second stage of the compressor was kept in the operation conditions, of which the suction pressure and temperature were 5 MPa and 20 °C, and the discharge pressure was 22 MPa. A vortex volume flowmeter was used to measure the actual flow rate, and six K-type thermocouples were installed in the gas head cover to measure the temperature at several special points. The uncertainty of the K-type thermocouple is ±1.5 °C. To measure the temperature of the internal temperature of the gas head cover, deep holes with a diameter of 5 mm and a depth of 140 mm were machined on it, and the thermocouple temperature measuring heads were inserted into the bottom of the holes and filled with thermal conductive silicone grease. The thermal conductivity of the silicone grease is larger than 6 W·(m^{-1}·K^{-1}). Several very fine K-type thermocouple wires were bonded at the surface of the discharge stepped holes and the discharge small holes by high-temperature-resistant glue.

Figure 9. Temperature test rig.

The temperature measuring points on the gas head cover are shown in Figure 10. Point 1 was located on the central wall of the discharge small holes. Point 2 was located in the middle of the suction and discharge stepped holes, 140 mm deep from the top surface of the gas head cover. Point 3 was located on the other side of the exhaust hole, which is symmetrical with point 1 about the discharge stepped hole. Point 4 was located on the inner wall surface of the discharge stepped hole, on the side away from the suction hole, and is 115 mm deep from the top surface of the gas head cover. Point 5 was also located on the inner wall surface of the discharge stepped hole, 60 mm deep. Point 6 was located at the same depth as points 2 and 3, and was 150 mm away from the central.

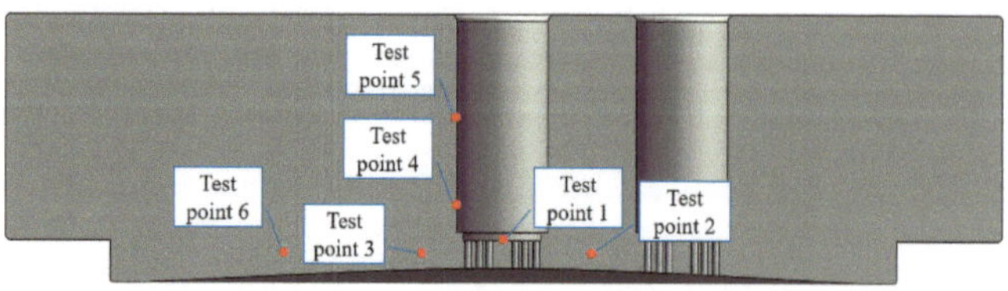

Figure 10. Positions of the temperature measurement points.

5. Results and Discussion

5.1. Finite Element Analysis Results and Comparison with Experimental Values

Figure 11 illustrates the temperature distribution in the cross-section along the centerline of the suction and discharge holes and the surface of the gas head cover. The comparison between the measured temperature and the simulated temperature of each measuring point is shown in Table 2. There are many reasons for experiment and simulation analysis to cause the deviation between the measured value and the simulated value. When measuring the surface temperature, e.g., Point 1, Point 4, and Point 5, one side of the temperature measuring element is attached to the surface of the gas head cover, and the other side is in contact with the high-temperature gas so that there is an error between the measured temperature and the actual surface temperature. The measured temperature values are higher than the actual surface temperature values. Especially, Point 1 is located in the centre of the discharge holes, where the surface temperature is close to the discharge temperature. Thus, the error due to the influence of high-temperature gas on the temperature-measuring element is smaller, and the deviation at this point is smaller. When measuring the internal temperature, e.g., Point 2, Point 3, and Point 6, the thermal conductivity of the filled silicone grease is smaller than that of the gas head cover itself, and the contact surface between the two will also have contact thermal resistance, so the measured values are slightly lower than the actual values there. In addition, the gas cavity of the diaphragm compressor is a flat flexing space, and the thickness and diameter of the space are also constantly changing during operation. The heat transfer coefficient between the gas and the surface of the gas cavity is also dynamic and non-uniform. The heat transfer coefficient between the gas and the gas cavity surface is simplified to be uniform to calculate in this study, which also brings errors to the simulation results. But, in general, the deviation between the measured temperature and the simulated value of each measuring point is less than 9.1%, which is within a reasonable range. The simulation results can reflect the actual temperature distribution characteristics of the compressor well. This indicates that this simulation analysis method is feasible and can be used to study the effect of different methods of optimizing heat transfer.

Figure 11. Temperature distribution of the gas head cover.

Table 2. Comparison of measured temperature and simulation results.

Positions	Point 1	Point 2	Point 3	Point 4	Point 5	Point 6
Measured temperature (°C)	165.2	123.5	132.3	146.3	134.2	124.6
Simulated temperature (°C)	162	134.8	142.5	138.3	125.1	135.4
Deviation	1.8%	9.1%	7.7%	5.5%	6.8%	8.7%

The temperature distribution graph declares that the maximum temperature of the gas head cover is 162 °C, which is located in the centre of the discharge small holes. The high-temperature gas continues to flow to the discharge nozzle through the small holes, transferring a large amount of heat to the surface of the holes. However, the thermal conductivity of the gas head cover is very low, and the heat around the discharge holes is too late to conduct to the surroundings, resulting in heat accumulation in the region of the discharge holes. This makes the temperature around the exhaust hole too high and the thermal deformation too large, which further aggravates the local stress of the diaphragm and reduces the life of the compressor. The temperature distribution graph also displays that the region of the discharge holes is the core high-temperature zone of the diaphragm compressor, and the temperature exceeds 150 °C. From the discharge holes to the surroundings, the temperature gradient decreases. The temperature of the part of the outer surface of the gas head cover near the discharge stepped hole also exceeds 100 °C. These revelations demonstrate that the problem of the high temperature of the gas head cover can be improved by two methods: augmented heat transfer of the core high-temperature zone around the discharge holes and augmented heat transfer of the outer surface of the gas head cover. But, it can be predicted that the former effect will be better.

5.2. *Influence of the Outer Surface Heat Transfer Coefficient on the Temperature Field of the Gas Head Cover*

The heat transfer on the outer surface was simplified in this study, and the convective heat transfer coefficient was considered to be a constant value, which would bring certain errors to the simulation results. To verify the influence of this error on the results, the sensitivity analysis of the natural convection heat transfer coefficient of the outer surface was conducted. The temperatures at several special points when the natural convection heat transfer coefficients are 5, 10, 15, 20, and 25, are shown in Figure 12. It declares that the highest temperature hardly changes with the change of the natural convection conversion coefficient, and point 1 is located in the core high-temperature zone; the temperature at point 1 is approximately equal to the maximum temperature. Point 2 and Point 3 are 55 mm away from the centre of the discharge holes, and the temperature is slightly reduced, but the reduction value is within 5 °C. The results indicate that the error of the natural convection heat transfer coefficient has little effect on the simulation results, which is acceptable. The heat inside the gas head cover can only be released to the environment through heat conduction and convective heat transfer on the surface. The thermal conductivity of stainless steel is low, and the thermal resistance of natural convection on the surface is smaller than that of thermal conduction, so the change of natural convection heat transfer coefficient has no significant effect on heat dissipation. The heat conduction resistance increases with distance from the outer surface of the gas head cover, whereas the impact of the surface convective heat transfer coefficient decreases.

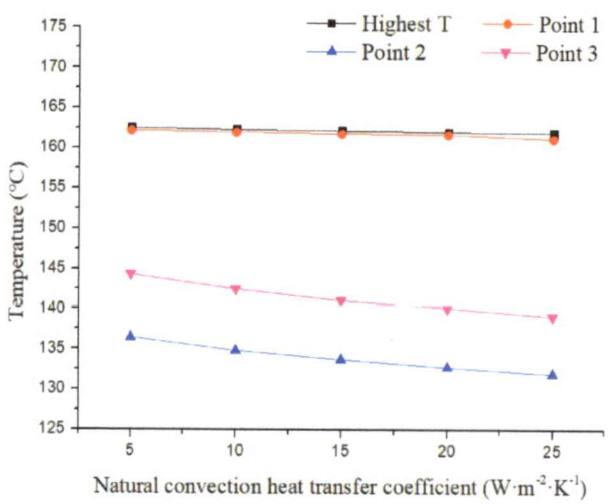

Figure 12. Temperature with different natural convection heat transfer coefficients.

5.3. *Effect of Augmented Heat Transfer of the Outer Surface and Core High-Temperature Zone*

To study the effect of augmented heat transfer of the outer surface and the core high-temperature zone, simulation analysis and experimental verification of the two methods were carried out. Two gas head covers were fabricated, one with water grooves machined on the outer surface and the other with an annular water groove machined along the discharge stepped hole, as shown in Figures 13 and 14.

Figure 13. Structure of augmented heat transfer of the outer surface.

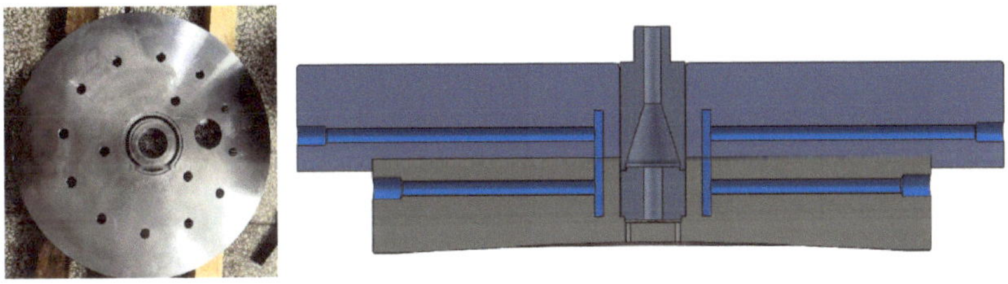

Figure 14. Structure of augmented heat transfer of the core high-temperature zone.

Figure 15 illustrates the temperature distribution of the gas head cover with augmented heat transfer of the outer surface. The temperature in the area close to the water grooves significantly reduces, and the temperature of the upper part of the gas head cover is lower than 100 °C. However, the highest temperature at the discharge holes still reaches 160.4 °C, which is only 1.9 °C lower than the temperature without heat transfer enhancement. Although the internal temperature at most parts reduces significantly, the method of augmented surface heat transfer has a limited effect on reducing the highest temperature.

Figure 15. Temperature distribution with augmented heat transfer of outer surface.

Figure 16 illustrates the temperature distribution of the gas head cover with augmented heat transfer of the core high-temperature zone. The highest temperature is reduced to 148.2 °C, which is 14.1 °C lower than that without heat transfer enhancement, and the area of the core high-temperature zone is also greatly reduced. The method of augmented heat transfer with cooling water around the discharge holes can effectively reduce the temperature of the gas head cover and solve the problem of the local excessive temperature of diaphragm compressors.

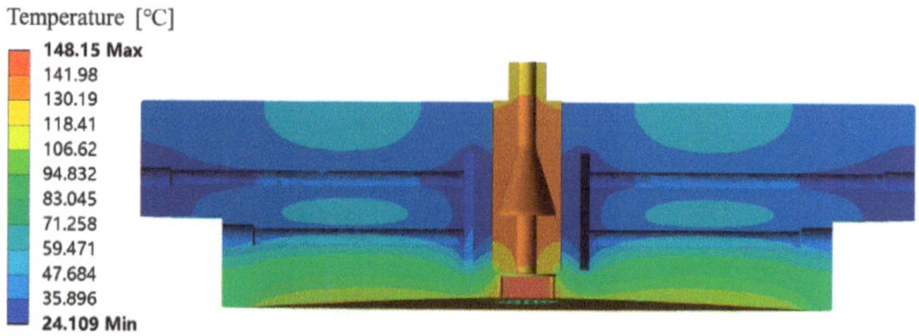

Figure 16. Temperature distribution with augmented heat transfer of the core high-temperature zone.

Table 3 lists the measured discharge temperature, as well as simulated and measured temperatures of several special test points under different gas head cover structures. It can be seen that the discharge temperature is also distinctly reduced with augmented heat transfer of the core high-temperature zone, and the temperature of each point is greatly reduced, which is of great significance for reducing thermal deformation of the gas head cover and improving the life of diaphragm compressors. The results also show that the simulated and measured temperatures are in good agreement for different structures of the head cover. This means that this temperature field simulation method can be used to evaluate the heat dissipation effect of the head cover structure when designing a diaphragm compressor.

Table 3. Temperatures under different gas head cover structures.

Structures	No Heat Transfer Enhancement	Augmented Heat Transfer of Outer Surface	Augmented Heat Transfer of Core High-Temperature Zone
Measured discharge temperature (°C)	171.5	169.2	155.4
Simulated highest temperature (°C)	162.2	160.4	148.2
Point 1 Simulated/Measured (°C)	162/165.2	160.1/164.7	147.9/147.2
Point 2 Simulated/Measured (°C)	134.8/123.5	121.9/112.5	88.6/82.6
Point 3 Simulated/Measured (°C)	142.5/132.3	127.4/121.8	93.9/89.6
Point 4 Simulated/Measured (°C)	138.3/146.3	122.2/131.1	46.8/50.2
Point 5 Simulated/Measured (°C)	125.1/134.2	72.7/75.1	37.9/40.3
Point 6 Simulated/Measured (°C)	135.4/124.6	117.7/105.5	119.6/111.9

This study also demonstrates that the thermal resistance of the gas head cover is the primary factor influencing the heat dissipation. Only through the heat conduction of the gas head cover can the heat inside the gas head cover exchange heat with the cooling water when the water is on the outer surface. The primary part of thermal resistance, heat conduction via the gas head cover, has not been much addressed. When the cooling water is close to the middle high-temperature area, the thermal resistance of the heat transfer process is greatly reduced, and the heat transfer effect is greatly improved.

6. Conclusions

This paper analyzed the heat transfer boundary of the gas head cover of hydrogen diaphragm compressors, based on which the simulation analysis was carried out to obtain the temperature distribution of the gas head cover, and a temperature test rig of a diaphragm compressor was built to measure the temperature of the gas head cover and verify the accuracy of the simulation calculation model. The main conclusions are as follows.

- The temperature field simulation analysis model established in this paper can calculate and has high accuracy to analyze the temperature distribution characteristics of the diaphragm compressor gas head cover of hydrogen refueling stations. The deviation between the measured temperature and the simulated value of each special measuring point is less than 9.1%.
- The region of the discharge holes is the core high-temperature zone of the diaphragm compressor, and the temperature exceeds 150 °C under the mother hydrogen refueling station conditions with suction pressure of 5 MPa and discharge pressure of 22 Mpa. From the discharge holes to the surroundings, the temperature gradient decreases. This is so because the central part has the maximum thermal resistance to the outside air and the highest temperature of the gas in contact with it.
- The temperature field simulation results of the two enhanced heat exchange head cover structures with different enhancing heat transfer methods are in good agreement with the measured values. This simulation method can be used to evaluate the heat dissipation effect of the head cover structure.
- The highest temperature and discharge are reduced by 14.1 °C and 16.1 °C with colling water around the discharge holes, respectively. This method successfully lowers the thermal resistance of heat dissipation in the middle high-temperature area, and significantly lowers the temperature, which can effectively solve the problem of the local excessive temperature of diaphragm compressors and enhance the reliability of diaphragm compressors for hydrogen refueling stations.

Author Contributions: Conceptualization, S.R.; Methodology, S.R.; Software, S.R.; Validation, S.R.; Investigation, J.Z.; Resources, X.J. and X.P.; Data curation, J.Z. and D.X.; Writing—original draft, S.R.; Writing—review & editing, X.J. and X.P.; Project administration, X.J. and X.P.; Funding acquisition, X.J. All authors have read and agreed to the published version of the manuscript.

Funding: This paper is supported by Inner Mongolia Major Science and Technology Major Project: 2020ZD0022 and National Natural Science Foundation of China Grant No. 51876155.

Data Availability Statement: The data is unavailable due to privacy or ethical restrictions.

Conflicts of Interest: The authors declare that they have no known competing financial interests or personal relationships that could have appeared to influence the work reported in this paper.

Nomenclature

T_d	Discharge temperature, K
T_s	Suction temperature, K
p_d	Discharge pressure, MPa
p_s	Suction pressure, MPa
K_T	Temperature adiabatic index
T_a	Average temperature, K
h	Convective heat transfer coefficient, $W \cdot m^{-2} \cdot K^{-1}$
k	Rate of decrease of heat transfer coefficient
R	Radius of the gas cavity
r	Radius
ε	Blackness
σ	Stefan-Boltzmann constant
T_w	Temperature of the outer surface, K
T_e	Temperature of the environment, K
h_r	Heat transfer coefficient of heat radiation, $W \cdot m^{-2} \cdot K^{-1}$
h_c	Heat transfer coefficient of air natural convective, $W \cdot m^{-2} \cdot K^{-1}$
u_s	Flow velocity of the gas flowing through the suction holes, m/s
u_d	Flow velocity of the gas flowing through the discharge holes, m/s

Q_o	Volume flow rate in standard conditions, m^3/h
p_o	Pressure in standard conditions, MPa
T_0	Temperature in standard conditions, K
Z_s	Compressibility factors in suction conditions
Z_d	Compressibility factors in discharge conditions
A_s	Total sectional areas of suction holes, m^2
A_d	Total sectional areas of discharge holes, m^2
Re	Reynolds numbers
v	Kinematic viscosity
d	Diameter of the small holes, m^2
Nu	Nusselt number
Pr	Prandtl number
λ	Thermal conductivity
n	Exponent of the Prandtl number
Φ_r	Heat transfer capacity of thermal radiation
Φ_c	Heat transfer capacity of heat conduction
Φ_t	Total heat transfer capacity
l	Length of the nozzle, m
λ_{air}	Thermal conductivity of air

References

1. Zheng, F.; Zhou, X.; Rahat, B.; Rubbaniy, G. Carbon neutrality target for leading exporting countries: On the role of economic complexity index and renewable energy electricity. *J. Environ. Manag.* **2021**, *299*, 113558.
2. Qin, L.; Kirikkaleli, D.; Hou, Y.; Miao, X.; Tufail, M. Carbon neutrality target for G7 economies: Examining the role of environmental policy, green innovation and composite risk index. *J. Environ. Manag.* **2021**, *295*, 113119.
3. Espegren, K.; Damman, S.; Pisciella, P.; Graabak, I.; Tomasgard, A. The role of hydrogen in the transition from a petroleum economy to a low-carbon society. *Int. J. Hydrogen Energy* **2021**, *46*, 23125–23138.
4. Muradov, N.Z.; Veziroğlu, T.N. "Green" path from fossil-based to hydrogen economy: An overview of carbon-neutral technologies. *Int. J. Hydrogen Energy* **2008**, *33*, 6804–6839.
5. Winter, C. Hydrogen energy—Abundant, efficient, clean: A debate over the energy-system-of-change. *Int. J. Hydrogen Energy* **2009**, *34*, S1–S52.
6. Acar, C.; Dincer, I. The potential role of hydrogen as a sustainable transportation fuel to combat global warming. *Int. J. Hydrogen Energy* **2020**, *45*, 3396–3406.
7. Liu, F.; Zhao, F.; Liu, Z.; Hao, H. The impact of fuel cell vehicle deployment on road transport greenhouse gas emissions: The China case. *Int. J. Hydrogen Energy* **2018**, *43*, 22604–22621.
8. Sherif, S.A.; Goswami, D.Y.; Stefanakos, E.L.; Steinfeld, A. (Eds.) *Handbook of Hydrogen Energy*; CRC Press: Boca Raton, FL, USA, 2014.
9. Tarhan, C.; Çil, M.A. A study on hydrogen, the clean energy of the future: Hydrogen storage methods. *J. Energy Storage* **2021**, *40*, 102676.
10. Kikukawa, S.; Yamaga, F.; Mitsuhashi, H. Risk assessment of Hydrogen fueling stations for 70 MPa FCVs. *Int. J. Hydrogen Energy* **2008**, *33*, 7129–7136.
11. Chalk, S.G.; Miller, J.F.; Wagner, F.W. Challenges for fuel cells in transport applications. *J. Power Source* **2000**, *86*, 40–51.
12. Bhogilla, S.S.; Niyas, H. Design of a hydrogen compressor for hydrogen fueling stations. *Int. J. Hydrogen Energy* **2019**, *44*, 29329–29337. [CrossRef]
13. Weinert, J.X. *A Near-Term Economic Analysis of Hydrogen Fueling Stations*; Working Paper Series; Institute of Transportation Studies: Berkeley, CA, USA, 2005.
14. Sdanghi, G.; Maranzana, G.; Celzard, A.; Fierro, V. Review of the current technologies and performances of hydrogen compression for stationary and automotive applications. *Renew. Sustain. Energy Rev.* **2019**, *102*, 150–170.
15. Tian, Z.; Lv, H.; Zhou, W.; Zhang, C.; He, P. Review on equipment configuration and operation process optimization of hydrogen refueling station. *Int. J. Hydrogen Energy* **2022**, *47*, 3033–3053.
16. Kermani, N.A.; Petrushina, I.; Nikiforov, A.; Jensen, J.O.; Rokni, M. Corrosion behavior of construction materials for ionic liquid hydrogen compressor. *Int. J. Hydrogen Energy* **2016**, *41*, 16688–16695.
17. Reddi, K.; Elgowainy, A.; Sutherland, E. Hydrogen refueling station compression and storage optimization with tube-trailer deliveries. *Int. J. Hydrogen Energy* **2014**, *39*, 19169–19181. [CrossRef]
18. Altukhov, S.M.; Kuznetsov, V.D.; Levin, G.A. Calculation of the volumetric characteristics of diaphragm compressors. *Chem. Pet. Eng.* **1971**, *7*, 671–675. [CrossRef]
19. Lei, J.; Lin, Z.; Qian, Z.; Han, G.; Peng, L. Analysis of Diaphragm Compressor Exhausts Volume Decrease. In Proceedings of the 2013 Fifth International Conference on Measuring Technology and Mechatronics Automation, Hong Kong, China, 16–17 January 2013; pp. 233–235.

20. Park, H.W.; Shin, Y.I.; Lee, Y.J.; Song, J.H.; Jeon, C.H. A Numerical Analysis on a Dependence of Hydrogen Diaphragm Compressor Performance on Hydraulic Oil Conditions. *Trans. Korean Hydrog. New Energy Soc.* **2009**, *20*, 471–478.
21. Jia, X.; Zhao, Y.; Chen, J.; Peng, X. Research on the flowrate and diaphragm movement in a diaphragm compressor for a hydrogen refueling station. *Int. J. Hydrogen Energy* **2016**, *41*, 14842–14851.
22. Ren, S.; Jia, X.; Jiang, J.; Zhang, S.; Zhao, B.; Peng, X. Effect of hydraulic oil compressibility on the volumetric efficiency of a diaphragm compressor for hydrogen refueling stations. *Int. J. Hydrogen Energy* **2022**, *47*, 15224–15235.
23. Li, X.; Chen, J.; Wang, Z.; Jia, X.; Peng, X. A non-destructive fault diagnosis method for a diaphragm compressor in the hydrogen refueling station. *Int. J. Hydrogen Energy* **2019**, *44*, 24301–24311.
24. Antonova, L.I.; Kalashnikov, V.F.; Degtyareva, T.S. Calculation of the reliability of the parametric and functional elements of diaphragm compressors. *Chem. Pet. Eng.* **1987**, *23*, 153–155.
25. Lu, W.; Wang, X.; Cong, Y. Analysis on factors influencing diaphragm life of diaphragm compressors. *China Sci. Technol. Overv.* **2019**, *4*, 93–94.
26. Jia, X.; Chen, J.; Wu, H.; Peng, X. Study on the diaphragm fracture in a diaphragm compressor for a hydrogen refueling station. *Int. J. Hydrogen Energy* **2016**, *41*, 6412–6421.
27. Hu, Y.; Xu, X.; Wang, W. A new cavity profile for a diaphragm compressor used in hydrogen fueling stations. *Int. J. Hydrogen Energy* **2017**, *42*, 24458–24469.
28. Li, J.Y.; Chen, J.H.; Zhao, H.L.; Wang, S.Y.; Jia, X.H. Theoretical analysis of diaphragm fracture in diaphragm compresssor. *Eng. Mech.* **2015**, *32*, 192–197.
29. Li, J.; Jia, X.; Wu, Z.; Peng, X. The cavity profile of a diaphragm compressor for a hydrogen refueling station. *Int. J. Hydrogen Energy* **2014**, *39*, 3926–3935.
30. Li, J.; Lin, L.; Jia, X.; Peng, X. A new generatrix of the cavity profile of a diaphragm compressor. *Proc. Inst. Mech. Eng. Part C J. Mech. Eng. Sci.* **2014**, *228*, 1754–1766.
31. Li, J.Y.; Jia, X.H.; Jiang, R.M.; Peng, X.Y. Optimization Design of Diaphragm Compressor Cavity Contour with New Generatrix. *Compress. Technol.* **2013**, 6–10. [CrossRef]
32. Li, J.; Zhao, X.; Jia, X.; Peng, X. Design and experimental verification for diaphragm compressor cavity profile by two exponential terms generatrix. *J. Xi'an Jiaotong Univ.* **2014**, *48*, 127–132.
33. Wang, T.; Jia, X.; Li, X.; Ren, S.; Peng, X. Thermal-structural coupled analysis and improvement of the diaphragm compressor cylinder head for a hydrogen refueling station. *Int. J. Hydrogen Energy* **2020**, *45*, 809–821.
34. Wang, T.; Tang, Z.; Jia, X. Study on the stress and deformation of a diaphragm compressor cylinder head under extreme conditions. *IOP Conf. Ser. Mater. Sci. Eng.* **2019**, *604*, 12029. [CrossRef]
35. Arasteh, H.; Mashayekhi, R.; Ghaneifar, M.; Toghraie, D.; Afrand, M. Heat transfer enhancement in a counter-flow sinusoidal parallel-plate heat exchanger partially filled with porous media using metal foam in the channels' divergent sections. *J. Therm. Anal. Calorim.* **2020**, *141*, 1669–1685.
36. El-Shorbagy, M.A.; Eslami, F.; Ibrahim, M.; Barnoon, P.; Xia, W.; Toghraie, D. Numerical investigation of mixed convection of nanofluid flow in a trapezoidal channel with different aspect ratios in the presence of porous medium. *Case Stud. Therm. Eng.* **2021**, *25*, 100977.
37. Ahmadi, G.; Toghraie, D.; Akbari, O.A. Technical and environmental analysis of repowering the existing CHP system in a petrochemical plant: A case study. *Energy* **2018**, *159*, 937–949.
38. Yang, S.; Tao, W. *Heat Transfer*; Higher Education Press: Beijing, China, 2006; pp. 246–249.
39. Barthélémy, H. Effects of pressure and purity on the hydrogen embrittlement of steels. *Int. J. Hydrogen Energy* **2011**, *36*, 2750–2758. [CrossRef]
40. Shin, H.; Custodio, N.A.; Baek, U. Numerical analysis for characterizing hydrogen embrittlement behaviors induced in STS316L stainless steel using an in-situ small-punch test. *Theor. Appl. Fract. Mech.* **2021**, *116*, 103139.

processes

MDPI

Article

Investigation of the Interface Effects and Frosting Mechanism of Nanoporous Alumina Sheets

Song He [1,*], Heyun Liu [1], Yuan Zhang [2], Haili Liu [1] and Wang Chen [1]

1 School of Energy and Electromechanical Engineering, Hunan University of Humanities, Science and Technology, Loudi 417000, China; 2211@huhst.edu.cn (H.L.); 2770@huhst.edu.cn (H.L.); 3323@huhst.edu.cn (W.C.)

2 School of Graduate Education Teaching, Hunan University of Humanities, Science and Technology, Loudi 417000, China; 3527@huhst.edu.cn

* Correspondence: 3504@huhst.edu.cn

Abstract: Nanoporous alumina sheets can inhibit the growth of the frost layer in a low-temperature environment, which has been widely used in air-conditioning heat exchangers. In this study, nanoporous alumina sheets with pore diameters of 30 nm, 100 nm, 200 nm, 300 nm, and 400 nm were prepared by using the anodic oxidation method with the conventional polished aluminum sheet as the reference. A comprehensive and in-depth analysis of the frosting mechanism has been proposed based on the contact angle, specific surface area, and fractal dimension. It was found that compared with the polished aluminum sheet, the nanoporous alumina sheets had good anti-frost properties. Due to its special interface effects, the porous alumina sheet with a 100 nm pore diameter had strong anti-frost performance under low temperatures and high humidity. In an environment with low surface temperature and high relative humidity, it is recommended to use hydrophilic aluminum fins with large specific areas and small fractal dimensions for the heat exchange fins of air source heat pump air conditioning systems.

Keywords: nanoporous alumina sheet; interface effect; frosting; mechanism

Citation: He, S.; Liu, H.; Zhang, Y.; Liu, H.; Chen, W. Investigation of the Interface Effects and Frosting Mechanism of Nanoporous Alumina Sheets. *Processes* **2023**, *11*, 2019. https://doi.org/10.3390/pr11072019

Academic Editors: Udo Fritsching, Chenyu Wu, Zhongkai Yi and Chenhui Lin

Received: 29 May 2023
Revised: 27 June 2023
Accepted: 4 July 2023
Published: 6 July 2023

1. Introduction

Air conditioning systems are widely used to adjust the comfort of a building environment. While enjoying a comfortable indoor environment, people are also worried about the high energy consumption of buildings and serious environmental pollution. Studies have shown that building energy consumption accounts for 30–40% of the total social energy consumption [1]. Moreover, air conditioning energy consumption accounts for 40–60% of building energy consumption [2,3]. Air source heat pumps constitute the most widely used air conditioning equipment in domestic residential buildings due to their energy saving, environmental friendliness, short investment return period, and low operating cost. However, in severely cold winter, when the surface temperature of the fins of the outdoor heat exchanger of the air source heat pump air conditioning system is lower than the ambient dew point temperature and is lower than the freezing temperature of the condensate water, the surface of the fins will be frosted [4], which reduces the heat transfer efficiency of the heat exchanger by 50–75% [5]. This is energy waste and environmentally unfriendly. Therefore, the optimization of air source heat pump air conditioning systems and the improvement of anti-frosting performance is a hotspot of current research, attracting many researchers' attention [6].

Inspired by the superhydrophobic "lotus effect" [7], a passive defrosting method was proposed by creating a surface that can prevented the condensation/frosting of water molecules on the surface or promoted the shedding of condensate/frost on the cold surface [8]. The passive method requires low energy consumption and is environmentally friendly. Both hydrophilic and hydrophobic fins on the heat exchanger meet the requirements of passive defrosting. Liu et al. [9] demonstrated the delayed formation of the frosting layer on hydrophilic heat exchange fins by at least 15 min compared to the ordinary heat exchange fins with a reduced amount of frost by 40%. Okoroafor et al. [10] found that due to the reduced water adsorption, the aluminum sheets coated with hydrophilic layers showed good anti-frosting properties in humid air. Their frosting rate and the thicknesses of the frost layers were 10–30% lower than that of an ordinary aluminum sheet. Lee et al. [11] compared the frost formation on the surfaces with different dynamic contact angles (DCA) of 23° and 88°. The results showed that low-DCA surfaces had lower frost thickness and higher frost density than high-DCA surfaces. Yang et al. [12] developed a nanoporous hydrophilic aluminum sheet with an improved anti-frosting performance at −15 °C by reducing the frosting rate by a factor of 3. Meanwhile, Wang et al. [13] found that the superhydrophobic surface also had good anti-frosting performance under high humidity conditions. A superhydrophobic aluminum surface with hierarchical micro–nano structures was created via a combination of chemical etching and sandblasting [14]. The frosting delay time was more than 10 times that of the bare substrate, and the frosting area was significantly smaller. Lei et al. [15] used the spray coating method to prepare a superhydrophobic coating based on silica nanoparticles and room-temperature vulcanized silicone rubber. The experimental results showed that the superhydrophobic coating had a strong ability to prevent frost growth. Fan et al. [16] prepared a transparent hydrophobic surface by using a sol–gel method with the frost formation time increased by 4.5 times that on the unmodified surface at a propagation speed of 47.5 μm/s, which was 29.6% lower than that of the unmodified surface.

Hydrophilic/hydrophobic surfaces produced by surface coating have some shortcomings. The surface coating could be structurally fragile with poor adhesion, easily falling off, and showing poor durability [17]. The preparation of a nanoporous surface via anodization method can avoid such shortcomings because this way, the nanoporous surface is directly formed, giving it better durability. Wilson et al. [18] found that nanoporous surface coating could reduce the nucleation temperature of condensed water, and the freezing property of the coating surface was not damaged after 150 cycles of freezing/melting. Kim et al. [19] conducted experimental tests on a nanoporous surface and found that it can promote the condensation of water droplets and the falling off of the frost layer from the surface under the action of its gravity, which makes up for the limitations of the hydrophobic surface in a high-humidity environment.

The anti-frosting properties of nanoporous alumina sheets have been studied, but their anti-frosting mechanism is seldom studied. To further study the mechanism of surface frosting and provide a reference for the subsequent preparation of anti-frosting surfaces, many researchers have conducted studies based on the contact angle and surface energy of a surface. Huang et al. [20] prepared a surface with a contact angle of 96.2°~154.9° by using solution immersion. Under natural convection conditions, a comparative experiment of frosting on the surface of ordinary copper was carried out. The results showed that the delayed frosting time on the surface was directly proportional to the surface contact angle. The larger the contact angle was, the better the surface anti-frosting performance was. Piucco et al. [21] found that the initial frosting area of the surface was inversely proportional to the surface contact angle. As the contact angle increased, a lower surface temperature was required to produce frost crystals. Mangini et al. [22] observed the frost layer morphology on a surface with different contact angles by using infrared thermal imaging cameras. They found that a dense ice layer was formed on the surface with a small contact angle while a sparsely distributed ice layer was formed on the surface with a large contact angle. This indicated that the contact angle would affect the shape and density of the surface frost layer and have a certain effect on the long-term frost formation on the surface. Na and Webb et al. [23] found that a low-energy surface required a much higher supersaturation degree for frost nucleation than a high-energy surface. Liu et al. [24]

proposed that the retarding frost nucleation on a low-energy surface is due to a smaller area of contact with the condensed water nuclei than on high-energy surfaces.

It is generally considered that frosting is a continuous problem and cannot be limited to the initial stage of frosting. However, the contact angle and surface energy mentioned above can only be used to study the initial frosting performance at the interface. The effects on the growth of the frost layer also need to be investigated, and these have important effects on the frosting process [25]. The growth structure of the frost layer directly affects the density of the frost layer, the effective diffusion coefficient of water vapor in the frost layer, and the thermal conductivity of the frost layer [26]. These three physical parameters will adversely affect the frost structure. Due to the complexity of frosting, most current studies are still limited to experimental tests and simple theoretical analyses. Limited studies have been conducted on the mechanism of frosting on surfaces.

In this study, nanoporous alumina sheets were prepared by using the anodic oxidation method with averaged pore diameters of 30 nm, 100 nm, 200 nm, 300 nm, and 400 nm. The mechanism of frosting on the nanoporous alumina sheet surfaces was explored. Focusing on the frost layer morphology at the initial frosting stage of the nanoporous surface, the experimental results were analyzed for Gibbs free energy, specific surface area, and fractal dimensions, and the theoretical morphology was compared with the actual morphology through visual observation. The subsequent research will further quantitatively analyze the frosting properties of a nanoporous surface by combining the thickness and density of the frosting layer. This research will help analyze the mechanistic factors that affect surface frosting with provided performance parameters for preparing surfaces with high anti-frosting properties. The results promote applying nanoporous material as the anti-frost material in air source heat pumps with a potential energy-saving technology.

2. Sample Preparation and Characterization

The nanoporous alumina sheets were prepared by using an anodization process. The pore diameters could be manipulated by controlling the intensity of the oxidation current and the hole expansion treatment. The surface morphology, contact angle, surface energy, and fractal dimension were observed and measured.

2.1. Preparation of the Nanoporous Alumina Sheets

The preparation process mainly involved 4 steps. Firstly, aluminum sheets (99.99% purity) were ultrasonically cleaned in the acetone/ethanol solution for 10 min to remove the residual grease before being dried and used as substrates. The samples were immersed in the 1 M sodium hydroxide solution for 5 min to remove the surface oxide film. After that, the substrate was electrochemically polished in a 400 mL 1:9 ratio perchloric acid in ethanol solution for 2 min, using graphite as the cathode and an aluminum sample as the anode. Finally, the electrolyte was replaced with 0.3 M oxalic acid solution for the anodization to form nanoporous material under a DC bias. It should be mentioned that after each step was finished, the substrate was thoroughly washed with deionized water.

A nanoporous surface with different apertures could be obtained by changing the current amplitude, and the relationship between the surface structure parameters and the anodization current is presented in Table 1 ('diameter' refers to the diameter of a nanopore, and 'spacing' refers to the distance between the centers of two adjacent nanopores). The nanoporous samples with 300 nm and 400 nm pore diameters (Samples 5 and 6) were obtained using the hole expansion treatment by immersing the aluminum sheet with a 200 nm pore diameter in 5 wt% phosphoric acid solution for 30 min and 60 min. Alongside this, the polished aluminum sheet (simple 1) was prepared for comparison.

Table 1. Surface structure parameters and fabrication craft.

Simple	Diameter	Spacing	Depth	Current	Size
1	0 nm	0 nm	0 µm	0 A	
2	30 nm	65 nm		0.2 A	
3	100 nm	100 nm		0.5 A	$20 \times 20 \times 0.2$ mm
4	200 nm	450 nm	60 ± 5 µm	1.2 A	
5	300 nm	450 nm		1.2 A	
6	400 nm	450 nm		1.2 A	

2.2. Surface Topography

A scanning electron microscope (SEM, S-3400N, Hitachi, Tokyo, Japan) was used to study the surface morphology, as shown in Figure 1. Uniformly distributed nanopore structures with relatively uniform pore sizes were observed for all samples. Sample 2 has a higher density of nanopores than sample 3. Samples 4, 5, and 6 have similar hole spacings, while the pore diameters are increased accordingly, with the wall gradually becoming thinner. The images below have been enlarged 50,000 times.

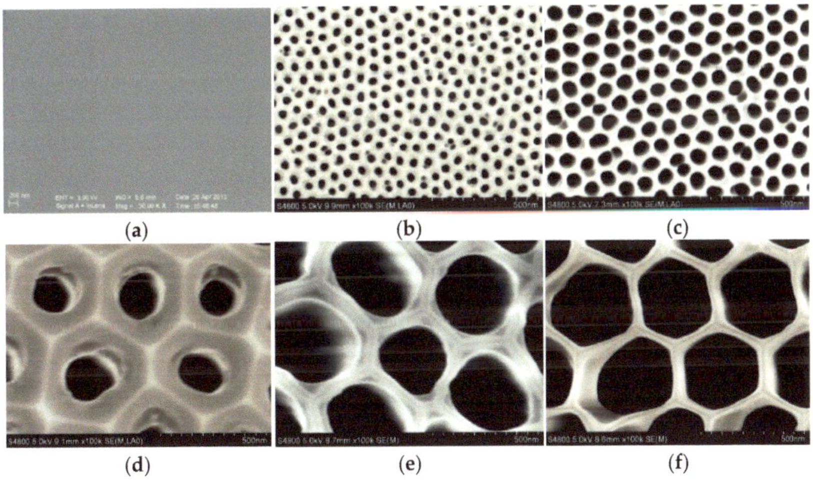

Figure 1. The SEM images of the nanoporous samples: (**a**) sample 1, (**b**) sample 2, (**c**) sample 3, (**d**) sample 4, (**e**) sample 5, and (**f**) sample 6.

The number of nanopores and the specific surface area for each sample could be calculated according to its pore diameter, depth, and spacing (the number of nanopores on the abscissa can be obtained by dividing the length of the surface by the spacing of the pores, and the same method obtains the number of nanopores in the ordinate; their product is the total number of nanopores. Specific surface area refers to the ratio of the total microscopic area of the surface of a nanoporous alumina sheet to the macroscopic area). The calculation results are listed in Table 2.

Table 2. The parameters of the prepared samples.

Sample	2	3	4	5	6
Number of nanopores	9.5×10^{10}	4×10^{10}	1.98×10^{9}	1.98×10^{9}	1.98×10^{9}
Specific surface area (m^2/m^2)	1342.5	1697.5	187.5	282.5	375

2.3. Contact Angle and Surface Energy

Static contact angle and surface energy were measured using an optical contact angle measuring instrument (ZJ-7000, Shenzhen, China), with the contact angle range being 0–180°, accuracy ±0.1°, surface energy range 0–1000 mN/m, and accuracy ±0.01 mN/m. It should be mentioned that surface energy is defined as the extra energy generated by the surface of a material relative to its interior. Water droplets are spherical on surfaces with low surface energy because low surface energy is not sufficient to break the surface tension of water droplets. The results are shown in Figure 2 and Table 3.

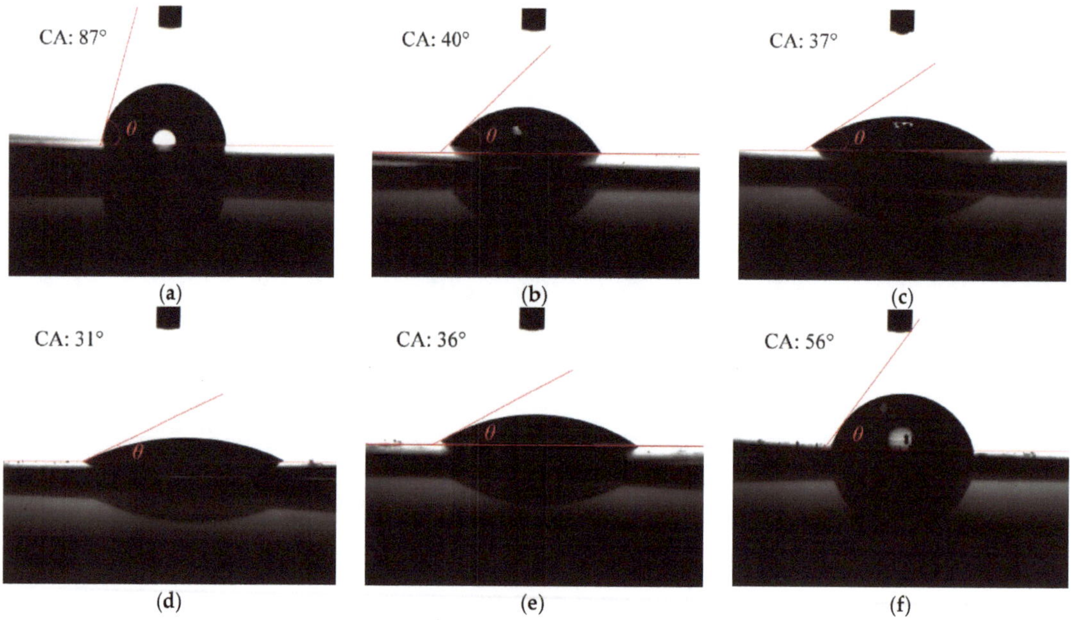

Figure 2. The static contact angles of the prepared samples: (**a**) sample 1, (**b**) sample 2, (**c**) sample 3, (**d**) sample 4, (**e**) sample 5, and (**f**) sample 6.

Table 3. The parameters of the prepared aluminum sheets.

Sample	1	2	3	4	5	6
Contact angle (degree)	87	40	37	31	36	56
Surface energy (mN/m)	50	142	147	157	149	111

2.4. Surface Fractal Dimension

The fractal dimension reflects the effectiveness of the space occupied by a complex form, which is a measure of the irregularity of a complex form. It can be used to analyze the space occupancy efficiency of nanoporous surfaces, i.e., how many nucleation active points water molecules have on their surfaces.

The fractal dimension was analyzed using Photoshop, Matlab, and the Fraclab toolbox from the SEM images. The Photoshop software (Photoshop CC) was used to convert the SEM images of the nanoporous alumina sheets into grayscale images, which were then converted into binary images represented by binary numbers. After that, the Matlab software (Matlab 2020) was used to convert the binary data into Fraclab-identifiable double-precision numerical data. Finally, the Fraclab toolbox was used to calculate the box dimension to obtain the fractal dimensions of the surfaces of the nanoporous alumina sheets. The surface fractal dimensions of nanoporous alumina sheets are shown in Table 4.

Table 4. The surface fractal dimension of nanoporous alumina sheets.

Sample	2	3	4	5	6
Fractal dimension	2.8788	2.8084	2.8643	2.7684	2.8247

3. Theoretical Analysis

The heat transfer relationship between the water vapor and the solid surface and the morphological characteristics of the frost layer affect the frosting rate and frosting quantity on a solid surface. The frosting mechanism of the prepared nanoporous alumina sheets will be analyzed from their surface heat, mass transfer, and interface effects—that is, using surface contact angles, fractal dimensions, and specific surface areas.

3.1. Thermodynamitical Analysis

The frosting nucleus is formed from a vapor embryo through condensing water molecules. Although the saturated air is in direct contact with a cold surface whose temperature is below 0 °C, the nucleation begins only after overcoming the energy barrier.

The model representing the homogeneous and heterogenous nucleation on a surface is shown in Figure 3a,b [21] (Reproduced with permission from Hermes, *Experimental Thermal and Fluid Science*, published by ELSEVIER, 2008). The vapor transformation into nucleation is accompanied by a transition from a saturated state to a supersaturated state. The following equation shows the free energy difference, ΔG, between the supersaturated and saturated states [23,27].

$$\Delta G = V\Delta g_v + \gamma_{se}A_{se} + \gamma_{ew}A_{ew} - \gamma_{sw}A_{ew} \tag{1}$$

Here, V is the volume. A_{ew} and γ_{ew} are the surface area and surface energy at the embryo–wall interface. A_{se} and γ_{se} correspond to the surface area and surface energy at the surroundings–embryo interface. γ_{sw} is the surface energy at the surroundings–wall interface. Δg_v is the volume-specific Gibbs free energy deviation between the supersaturated and saturated states, which is described in Equation (2) [21].

$$\Delta g_v = \Delta h_{lat} - TS = \frac{\rho RT}{M}\ln\left(\frac{\omega_s}{\omega_{sat,e}}\right) \tag{2}$$

Here, ω_s and $\omega_{sat,e}$ are the humidities of the surrounding air and the air at the embryo surface, respectively. T is the embryo temperature and R is the ideal gas constant. ρ and M are the embryo density and molar mass, respectively. Hence, the total free energy difference can be expressed in Equation (3).

$$\Delta G = \frac{\rho VRT}{M}\ln\left(\frac{\omega_s}{\omega_{sat,e}}\right) + \gamma_{se}A_{se} + \gamma_{ew}A_{ew} - \gamma_{sw}A_{ew} \tag{3}$$

Fletcher [28] assumed that the shape of the parent phase for nucleation is a spherical segment as shown in Figure 3b. The embryo size significantly affects the Gibbs free energy deviation. As Bai [29] mentioned, at a critical radius, r', the Gibbs free energy to be overcome for the initiate nucleation will be minimized. Using the Fletcher model, the critical Gibbs free energy deviation ($\Delta G'$) can be obtained at the critical embryo size (r').

$$\Delta G'_{tot} = \frac{4\pi}{3}\frac{{\gamma_{se}}^3}{\left[\frac{\rho RT}{M}\ln\left(\frac{\omega_{sat,e}}{\omega_s}\right)\right]^2}(1-\cos\theta)^2(2+\cos\theta) \tag{4}$$

$$r' = 2\gamma_{se}\left[\frac{\rho RT}{M}\ln\left(\frac{\omega_s}{\omega_{sat,e}}\right)\right]^{-1} \tag{5}$$

Becker and Doring [30] proposed the following equation for the embryo formation rate on a unit surface area for heterogeneous nucleation:

$$I = I_0 \exp(-\frac{\Delta G'}{kT_w}) \tag{6}$$

Here, I is the embryo formation rate at temperature T_w. $\Delta G'$ is scally by Volmer and Flood [31]. $I_0 = 10^{25}$ embryo/(cm^2 s) is the kinetic constant and k is the Boltzmann constant (1.381×10^{-23} J/K) [28].

It could be concluded from Equations (4)–(6) that the nucleation rate of frost crystals is inversely proportional to the surface contact angle in a certain thermal system.

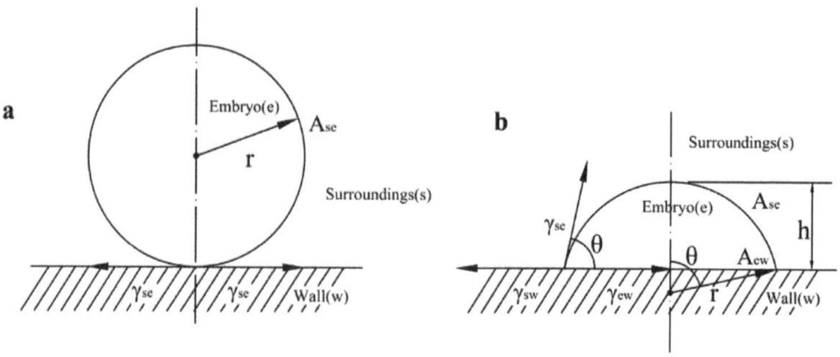

Figure 3. Schematic representation of (**a**) homogeneous and (**b**) heterogeneous nucleation [21].

3.2. Fractal Dimension Analysis

Both the nucleation rate and the macroscopic morphology of the frost layer can affect the formation of frost on cold surfaces. The fractal dimension is a quantitative reaction of self-similarity and can reflect the complexity of surface microstructure [32]. The larger the fractal dimension of the surface is, the more complex the surface structure will be, with a more active surface containing more edges and angles and thus increasing the number of attachment points for the initial liquid nuclei. Hou [33] and Mei [34] have discussed the effects of a cold surface having different fractal dimensions on surface condensation and frosting. They realized that differentiation in frost layer morphology significantly affects the heat transfer between the frost crystals, the cold surface, and the surrounding environment. Ding et al. [35] proposed that the smaller the fractal dimension of the surface was, the more regular the surface morphology would be and the fewer active points the initial liquid nucleus would need to survive, leading to a smaller number of condensing droplets and thus reducing the surface frost crystal coverage.

3.3. Surface Adsorption

The essence of frosting is that water molecules contact a cold surface and the liquid phase turns into frost crystals exothermically. Free collisions of water molecules and active adsorption on the surface are involved. Physisorption is the main adsorption mechanism at low temperatures, involving the Van der Waals forces between molecules. Studies have shown that the Van der Waals force can be affected by the specific surface area of a surface. The larger the surface area is, the greater the Van der Waals force is [36]. The active adsorption of water molecules on the surface will increase the probability of water molecules contacting the surface. Strong, active adsorption will enhance the initial frosting on the surface. However, the long-term frosting on the surface needs to be further analyzed because the strong adsorption force will also affect the shedding of the frost layer and the heat transfer efficiency of the surface.

4. Experiment and Discussion

The frosting morphologies of nanoporous alumina sheets were observed and analyzed. The influence of the interface effects (contact angle, specific surface area, and fractal dimension) of nanoporous surfaces on their frosting properties was analyzed.

4.1. Experimental Test

The experimental test setup, shown in Figure 4, included a computer, microscope, and semiconductor cooler. The frost topographies of the fronts and sides of the aluminum sheets were photographed. Frontal shooting was performed with a wind speed of 0.1 m/s, an ambient temperature of 28 °C, a relative humidity of 62%, and a surface temperature of −10 °C. Lateral shooting was performed with a wind speed of 0.1 m/s, an ambient temperature of 16 °C, a relative humidity of 58%, and a surface temperature of −15 °C.

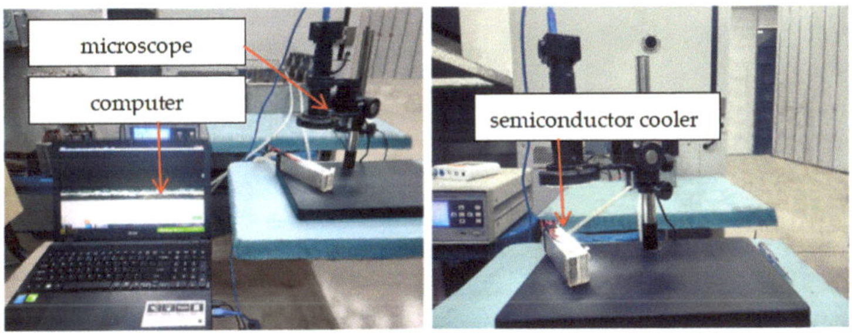

Figure 4. Frost morphology recording device on aluminum surface.

The microscope and computer were connected with a CCD, the frosting process on the surface of each aluminum sheet was transmitted to the computer through a video cable image, and the pictures of different aluminum sheets were taken at the same multiple. The experimental results are shown in Section 4.1.

The following phenomena can be obtained by analyzing Figure **??**.

On the surface of the polished aluminum sheet, the water molecules first condensed into hemispherical water droplets, next, they gradually froze, and then, frost crystals appeared on the frozen water droplets. On the surface of the 30 nm pore diameter alumina sheet, the water molecules condensed into droplets without obvious shapes, the spreading area was large, the heights of the droplets were smaller than the heights of the water droplets condensed on the surface of the polished aluminum sheet, and a dense frost layer was generated on the surface of the droplets after freezing. The surface of the 100 nm pore diameter alumina sheet first condensed into a water film and then froze, forming frost crystals on the surface of the frozen water film. The frosting processes of the 200 nm, 300 nm, and 400 nm pore diameter alumina sheets were similar, and frost crystals were generated directly on their surfaces. The amount of frost on the surface of each aluminum sheet in the early stage of frosting could be preliminarily judged thus: polished >30 nm > 100 nm; 200 nm > 400 nm > 300 nm.

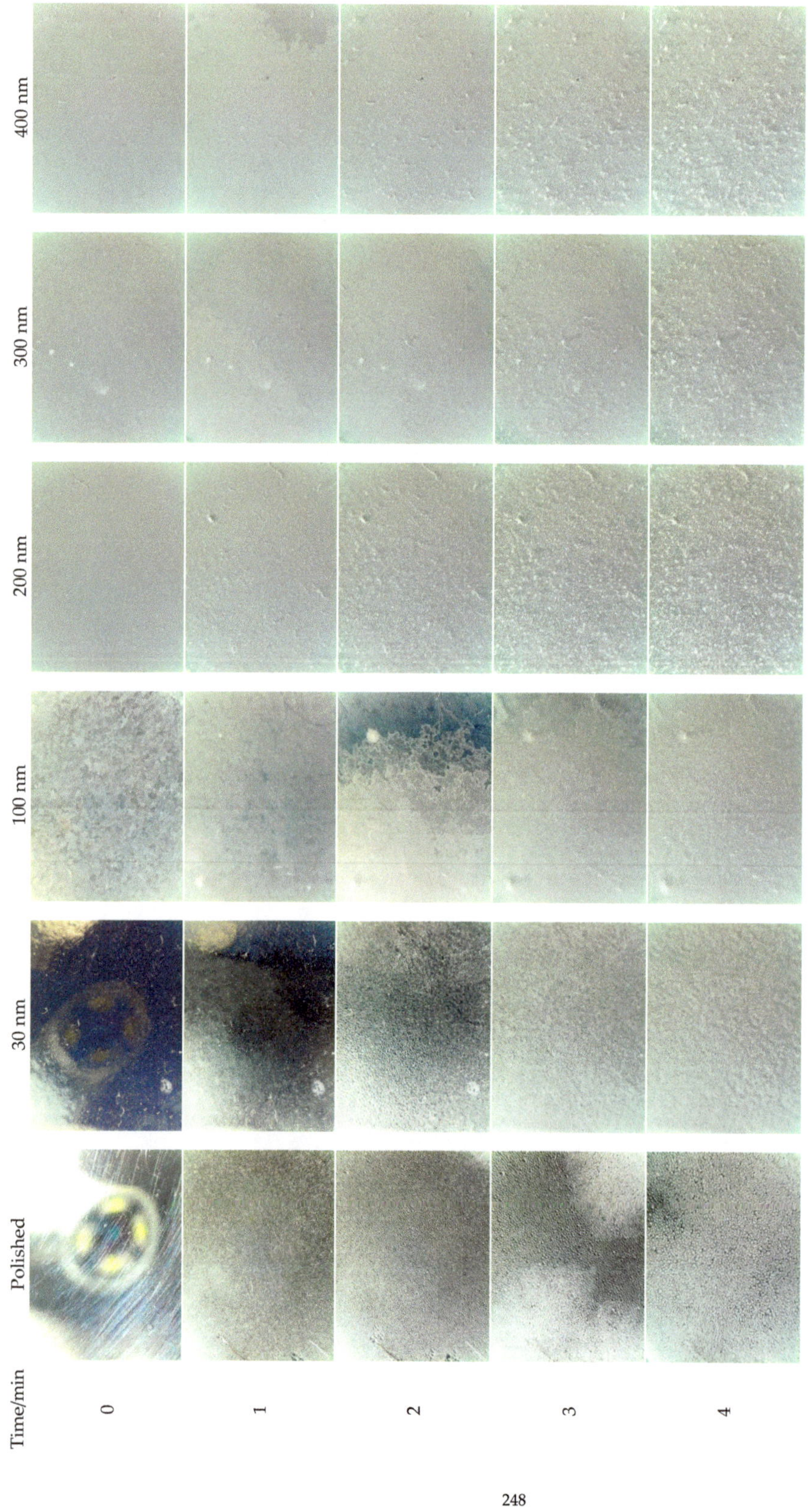

Figure 5. *Cont.*

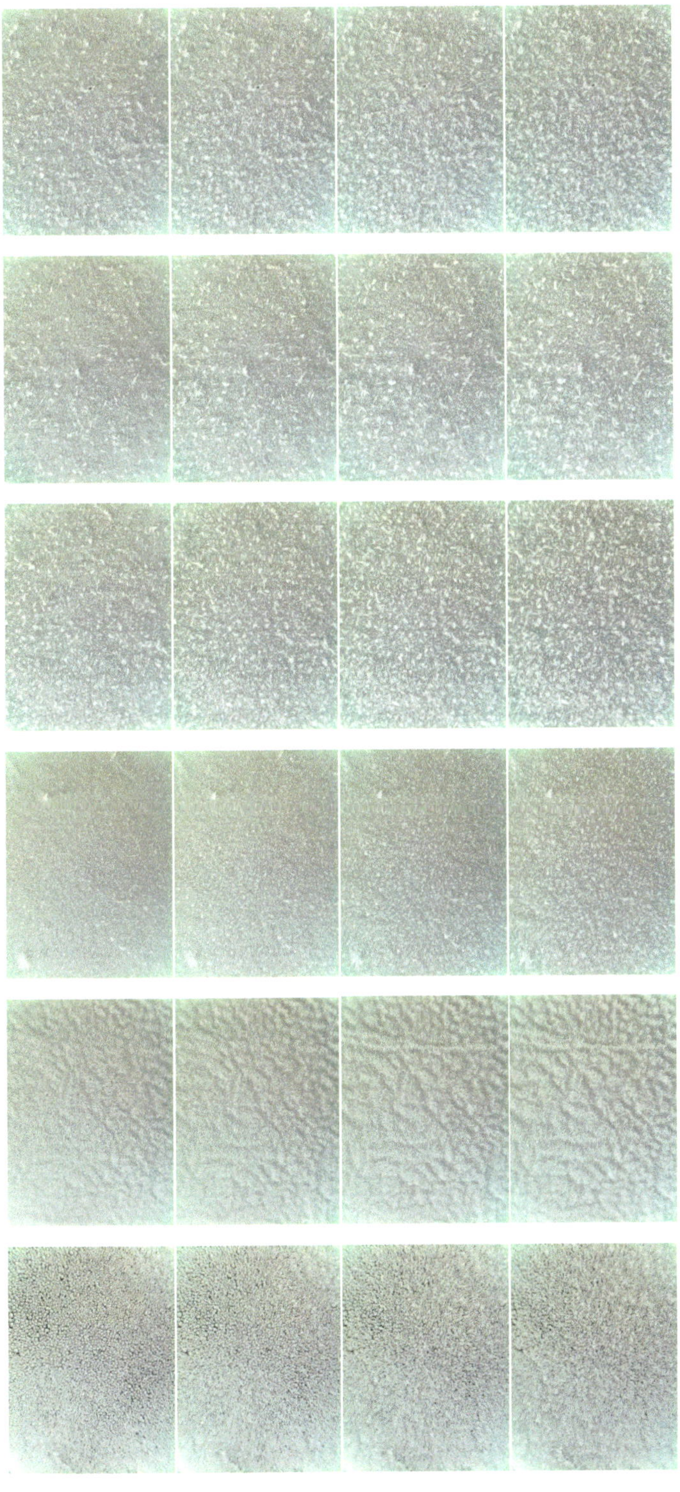

5

6

7

8

Figure 5. The frosting process on the fronts of the aluminum sheet surfaces (the scale bar is 1:20).

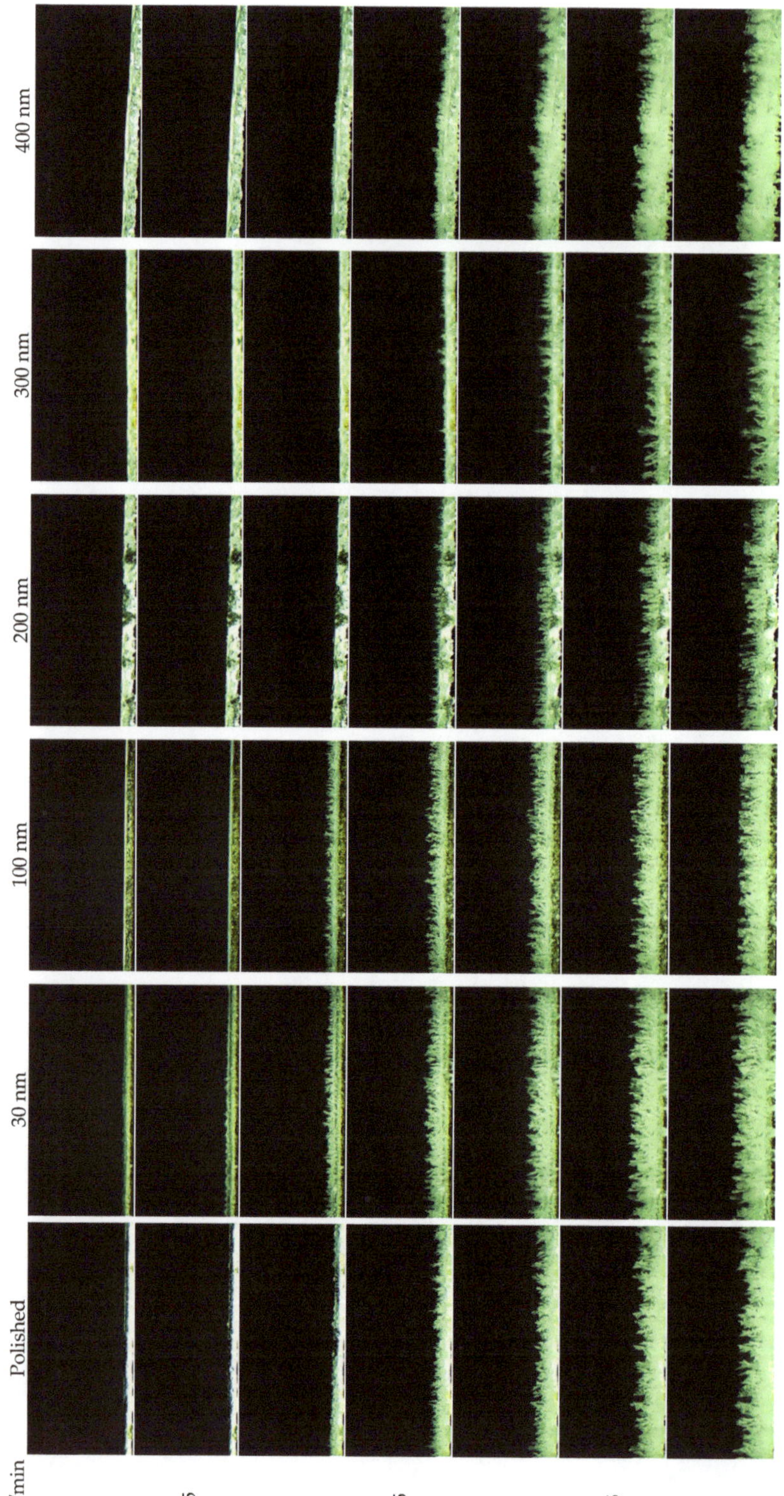

Figure 6. The frosting process on the sides of the aluminum sheet surfaces (the scale bar is 1:20).

From the results in Figure **??**, the following conclusions could be obtained.

When ice was formed on the surface of each polished aluminum sheet, dendritic frost crystals continued to grow on the surface of the ice block, and the dendritic frost crystals increased the contact area with air and promoted the subsequent frosting process. The surface frost crystals of the 30 nm pore diameter alumina sheet were tree-shaped. The frost crystals grown on the surface of the 100 nm and 200 nm pore diameter alumina sheets were needle-like, and the frost crystal density on the surface of the 100 nm pore diameter alumina sheet was greater than that on the 200 nm pore diameter alumina sheet; from the results of the long-term experiment, it was found that the dense needle-like frost crystals would play a role in heat insulation and hinder the subsequent frosting process of water molecules. The dendritic frost crystals on the surface of the 300 nm pore diameter alumina sheet reduced the obstruction of heat exchange with water molecules and enhanced the subsequent frosting process. A flat frost layer on the surface of the 400 nm alumina sheet would hinder the subsequent frosting process.

4.2. *Interface Effect of Nanoporous Alumina Sheets*

The effects of contact angle, specific surface area, and fractal dimension on frost morphology, supercooling degree, and embryo formation rate in the alumina sheet surfaces were analyzed. The interface parameters of different nanoporous alumina sheets are shown in Figures 7–9 (the data in Figures 7–9 corresponds to Tables 2–4. Each contact angle is measured by using the optical contact-angle-measuring instrument at 3 random points on the surface and selecting their average value. The specific surface area is calculated by combining the surface pore size and pore spacing. The fractal dimension is calculated, by using the MATLAB software, based on the surface topography).

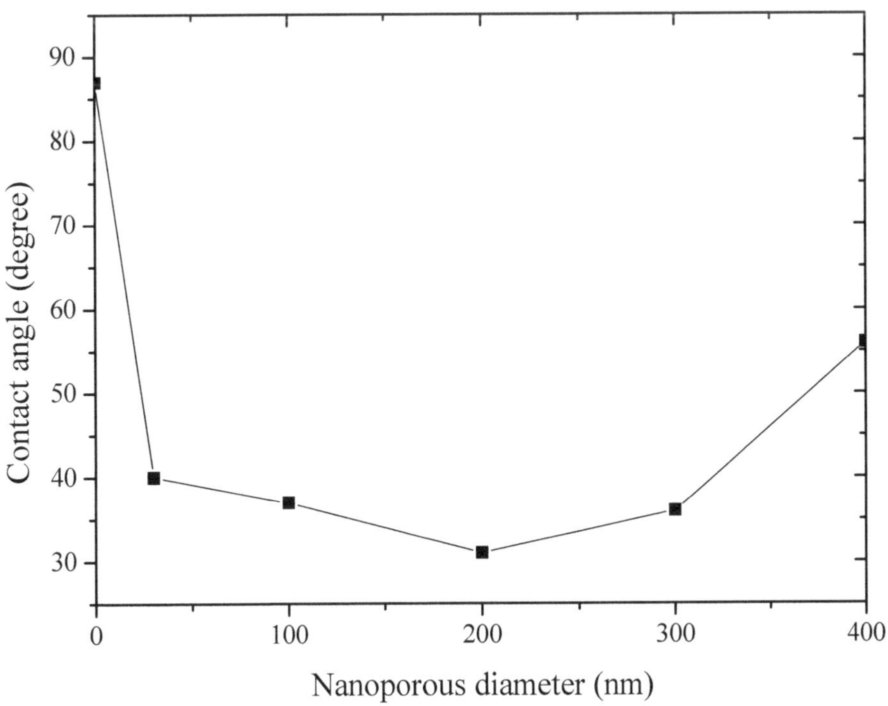

Figure 7. Contact angle value of each aluminum sheet.

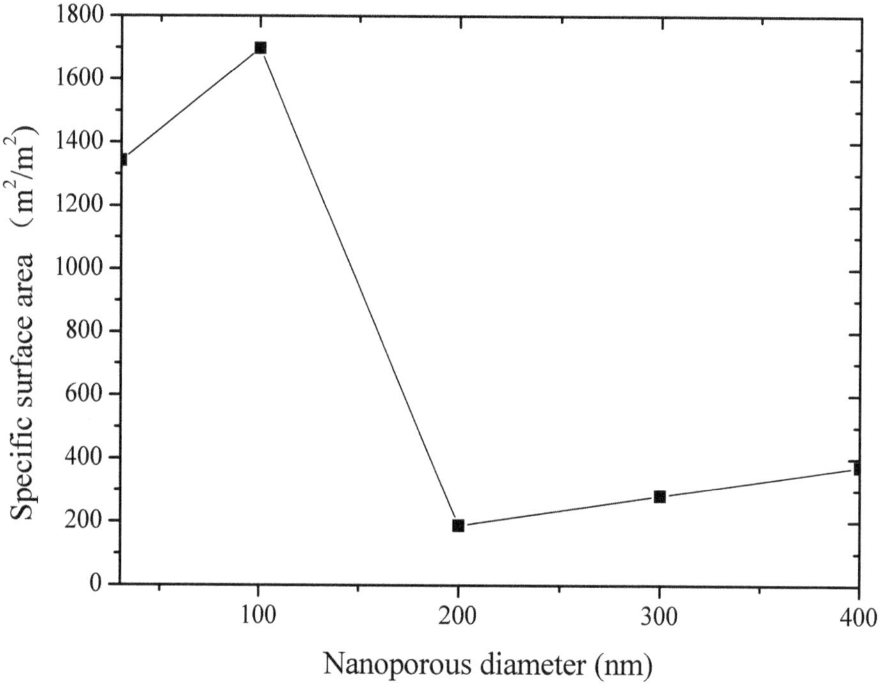

Figure 8. Specific surface area value of each aluminum sheet.

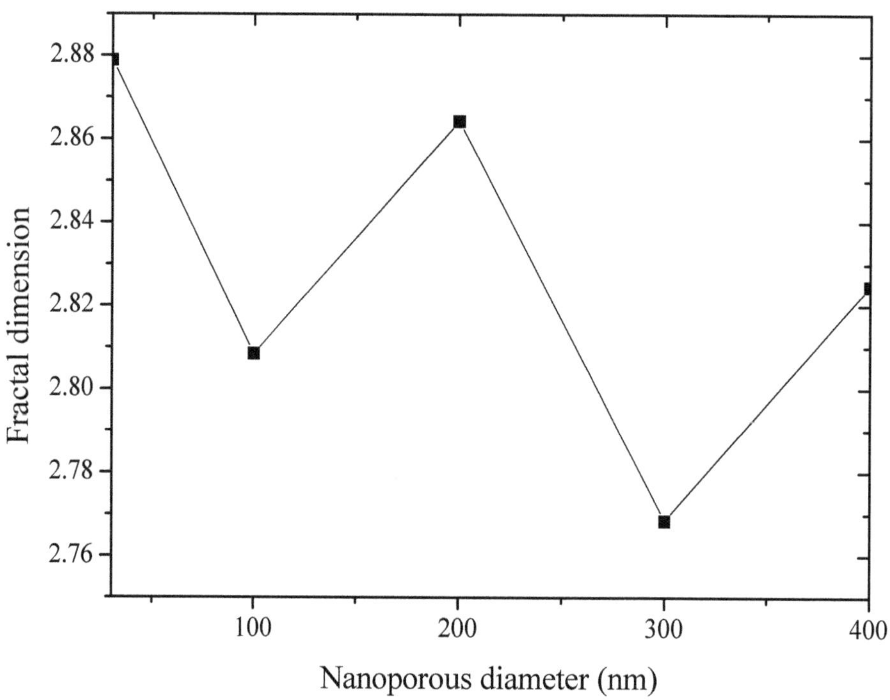

Figure 9. Fractal dimension value of each aluminum sheet.

The initial frosting phenomenon was analyzed with the interface parameters of the nanoporous alumina sheets. Morphology models representing the formations of initial frost crystals on different aluminum sheet surfaces were built, as shown in Figure 10, based on the photographs at the beginning of the frosting in Figure 11.

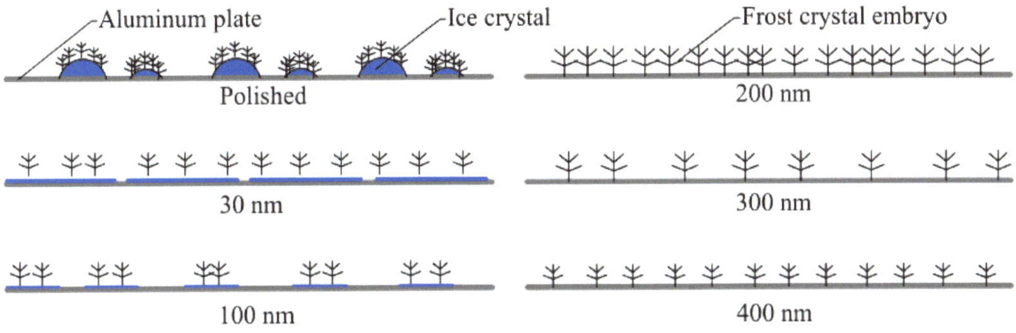

Figure 10. Morphology model of initial frost crystals on aluminum sheet surfaces.

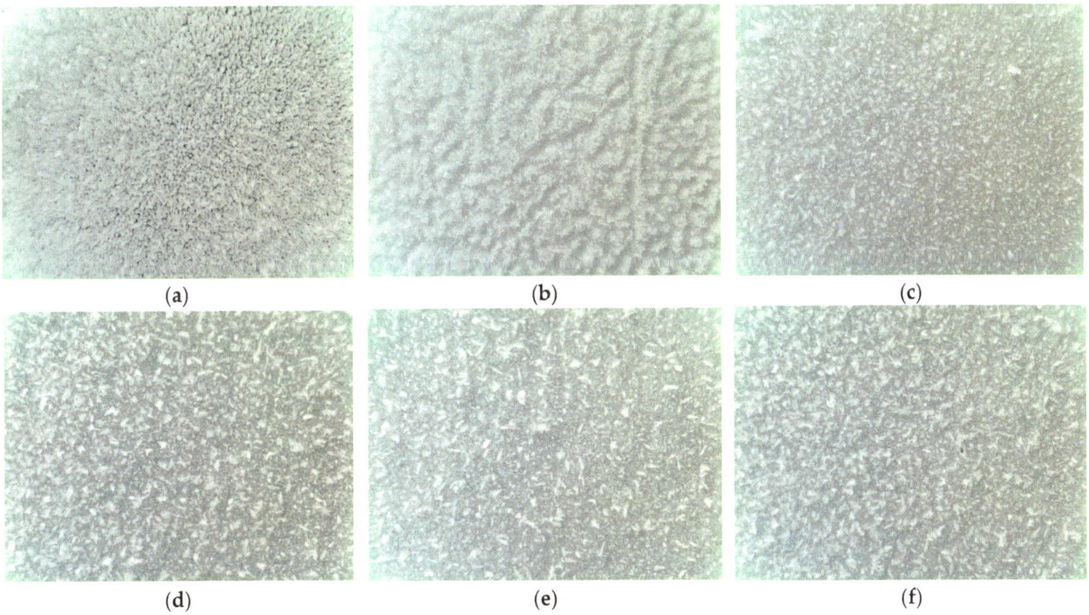

Figure 11. Initial frost morphology images of aluminum sheets: (**a**) polished, (**b**) 30 nm, (**c**) 100 nm, (**d**) 200 nm, (**e**) 300 nm, and (**f**) 400 nm.

(1) The surface contact angle of the polished aluminum sheet was much larger than those of the other prepared aluminum sheets, resulting in a larger Gibbs free energy barrier and a longer phase transition time. More water molecules would condense on its surface before they froze. After freezing, the water molecules in the air would subject phase changes on the ice crystals exothermally, generating frost crystals.

(2) The specific surface areas of the 30 nm pore diameter alumina sheet and 100 nm pore diameter alumina sheet were similar and much larger than those of the other aluminum sheets. At the beginning of the frosting process, water molecules in the moist air would be trapped with a larger molecular adsorption force. This process would be very rapid,

and the captured water molecules would not have time to transform into ice, forming a liquid water film first. Since the surface fractal dimension (i.e., active surface points) of the 30 nm pore diameter alumina sheet was larger than that of the 100 nm pore diameter alumina sheet, the spreading area of the water film on the 30 nm pore diameter alumina sheet would be larger. Since the contact angle on the surface of the 30 nm pore diameter alumina sheet was greater than that on the 100 nm alumina sheet, the phase transition was more difficult than that on the 100 nm pore diameter alumina sheet, leading to a thicker liquid water layer.

(3) The specific surface areas of the 200, 300, and 400 nm pore diameter alumina sheets were similarly low. The adsorption of water molecules depended on the probability of free collision between water molecules and the nanoporous alumina sheet surfaces. Therefore, after contacting these nanoporous alumina sheet surfaces, the adsorbed water molecules would directly form ice crystals rather than liquid. Their fractal dimensions followed the order of 200 nm > 400 nm > 300 nm, and so, the number of formed frost embryos would also show 200 nm > 400 nm > 300 nm. The contact angle of the surface of the 400 nm pore diameter alumina sheet was greater than those of the 200 nm and 300 nm pore diameter alumina sheets. Hence, the phase transition of water molecules on the 400 nm pore diameter surface was more difficult and slower. Therefore, the height of the frost crystal embryos would be smaller than that of the 200 nm and 300 nm pore diameter alumina sheets. Since the contact angles of the 200 nm and 300 nm pore diameter alumina sheets were similar, their frost crystal embryo heights were also comparable.

By comparing Figures 10 and 11, it can be seen that the frost crystal morphology model can reflect the initial frost crystal morphology on the surface of an aluminum sheet.

Analyzing the frost crystal morphology model, one could draw the following conclusions:

(1) The polished aluminum sheet has the best thermal conductivity. The contact area of the surface ice crystal with air is large. Therefore, the polished aluminum sheet has the highest average frosting rate when compared to the other aluminum sheets.
(2) The formed frost crystals on the 100 nm pore diameter alumina sheet are distributed relatively and sparsely due to the sparsely distributed surface-active points. During the initial stage of frosting, the ice crystals play the role of overhead insulation on the cold surface to prevent frost formation. Hence, the 100 nm pore diameter alumina sheet offers good anti-frosting performance even in high humidity environments.

5. Conclusions

In this study, nanoporous alumina sheets with pore diameters of 30, 100, 200, 300, and 400 nm were prepared by using the anodic oxidation method with a polished aluminum sheet as a reference. The mechanism of surface frosting was established based on the contact angles, specific surface areas, and fractal dimensions. Based on the influence of mechanism parameters on frost performance, the frost morphology model in the early stage of surface frosting was established, and the reliability of the frost morphology model was verified by comparing it with the actual observation; moreover, the surface performance parameters of high frost resistance were introduced to the frost morphology model to provide a reference for the preparation of high-frost-resistance surfaces.

The alumina sheet with a 100 nm pore diameter has strong anti-frost performance at low temperatures in a high humidity environment due to its hydrophilic surface with a large specific area and small fractal dimension. Sparsely distributed ice crystals appear on its surface first, and then, frost crystals grow on the ice crystals. The ice crystals form an insulation layer limiting the growth of frost crystals.

Through theoretical analysis, in the environment of low surface temperature and high relative humidity, it is recommended to use hydrophilic aluminum fins with a large specific area and small fractal dimension for the heat exchange fins of air source heat pump air conditioning systems.

Author Contributions: Methodology, H.L. (Heyun Liu); Software, Y.Z.; Validation, H.L. (Heyun Liu) and W.C.; Formal analysis, H.L. (Heyun Liu) and Y.Z.; Investigation, S.H.; Resources, H.L. (Haili Liu); Data curation, Y.Z.; Writing—original draft, S.H.; Visualization, W.C.; Supervision, H.L. (Haili Liu); Project administration, H.L. (Haili Liu); Funding acquisition, S.H. All authors have read and agreed to the published version of the manuscript.

Funding: This work was financially supported by the Loudi Xiaohe Talent Project [202211]. And The APC was funded by [Huhst PhD Research Start-up Fund].

Data Availability Statement: All relevant data are within the paper.

Conflicts of Interest: The authors declare no conflict of interest.

Nomenclature

A area (m^2)
G Gibbs free energy (J)
G' critical Gibbs free energy (J)
G unit volume Gibbs free energy (J)
H enthalpy (J)
I embryo formation rate (embryo cm^{-2} s^{-1})
K Boltzmann constant
M molar mass (kg mol^{-1})
R ideal gas constant (J mol^{-1} K^{-1})
R radius (m)
r' critical radius (m)
S entropy (J)
T temperature (K)
V volume (m^3)
Γ surface energy (J m^{-2})
P embryo density (kg m^3)
Θ contact angle (degrees)
Ω humidity ratio (kg_v $kg_a{}^{-1}$)
Δ deviation

Subscripts

E embryo
Ew interface embryo–wall
Lat latent
S surroundings
Sat saturation
Se interface embryo–surroundings
Sw interface surroundings–wall
Tot total
W wall

References

1. Song, M.J.; Mao, N.; Xu, Y.J.; Deng, S.M. Challenges in, and the development of building energy saving techniques, illustrated with the example of an air source heat pump. *Therm. Sci. Eng. Prog.* **2019**, *10*, 337–356. [CrossRef]
2. Ma, Z.J.; Song, J.L.; Zhang, J.L. Energy consumption prediction of air conditioning systems in buildings by selecting similar days based on combined weights. *Energy Build.* **2017**, *151*, 157–166. [CrossRef]
3. Fekadu, G.; Subudhi, S. Renewable energy for liquid desiccants air conditioning system: A review. *Renew. Sustain. Energy Rev.* **2018**, *93*, 364–379. [CrossRef]
4. Wang, F.; Liang, C.H.; Zhang, X.S. Research of anti-frosting technology in refrigeration and air conditioning fields: A review. *Renew. Sustain. Energy Rev.* **2018**, *81*, 707–722. [CrossRef]
5. Badri, D.; Toublanc, C.; Rouaud, O.; Havet, M. Review on frosting, defrosting and frost management techniques in industrial food freezers. *Renew. Sustain. Energy Rev.* **2021**, *151*, 111545. [CrossRef]
6. Zhang, L.; Song, M.J.; Shen, J.; Zhang, X.; Xu, Y.J.; Hu, Y.X. Impacts of initial cooling rate on local frosting characteristics of horizontal cold plate surface with edge effect considered. *Int. Commun. Heat Mass* **2023**, *143*, 106654. [CrossRef]

7. Barthlott, W.; Neinhuis, C. Purity of the sacred lotus, or escape from contamination in biological surfaces. *Planta* **1997**, *202*, 1–8. [CrossRef]
8. Yang, S.Y.; Wu, C.Y.; Zhao, G.L.; Sun, J.; Yao, X.; Ma, X.H.; Wang, Z.K. Condensation frosting and passive anti-frosting. *Cell Rep. Phys. Sci.* **2021**, *2*, 100474. [CrossRef]
9. Liu, Z.L.; Wang, H.Y.; Zhang, X.H.; Meng, S.; Ma, C.F. An experimental study on minimizing frost deposition on a cold surface under natural convection conditions by use of a novel anti-frosting paint. Part II. Long-term performance, frost layer observation and mechanism analysis. *Int. J. Refrig.* **2006**, *29*, 236–242. [CrossRef]
10. Okoroafor, E.U.; Newborough, M. Minimising frost growth on cold surfaces exposed to humid air by means of crosslinked hydrophilic polymeric coatings. *Appl. Therm. Eng.* **2000**, *20*, 736–758. [CrossRef]
11. Lee, H.; Shin, J.; Ha, S.; Choi, B.; Lee, J. Frost formation on a plate with different surface hydrophilicity. *Int. J. Heat Mass Transf.* **2004**, *47*, 4881–4893. [CrossRef]
12. Yang, W.; Zeng, B.; Zhang, Y.; He, S.; Zhao, X. Frosting performance of a nanoporous hydrophilic aluminum surface. *Energies* **2018**, *11*, 3483. [CrossRef]
13. Wang, F.; Tang, R.; Wang, Z.; Yang, W. Experimental study on anti-frosting performance of superhydrophobic surface under high humidity conditions. *Appl. Therm. Eng.* **2022**, *217*, 119193. [CrossRef]
14. Wang, G.Y.; Shen, Y.Z.; Tao, J.; Luo, X.Y.; Jin, M.M.; Xie, Y.H.; Li, Z.Z.; Guo, S.M. Facilely constructing micro-nanostructure superhydrophobic aluminum surface with robust ice-phobicity and corrosion resistance. *Surf. Coat. Technol.* **2017**, *329*, 224–231. [CrossRef]
15. Lei, S.; Wang, F.; Fang, X.; Ou, J.; Li, W. Icing behavior of water droplets impinging on cold superhydrophobic surface. *Surf. Coat. Technol.* **2019**, *363*, 362–368. [CrossRef]
16. Fan, P.Y.; Li, Y.X.; Liu, Z.L.; Yu, F.J.; Chen, Y.L.; Li, Y. Preparation and performance of nanoparticles-based anti-frosting transparent hydrophobic surfaces. *Int. J. Refrig.* **2021**, *130*, 404–412. [CrossRef]
17. Diamantino, T.C.; Gonçalves, R.; Nunes, A.; Páscoa, S.; Carvalho, M.J. Durability of different selective solar absorber coatings in environments with different corrosivity. *Sol. Energy Mater. Sol. Cells* **2017**, *166*, 26–38. [CrossRef]
18. Wilson, P.W.; Lu, W.; Xu, H.J.; Kim, P.; Kreder, M.J.; Alvarengad, J.; Aizenbergb, J. Inhibition of ice nucleation by slippery liquid-infused porous surfaces. *Appl. Surf. Sci.* **2013**, *15*, 581–584. [CrossRef]
19. Kim, P.; Wong, T.S.; Alvarenga, J.; Kreder, M.J.; Adorno-Martinez, W.E.; Aizenberg, J. Liquid-infused nanostructured surfaces with extreme anti-ice and anti-frost performance. *ACS Nano* **2012**, *6*, 6569–6576. [CrossRef]
20. Huang, L.; Liu, Z.; Liu, Y.; Guo, Y.; Wang, L. Effect of contact angle on water droplet freezing process on a cold flat surface. *Exp. Therm. Fluid Sci.* **2012**, *40*, 73–80. [CrossRef]
21. Piucco, R.O.; Hermes, C.J.L.; Melo, C.; Barbosa, J.R. A study of frost nucleation on flat surfaces. *Exp. Therm. Fluid Sci.* **2008**, *32*, 1710–1715. [CrossRef]
22. Mangini, D.; Antonini, C.; Marengo, M.; Amirfazli, A. Runback ice formation mechanism on hydrophilic and superhydrophobic surfaces. *Cold Reg. Sci. Technol.* **2015**, *109*, 53–60. [CrossRef]
23. Na, B.; Webb, R.L. A fundamental understanding of factors that affecting frost nucleation. *Int. J. Heat Mass Transf.* **2003**, *46*, 3797–3808. [CrossRef]
24. Liu, Z.; Wang, H.; Zhang, X.; Meng, S.; Ma, C. An experimental study on minimizing frost deposition on a cold surface under natural convection conditions by use of a novel anti-frosting paint. Part I. Anti-frosting performance and comparison with the uncoated metallic surface. *Int. J. Refrig.* **2006**, *29*, 229–236. [CrossRef]
25. Liu, J.; Zhu, C.Q.; Liu, K.; Jiang, Y.; Song, Y.L.; Francisco, J.S.; Zeng, X.C.; Wang, J.J. Distinct ice patterns on solid surfaces with various wettabilities. *Proc. Natl. Acad. Sci. USA* **2017**, *114*, 11285–11290. [CrossRef] [PubMed]
26. Byun, S.J.; Jeong, H.J.; Son, H.; Kim, D.R.; Lee, K.S. Frost formation from general-low to ultra-low temperatures: A review. *Int. J. Heat Mass Transf.* **2022**, *195*, 123164. [CrossRef]
27. Jing, T.; Kim, Y.; Lee, S.; Kim, D.; Kim, J.; Hwang, W. Frosting and defrosting on rigid superhydrohobic surface. *Appl. Surf. Sci.* **2013**, *276*, 37–42. [CrossRef]
28. Fletcher, N.H. *The Chemical Physics of Ice*; Cambridge University Press: Cambridge, UK, 1970; Chapters 4–5.
29. Bai, G.Y.; Gao, D.; Liu, Z.; Zhou, X.; Wang, J.J. Probing the critical nucleus size for ice formation with graphene oxide nanosheets. *Nature* **2019**, *576*, 437–441. [CrossRef] [PubMed]
30. Becker, R.; Doring, W. Kinetische behandlung der keimbildung in übersättigten dämpfen. *Ann. Phys. Lpz.* **1935**, *24*, 719. [CrossRef]
31. Volmer, M.; Flood, H. Tröpfchenbildung in dämpfen. *Z. Phys. Chem. A* **1934**, *170*, 273. [CrossRef]
32. Mandelbrot, B.B. How long is the coast of britain? statistical self-similarity and fractional dimension. *Science* **1967**, *156*, 636–638. [CrossRef] [PubMed]
33. Hou, P.; Liang, C.; Yu, W.-P. Experimental study and fractal analysis of ice crystal structure at initial period of frost formation. *J. Appl. Sci.* **2007**, *25*, 193–197.
34. Mei, M.; Yu, B.; Cai, J.; Liang, L. A fractal analysis of dropwise condensation heat transfer. *Int. J. Heat Mass Transf.* **2009**, *52*, 4823–4828. [CrossRef]

35. Ding, Y.F.; Yin, S.; Liao, Y.D.; Wu, H.J. Frosting mechanism and suppression on nano/micro-structured hydrophobic surfaces. *CIESC J.* **2012**, *63*, 3213–3219.
36. Carrales, D.H.; Rodríguez, I., Leyva, R.; Mendoza, E.; Villela, D.E. Effect of surface area and physical-chemical properties of graphite and graphene-based materials on their adsorption capacity towards metronidazole and trimethoprim antibiotics in aqueous solution. *Chem. Eng. J.* **2020**, *402*, 126154.

MDPI AG
Grosspeteranlage 5
4052 Basel
Switzerland
Tel.: +41 61 683 77 34

Processes Editorial Office
E-mail: processes@mdpi.com
www.mdpi.com/journal/processes